T0332610

THE TAIWAN CRISIS: A SHOWCASE OF THE GLOBAL ARSENIC PROBLEM

ISGSD

*International Society of
Groundwater for
Sustainable Development*

Cover photo

The cover photo shows the Wushanting mud volcano in southwestern Taiwan, which is one of the 64 known active on-shore mud volcanoes of Taiwan. Several of them are located at the east of Chianan plain and their erupted muds and fluids contain high concentrations of arsenic, which could have been transported by predominantly fluvial processes to the coastal Chianan plain where it contributed to the high arsenic concentrations in the groundwater of the deep aquifers, which are a major factor that has caused the endemic blackfoot disease in SW Taiwan before 1990, especially being prevalent during the 1950s.

Arsenic in the Environment

Series Editors

Jochen Bundschuh
Institute of Applied Research, Karlsruhe University of Applied Sciences,
Karlsruhe, Germany
Royal Institute of Technology (KTH), Stockholm, Sweden

Prosun Bhattacharya
KTH-International Groundwater Arsenic Research Group, Department of Land
and Water Resources Engineering, Royal Institute of Technology (KTH),
Stockholm, Sweden

ISSN: 1876-6218

Volume 3

The Taiwan Crisis: a showcase of the global arsenic problem

Jiin-Shuh Jean
*Department of Earth Sciences, National Cheng Kung University,
Tainan City 701, Taiwan*

Jochen Bundschuh
*Institute of Applied Research, Karlsruhe University of Applied Sciences,
Karlsruhe, Germany; Royal Institute of Technology (KTH), Stockholm, Sweden*

Chien-Jen Chen
Genomics Research Center, Academia Sinica, Taipei City, Taiwan

How-Ran Guo
*Department of Environmental and Occupational Health, Medical College,
National Cheng Kung University, Tainan City, Taiwan*

Chen-Wuing Liu
*Department of Bioenvironmental Systems Engineering,
National Taiwan University, Taipei City, Taiwan*

Tsair-Fuh Lin
*Department of Environmental Engineering, National Cheng Kung University,
Tainan City, Taiwan*

Yen-Hua Chen
*Department of Earth Sciences, National Cheng-Kung University, Tainan City,
Taiwan*

CRC Press
Taylor & Francis Group
Boca Raton   London   New York   Leiden

CRC Press is an imprint of the
Taylor & Francis Group, an **informa** business

A BALKEMA BOOK

The digital elevation model PIA3395 (World: Figure 1.1) (http://photojournal.jpl.nasa.gov) is courtesy of NASA/JPL-Caltech.

The digital elevation model of Taiwan used in Figures 3.1, 3.9, 6.1, 6.2, 6.4–6.6, 6.8, 6.9, 6.11, 6.13, 6.14, 7.1, 8.1 and 8.5 was prepared from 40 × 40 m digital terrain model data of Taiwan and is courtesy of Chia-Hung Jen from National Kaohsiung Normal University, Taiwan.

CRC Press/Balkema is an imprint of the Taylor & Francis Group, an informa business

© 2010 Taylor & Francis Group, London, UK

Typeset by Vikatan Publishing Solutions (P) Ltd., Chennai, India
Printed and bound in Poland by Poligrafia Janusz Nowak, Poznán

Published by: CRC Press/Balkema
P.O. Box 447, 2300 AK Leiden, The Netherlands
e-mail: Pub.NL@taylorandfrancis.com
www.crcpress.com – www.taylorandfrancis.co.uk – www.balkema.nl

Library of Congress Cataloging-in-Publication Data

The Taiwan crisis : a showcase of the global arsenic problem / editors, Jiin-Shuh Jean ... [et al.].
    p. cm.—(Arsenic in the environment ; v. 3)
    Includes bibliographical references.
    ISBN 978-0-415-58510-1 (hardback : alk. paper)
1. Groundwater--Arsenic content--Taiwan. 2. Arsenic--Physiological effect--Taiwan.
3. Environmental toxicology--Taiwan. I. Jean, Jiin-Shuh. II. Title.

TD427.A77T35 2010
363.738'4--dc22   2010000331

ISBN: 978-0-415-58510-1 (Hbk)
ISBN: 978-0-203-84806-7 (eBook)

About the book series

Although arsenic has been known as a 'silent toxin' since ancient times, and the contamination of drinking water resources by geogenic arsenic was described in different locations around the world long ago — e.g., in Argentina in 1917— it was only two decades ago that it received overwhelming worldwide public attention. As a consequence of the biggest arsenic calamity in the world, which was detected more than twenty years back in West Bengal, India and other parts of southeast Asia. As a consequence, there has been an exponential rise in scientific interest that has triggered high quality research. Since then, arsenic contamination (predominantly of geogenic origin) of drinking water resources, soils, plants and air, the propagation of arsenic in the food chain, the chronic affects of arsenic ingestion by humans, and their toxicological and related public health consequences, have been described in many parts of the world, and every year, even more new countries or regions are discovered to have arsenic problems

Arsenic is found as a drinking water contaminant, in many regions all around the world, in both developing as well as industrialized countries. However, addressing the problem requires different approaches which take into account, the differential economic and social conditions in both country groups. It has been estimated that 200 million people worldwide are at risk from drinking water containing high concentrations of As, a number which is expected to further increase due to the recent lowering of the limits of arsenic concentration in drinking water to 10 μg L^{-1}, which has already been adopted by many countries, and some authorities are even considering decreasing this value further.

The book series "Arsenic in the Environment" is an inter- and multidisciplinary source of information, making an effort to link the occurrence of geogenic arsenic in different environments and the potential contamination of ground- and surface water, soil and air and their effect on the human society. The series fulfills the growing interest in the worldwide arsenic issue, which is being accompanied by stronger regulations on the permissible Maximum Contaminant Levels (MCL) of arsenic in drinking water and food, which are being adopted not only by the industrialized countries, but increasingly by developing countries.

The book series covers all fields of research concerning arsenic in the environment and aims to present an integrated approach from its occurrence in rocks and mobilization into the ground- and surface water, soil and air, its transport therein, and the pathways of arsenic introduction into the food chain including uptake by humans. Human arsenic exposure, arsenic bioavailability, metabolism and toxicology are treated together with related public health effects and risk assessments in order to better manage the contaminated land and aquatic environments and to reduce human arsenic exposure. Arsenic removal technologies and other methodologies to mitigate the arsenic problem are addressed not only from the technological perspective, but also from an economic and social point of view. Only such inter- and multidisciplinary approaches, will allow case-specific selection of optimal mitigation measures for each specific arsenic problem and provide the local population with arsenic safe drinking water, food, and air.

We have an ambition to make this book series an international, multi- and interdisciplinary source of knowledge and a platform for arsenic research oriented to the direct solution of problems with considerable social impact and relevance rather than simply focusing on cutting edge and breakthrough research in physical, chemical, toxicological and medical sciences. The book series will also form a consolidated source of information on the worldwide occurrences of arsenic, which otherwise is dispersed and often hard to access. It will also have role in increasing the awareness and knowledge of the arsenic problem among administrators, policy makers and company executives and improving international and bilateral cooperation on arsenic contamination and its effects.

Consequently, we see this book series as a comprehensive information base, which includes authored or edited books from world-leading scientists on their specific field of arsenic research, but also contains volumes with selected papers from international or regional congresses or other scientific events. Further, the abstracts presented during the homonymous biannual international congress series, which we organize in different parts of the world is being compiled in a stand-alone book series "Arsenic in the Environment—Proceedings" that would give short and crisp state of the art periodic updates of the contemporary trends in arsenic-related research. Both the series will be open for any person, scientific association, society or scientific network, for the submission of new book projects. Supported by a strong multi-disciplinary editorial board, book proposals and manuscripts are peer reviewed and evaluated.

Jochen Bundschuh
Prosun Bhattacharya
(*Series Editors*)

Editorial board

List of Contents

About the book series   VII

Editorial board   IX

Foreword   XIX

Authors' preface   XXI

About the authors   XXV

Acknowledgements   XXIX

1   Taiwan and the global arsenic problem   1

   1.1   General introduction to the arsenic problem   1

     1.1.1   Origin, release and occurrence of groundwater arsenic   1

     1.1.2   Geochemical arsenic mobility controls   2

     1.1.3   Other arsenic mobility controls   3

     1.1.4   Remediation of arsenic-contaminated sites   3

     1.1.5   Human exposure to arsenic and related health effects   5

   1.2   Arsenic: From history to Taiwan   5

     1.2.1   Arsenic discoveries in groundwater of Argentina   6

     1.2.2   Arsenic discoveries in groundwater of Mexico   6

     1.2.3   Arsenic discoveries in ground- and surface-water of Chile   7

     1.2.4   Arsenic discoveries in groundwater of Taiwan   7

   1.3   Arsenic: From Taiwan to the end of the 20th century   8

   1.4   Arsenic in the 21st century—Recognizing groundwater arsenic as a global problem   9

   1.5   Regulations of arsenic contents in drinking water and its impact on the exposed population   14

   1.6   Why was the "Taiwan signal" not immediately recognized worldwide?   16

   1.7   Why does arsenic continue to affect people worldwide?   16

   1.8   Demands for international cooperation and networking   17

2   Geological controls of arsenic concentrations in ground- and surface-waters—
    An overview of our worldwide state-of-the-art knowledge   19

    2.1   Arsenic in the earth's environments and introduction into ground-
          and surface-water resources   19

    2.2   Geogenic arsenic: Occurence and sources   20

          2.2.1   Arsenic in minerals and amorphous phases   20

          2.2.2   Arsenic in rocks   22

    2.3   Mechanisms of arsenic mobilization into aqueous
          environments: an overview   24

          2.3.1   Arsenic species in natural waters and reaction kinetics   25

          2.3.2   Arsenic release and mobility: solid-fluid interfacial processes 26

          2.3.3   Additional factors and processes influencing concentrations
                  of dissolved arsenic   27

          2.3.4   Arsenic transport in natural water   30

    2.4   Sulfide oxidation   30

          2.4.1   Mechanism and kinetics of arsenic mobilization
                  through sulfide oxidation   30

          2.4.2   Example: Arsenic mobilization by sulfide oxidation
                  in the near-neutral sandstone aquifer of northeastern
                  Wisconsin, USA   32

          2.4.3   Example: Franconian Upper Triassic sandstone aquifer, Germany   32

    2.5   Arsenic input due to leaching in geothermal reservoirs:
          the role of geothermal fluids   33

          2.5.1   Arsenic input from geothermal waters and other geothermal
                  manifestations   33

          2.5.2   Examples of arsenic input from geothermal waters   37

    2.6   The role of Fe, Mn, and Al oxides and oxyhydroxides
          as sources and sinks for dissolved arsenic   40

          2.6.1   Arsenic release by dissolution of metal oxyhydroxides   40

          2.6.2   Arsenic release/sequestration due to sorption
                  by Fe, Mn and Al oxides and oxyhydroxides   42

                  2.6.2.1   Influence of redox potential and pH
                            on adsorption capacity   43

                  2.6.2.2   Influence of competing ions on arsenic
                            adsorption capacity   44

                  2.6.2.3   Example: Chaco-Pampean plain, Argentina   45

                  2.6.2.4   Example: Molasse trough sand aquifer,
                            Southern Germany   46

    2.7   Adsorption processes and capacity of clay minerals   46

2.8   Precipitation/dissolution and sorption processes of calcite   46

2.9   Interactions between arsenic and humic substances   47

3   History of blackfoot disease   49

3.1   Prologue: A mysterious disease   49

3.2   Clinical characteristics of blackfoot disease   49

3.3   Pathological findings of blackfoot disease   56

3.4   Epidemiological characteristics of blackfoot disease   57

4   Cause of blackfoot disease: Arsenic in artesian well water   61

4.1   Types of wells in blackfoot disease-endemic area   61

4.2   Characteristics of well water in blackfoot disease endemic area   62

4.3   Arsenic levels in well water in Lanyang basin   64

4.4   Association between blackfoot disease and artesian well water   65

4.5   Arsenic in drinking water: The cause of blackfoot disease   66

4.6   Co-morbidity of unique arsenic-induced skin lesions and blackfoot disease   67

4.7   Host and environmental co-factors for blackfoot disease   69

4.8   Arsenic in drinking water and circulatory diseases
      other than blackfoot disease   69

4.9   Arsenic in drinking water and prevalence of diabetes and hypertension   70

4.10 Reduction in mortality of arsenic-induced diseases after implementation
      of public water supply system in the endemic area of blackfoot disease   71

5   Non-vascular health effects of arsenic in drinking water in Taiwan   73

5.1   Introduction   73

5.2   Skin cancer   73

5.3   Internal cancers   77

5.4   Eye diseases   82

5.5   Other health outcomes   82

5.6   Summary and conclusions   83

6   Arsenic sources, occurrences and mobility in surface water,
    groundwater and sediments   85

6.1   Introduction   85

6.2   Hydrogeology and sedimentology of arsenic in aquifers   85

6.2.1   Chianan plain   85

6.2.2   Lanyang plain (Yilan plain)   88

6.2.3   Guandu plain   91

6.3   Potential arsenic sources   92

6.3.1   Geogenic sources   92

6.3.1.1   Chianan plain   93

6.3.1.2   Lanyang (or Yilan) plain   96

6.3.2   Anthropogenic sources   99

6.3.2.1   Mining activity   99

6.3.2.2   Industrial activity   99

6.3.2.3   Agricultural activity   99

6.4   Arsenic distributions and mobility controls   101

6.4.1   Water chemistry in the Chianan and Lanyang plains   101

6.4.2   Arsenic in sediments   104

6.4.3   Mobilization and transport of arsenic   107

6.4.3.1   Arsenic speciation   107

6.4.3.2   Redox-mediated mobilization and transport of arsenic   107

6.4.3.3   Microbe-mediated mobilization and transport of arsenic   109

6.5   Arsenic in mud volcanoes and hot springs   109

6.6   Concluding remarks   113

7   Arsenic in soils and plants: accumulation and bioavailability   115

7.1   Accumulation and behavior of arsenic in soil   115

7.2   Bioaccumulation of arsenic in plants and crops   119

8   Potential threat of the use of arsenic-contaminated water in aquaculture   123

8.1   Introduction   123

8.2   Arsenic in aquacultural organisms   125

8.2.1   Tilapia   125

8.2.2   Milkfish   127

8.2.3   Mullet   128

8.2.4   Clam   128

8.2.5   Oyster   129

8.2.6   Arsenic levels in groundwater and farmed fish/shrimp in Lanyang plain   130

8.3   Arsenic methylation capability   131
8.4   Health risk assessment   132

9   Current solutions to arsenic-contaminated water   135
9.1   Introduction   135
9.2   Change of water source   135
9.3   Water treatment processes for centralized systems   136
       9.3.1   Precipitation methods   136
       9.3.2   Adsorption and ion exchange methods   137
       9.3.3   Membrane technology   138
9.4   Point-of-use and point-of-entry devices   139
9.5   Case study in southwestern Taiwan   141
9.6   Recommendations   143

10   Future areas of study and tasks for the Taiwan arsenic problem   145
10.1   Sources of arsenic and mobilization in groundwater   145
10.2   Human impact through the food chain   147
10.3   Health effects of arsenic in drinking water, treatment,
       risk assessment and prevention   148
10.4   Future treatment demands, including nanotechnology   149

References   151
Subject index   185
Locality index   197
Book series page   203

Foreword

Before 1990, blackfoot disease (BFD) was only found to be endemic in the coastal areas of southwestern Taiwan (most severe cases were in Beimen and Hsuechia townships of Tainan county and Budai and Yichu townships of Chiayi county). BFD was only colloquially named "black and dry snakes", a peripheral vascular gangrene disease. The early symptom of BFD patients was poor circulation at the extremeties of limbs and even a little pressure on these affected areas caused excruciating pain. As the conditions worsened, toes and even fingers would undergo apoptosis and necrosis so that amputations were eventually required.

BFD has long been considered to be caused by drinking groundwater that contains a large amount of arsenic. However, some literature indicated that BFD was only caused by coexposure to arsenic and fluorescent humic substances or organometallic complexes (binding of humic substances with arsenic). Although no new BFD cases have occurred since 1990, due to drinking tapwater instead of groundwater, the etiological factors that caused BFD remain unknown. Many countries in the world, such as Bangladesh, India, Vietnam, Cambodia, Thailand, China, USA and countries in Europe and Latin America have also been found to contain arsenic-contaminated groundwater. Interestingly, no BFD cases have yet been found in these areas. Such a finding requires further research to determine the causative factor(s) of BFD.

The authors outline the cause of the arsenic pollution and subsequent effects on health in Taiwan and propose solutions that could have worldwide implications for understanding and solving the arsenic problem elsewhere.

King – Ho Wang

王 金 河

Dr. King-Ho Wang
President of the Blackfoot Disease
Museum in Beimen, Tainan, Taiwan

Authors' preface

It was half a century ago when Taiwan became famous because of the detection of a new disease, which was named "blackfoot disease" (BFD) since it is characterized by its most striking clinical presentation in the late stage—the black discoloration caused by the gangrenous changes in the extremities, mostly in the feet. Blackfoot disease cases were known at the beginning of the 20th century and were first documented in 1954 in an endemic area of arsenic intoxication caused by drinking groundwater along the southwestern coast of Taiwan. Documented cases were most prevalent between 1950 and 1956 and detailed studies started in the 1960s. After 1956, some of the residents started to utilize public surface water supplies instead of the groundwater for drinking purposes and in 1970, virtually all of the residents drank tapwater instead of groundwater, so that the number of BFD cases was significantly reduced. Since 1990, virtually no new BFD cases have been identified; though arsenic is still found in the groundwater. Whereas BFD is generally believed to be caused by consuming arsenic-containing groundwater, specific health effects causing BFD depend on the presence of additional unknown compounds in the groundwater, the identity of which are still heavily debated and the subject of ongoing investigations. Moreover, although the epidemiological studies on the chronic arsenic toxicity have been extensively carried out in Taiwan, the geogenic sources and release processes of arsenic to the groundwater have not been comprehensively investigated and certainly deserve special attention.

The discovery of the arsenic problem in Taiwan in the 1950s occurred at about the same time as the detection of the arsenic problem in ground- and surface-waters in Chile and Mexico, whereas the arsenic problem of Argentina had already been described in the 1910s. At that time, the presence of geogenic arsenic in water resources, predominantly in groundwater, was thought to be a rare problem affecting only a very few areas. Consequently, the arsenic problem did not receive much attention, until the detection of arsenic in groundwater in Bengal delta and the related health effects in the population achieved worldwide interest. However, the world was still of the opinion that the arsenic problem affected only a few regions. This opinion was maintained until the end of the 20th century but the discovery of arsenic in the groundwater in more than 70 countries has changed this mindset and the arsenic problem is now recognized as a global problem, which requires global solutions.

The large number of studies related to the genesis of the aquifers with high arsenic concentrations in Taiwan, the geogenic processes of arsenic, the mechanism that caused BFD and the possible etiological agents of BFD have inspired us to write this book. It aims to summarize—as far as is possible—the contents of over 1000 published scientific papers and reports, many of them in Chinese and others inaccessible to the broad public in an easily understandable overview. Most of the older reports dealt with medical aspects (e.g., pathology, epidemiology), and it was not until the 1980s that the focus turned to the environmental aspects of groundwater and geological characteristics of the endemic BFD areas. Since then, the BFD issue has drawn the interest of environmental scientists and hydrogeologists from Taiwan and around the world.

There is a lack of and urgent need of easily understandable and accessible information in a comprehensive form on the experiences obtained during the last half century regarding the Taiwan arsenic problem. No doubt, such information does exist but is disseminated amongst many published scientific papers and is difficult for graduate students and researchers to collect in a short period of time. This is especially true in the case of students in developing countries, who face difficulties accessing such information. This, together with the absence of a comprehensive book that deals with all aspects of arsenic in Taiwan—including constraints on the mobility of arsenic in groundwater, its uptake from soil and water by plants, arsenic-propagation through the food chain, human health impacts, and the treatment of the arsenic-rich waters on an industrial scale as well as on a small scale suitable for rural areas—motivated us to write this book.

However we do not address the arsenic problem of Taiwan as an isolated case. In contrast we describe how it is imbedded in the global arsenic problem. This means that the Taiwan arsenic case experiences elucidated in this book can be applied worldwide. The book is a state-of-the-art overview of research on arsenic in Taiwan and is designed to:

- Create interest within areas of Taiwan affected by the presence of arseniferous aquifers.
- Draw the attention of the international scientific community.
- Increase awareness among researchers, administrators, policy makers, and company executives.
- Improve the international cooperation on the topic.

To fulfill this demand in Chapter 1, we give an overview of the global arsenic problem and the role which Taiwan plays within it. In Chapter 2 we give a state-of-the-art overview of the geological controls of arsenic concentrations in ground- and surface waters, which will help us to better understand the genesis of the arsenic-rich groundwater in Taiwan. Detailed discussions on arsenic sources, occurrences and mobility in surface water, groundwater and sediments in Taiwan are provided in Chapter 6. Chapters 3 to 5 are devoted to the toxicology and health effects of arsenic poisoning. Chapter 3 contains information on the history of the BFD, from the first detection of the disease at the beginning of the 20th century to the 1980s when the last newly occurred cases were reported. Chapter 4 deals with the causes of BFD, and Chapter 5 focuses on other health effects, principally skin cancers, internal cancers, and other diseases which occurred in addition to BFD. In Chapter 7 we address the occurrence of arsenic in soils and plants and discuss its accumulation and evaluate the hazards of rice grown on arsenic-rich soil for human consumption. Chapter 8 deals with the bioavailability of arsenic to fish and shellfish and the potential hazard of human arsenic uptake through the food chain is explored. In Chapter 9, various arsenic remediation methods for water treatment are discussed for both, industrial-scale water treatment plants and wellhead units. Finally in Chapter 10 we give a description of the research required to better understand the arsenic problem in Taiwan, and to remediate the problem where necessary (e.g. to improve the treatment of drinking water, especially in middle-sized communities and in rural areas, as well as solving the problem of using arsenic-contaminated water for farming fish and shellfish).

The comprehensive information on the Taiwanese arsenic problem in this book should make it a convenient source that aims to develop a strong human resource in both developing and developed countries to better understand and mitigate the arsenic problem in their own country. The book is for graduate students and researchers, and should prove useful not only for geologists, hydrogeologists, soil scientists, environmental scientists and engineers, toxicologists, health scientists, and persons related to agriculture, aquaculture, and food production, but also for those involved in environmental legislation and remediation of contaminated sites. It is a useful source of information for those involved in decision and policy making, for administrative leaders both in governments and in international bodies such as the United Nations family, the international and regional development banks, financial institutions, and donors that are concerned with technical and economic cooperation with developing countries.

As authors we expect that this multidisciplinary insight into the arsenic problem will serve the world community and help to address the arsenic problem in the over 200 areas in which the problem has so far been discovered and any areas where it may be detected in the future. We hope that this book will be useful for many people in helping society to effectively mitigate the arsenic problem worldwide, providing access to safe water and food everywhere and to everybody.

Jiin-Shuh Jean
Jochen Bundschuh
Chien-Jen Chen
How-Ran Guo
Chen-Wuing Liu
Tsair-Fuh Lin
Yen-Hua Chen
(*Authors*)

About the authors

Jiin-Shuh Jean (1952, Taiwan) finished his PhD degree in Hydrogeology, specializing in groundwater modeling, from Purdue University, West Lafayette, Indiana, USA. He is a full professor of Hydrogeology at the Department of Earth Sciences, National Cheng Kung University (NCKU), Tainan City, Taiwan. His current research interest is in arsenic mobilization and removal. His ongoing three-year grant project from the National Science Council of Taiwan aims to find out the etiological factor that caused blackfoot disease (BFD) which was prevalent in the coastal areas of SW Taiwan between 1955–1960. Why BFD occurred only in SW Taiwan and nowwere else in the world is also a major interest. He is now an associate editor of the Journal of Hydrology, Elsevier and a local chief organizer of the 3rd International Congress on Arsenic in the Environment, which was held on 17–21 May 2010 at NCKU, Tainan, Taiwan.

Jochen Bundschuh (1960, Germany) has been working in international academic and technical co-operation programs in different fields of water and renewable energy resources for more than 17 years for the German government. He was a long-term professor for the DAAD (German Academic Exchange Service) in Argentina and an expert of the German Agency of Technical Cooperation (GTZ) (1993–1998). From 2001 to 2008 he worked within the Integrated Expert Program of CIM (GTZ/BA) as an advisor to Costa Rica at the Instituto Costarricense de Electricidad (ICE). Here, he assisted the country in the evaluation and development of renewable energy resources. Since 2009 he has been teaching as a professor in the field of renewable energies, in particular gothermics, at the University of Applied Sciences in Karlsruhe (Germany) and is also a researcher at the Institute of Applied Research at the same university, where he works in geothermics and the groundwater arsenic issue. Research is performed in close cooperation with the Department of Earth Sciences at the National Cheng Kung University, Tainan, Taiwan, where he is a visiting professor. Prof. Bundschuh is an editor of the book series "Multiphysics Modeling" and "Arsenic in the Environment" and the principal organizer of the "Arsenic in the Environment" international congress series.

Chien-Jen Chen (1951, Taiwan) graduated from Johns Hopkins University with a Sc.D. in epidemiology and became a professor at the National Taiwan University. He has been appointed as the Director of the Graduate Institute of Epidemiology, and Dean of College of Public Health in the National Taiwan University. He has served in several government positions, including the Minister of the Department of Health and Minister of the National Science Council. He has dedicated himself to epidemiological research on chronic arsenic poisoning and virus-induced cancers. He has published 505 original, review and editorial articles in refereed journals, which have been cited more than 13,000 times. He has received many accolades and awards for his research achievements, including the Presidential Science Prize, the most prestigious science award in Taiwan. He was elected as an academician of the Academia Sinica in 1998. He was also elected as a member of the Academy of Sciences for the Developing World, and an honorary member of the Mongolian Academy of Science. He was elected as the Dr. DV Datta Memorial Orator by the Indian National Association for the Study of the Liver, and the Cutter Lecturer on Preventive Medicine by Harvard University. He also received the "Officier dans l'Ordre des Palmes Academiques" from the Ministry of Education in France, and the Science and Engineering Achievement Award from the Taiwan-American Foundation in the USA.

How-Ran Guo (1961, Taiwan) obtained his medical degree in 1988 at the Taipei Medical College (currently the Taipei Medical University), Taipei, Taiwan, where he started his research on arsenic by assessing the high incidence of bladder cancer in an endemic area of arsenic intoxication. He continued his study at the Harvard School of Public Health, Boston, Massachusetts, U.S.A. and earned an M.P.H. (1989), a M.S. in Epidemiology (1990), and a Sc.D. in Environmental Health, with a major in Occupational Medicine (1994). Currently, he serves as the Chair of the Department of Occupational and Environmental Medicine at the National Cheng Kung University as well as the Director of the Department of Occupational and Environmental Medicine at the National Cheng Kung University Hospital, Tainan, Taiwan. His ongoing two-year grant project from the National Science Council of Taiwan aims to describe the epidemiologic characteristics of skin cancer associated with arsenic. He is now an Associate Editor of both the Taiwan Journal of Public Health and the Journal of Occupational Safety and Health and a local chief organizer of the Joint 21st EPICOH and 38th Medichem Conference, which was held from 20–25 April 2010 in Taipei, Taiwan.

Chen-Wuing Liu (1955, Taiwan), PhD, is a professor in the Department of Bioenvironmental Systems Engineering at the National Taiwan University and since 2007 has been the President of the Taiwanese Soil and Groundwater Environmental Protection (TASGEP). His teaching covers contaminant hydrogeology and geochemistry, and his research focuses on subsurface flow, contaminant transport, and groundwater chemistry. Liu holds a PhD from the University of California at Berkeley in Hydrogeology.

Tsair-Fuh Lin (1963, Taiwan), PhD, is a distinguished professor in the Department of Environmental Engineering (DEnvE) at National Cheng Kung University (NCKU), Tainan City, Taiwan. Since 2009, he has served as the Department Chair of DEnvE at NCKU, Secretary-General of the Chinese Institute of Environmental Engineers in Taiwan, and Chair of the Specialist Group in Off Flavours in the Aquatic Environment, International Water Association. His research interests include adsorption processes, contaminant site remediation, and monitoring and control of cyanobacteria and metabolites in reservoirs. Lin holds a PhD from the University of California at Berkeley in Environmental Engineering.

Yen-Hua Chen (1976, Taiwan) received her B.S. and M.S from the Earth Sciences Department at National Cheng Kung University, Taiwan, in 1997 and 1999, respectively. She received a Ph.D. from the Material Sciences and Engineering Department of the National Tsing-Hua University, Taiwan, in 2006. She joined the Earth Sciences Department at National Cheng-Kung University as an Assistant Professor in 2007. Her current research interests include synthesizing nano-minerals and their application in the treatment of environmental pollution and the development of functional minerals for use as sensors or adsorbents.

Acknowledgements

This book would be incomplete without an expression of our sincere and deep sense of gratitude to Kirk Nordstrom at the US Geological Survey in Boulder (CO, USA) and Richard Wilson, formerly at Harvard University in Cambridge (MA, USA) for their careful reading and suggestions that greatly improved Chapters 1 and 2 on the global arsenic problem and the geological controls of arsenic mobilization into ground- and surface-waters. Their enormous input and dedication for their time-consuming efforts have significantly improved these overview chapters.

We thank Walter Klimecki at the Department of Pharmacology and Toxicology, University of Arizona (Tucson, AZ) and Miroslav Stýblo at the Department of Nutrition and The Center for Environmental Medicine, Asthma, and Lung Biology, University of North Carolina (Chapel Hill, NC) for their reviews and valuable suggestions which helped to improve the chapters related to health and toxicological issues (Chapters 3, 4 and 5).

Many thanks also to an anonymous reviewer for devoting valuable time to Chapter 6 on arsenic sources and mobility controls in Taiwan.

We express our gratitude to Thomas Rüde at the Institute of Hydrogeology, RWTH Aachen University (Germany), Anna Karczewska at the Institute of Soil Sciences and Environmental Protection, Wroclaw University of Environmental and Life Sciences (Wroclaw, Poland) and Andrew Meharg at the Institute of Biological and Environmental Sciences, University of Aberdeen, Aberdeen, who contributed significantly to improve Chapter 7 on arsenic in soils and plants. The latter also reviewed Chapter 8 dealing with the potential threat of the use of arsenic-contaminated water in aquaculture and his valuable comments improved this chapter substantially.

We thank Wolfgang H. Höll, formerly at the Institute for Technical Chemistry, Karlsruhe Institute of Technology (Karlsruhe, Germany) and Paul Sylvester at SolmeteX Inc. in Northborough (MA, USA) for their valuable reviews and comments, which significantly contributed to improve Chapter 9 on current treatment methods for As-contaminated water.

Thanks to the previously mentioned reviewers, who jointly provided valuable comments on Chapter 10, about the future areas of study and tasks for the Taiwan arsenic problem.

We are especially grateful to Paul Sylvester who overtook the huge effort to copyedit large parts of the book and his numerous comments which contributed to improve this publication.

We wish to express our sincere thanks to those whose efforts contributed to the high quality of the book.

CHAPTER 1

Taiwan and the global arsenic problem

Most of the problems of hydrologic arsenic contamination around the globe are the result of its mobilization and retention, which occur under natural conditions in a wide variety of natural environmental systems, under both oxidizing and reducing conditions. The contamination of drinking water wells in Taiwan by predominantly geogenic arsenic (As) constitutes a classical, well-studied and well-documented showcase. It is important historically because the first indications of public health problems were not followed up for 20 years and they were far worse than initially realized, and the studies both in terms of the effect on health and the understanding of the causes of the contamination make it a showcase which has stimulated activities elsewhere. Now it is time to see it as more than single standing issue but to analyze it in the context of the global As scenario. This will help us to better understand the occurrence and genesis of aquifers with high As concentrations in groundwater around the world, which are predominantly due to release from geogenic resources. This is despite some specific characteristics of the Taiwan case such as the occurrence of the endemic blackfoot disease (BFD), which is a result of a combination of some yet unconfirmed characteristics. However, worldwide other—not yet detected—sites with the same characteristics may exist. Therefore it is essential and of global interest to identify the causes of the BFD and to search for aquifer similar conditions, where this disease may be expected to occur.

Generally speaking, many of the findings and experiences from the Taiwan case study can be transferred to other parts of the world. This transference must include more than "hard scientific" results, but also political, social and economic ones all in all contributing to mitigate the As problem in the many countries, especially in the developing world, where the problem has not been solved so far, and where many people are still drinking As-contaminated water, many of them without their own knowledge as it happened half a century ago in Taiwan, too. Learning from positive and negative experiences obtained in Taiwan makes this case study to a showcase for people from all over the world who are involved in the As topic in the one or the other form. Seeing Taiwan within the global As scenario will (1) improve the understanding within Taiwan to better understand the As problem of their country and (2) to improve the understanding of the Taiwan case by the international scientific community; all in all increasing awareness among researchers, administrators, policy makers, and company executives and improve the international cooperation on that topic.

1.1 GENERAL INTRODUCTION TO THE ARSENIC PROBLEM

1.1.1 *Origin, release and occurrence of groundwater arsenic*

Arsenic is predominantly released from rocks with primary or secondary As or As-containing minerals due to physical, chemical or microbiological weathering into aqueous and other environments at many places of the world. There are numerous geogenic As sources, including over 200 As-bearing minerals and the As release and its mobility in the different environments is controlled by different geochemical, biogeochemical, geological, hydrogeological, geomorphological and climatic conditions and settings, which we will discuss in more detail in chapter 2. Correspondingly, the areas with aquifers with high As concentration in the groundwater can be classified according to specific similarities especially those regarding their geological, hydrogeochemical and climatic setting.

Thus, young volcanic rocks of Tertiary to recent age, and their weathering products form numerous sedimentary mostly oxidizing aquifers with high concentrations of As in groundwater. Examples are those along the circumpacific volcanic belt including areas such as the cordillera of the Americas; those of the Mediterranean area (e.g., Italy, Greece, Turkey, Canary islands, Morocco) and those of the eastern African rift system. However, in same of the cases, the weathering products derived from rocks of the circumvolcanic belt may be transported by eolic and fluvial processes over large distances and contribute to the sediments of aquifers far away from its original source, and contribute there to the genesis of high As groundwater (e.g., Chaco-Pampean plain of Argentina).

In mineralized areas, As is predominantly released under oxidizing conditions from sulfide minerals. Examples are the Andes and Andean Highlands, the Middle and North American cordillera, the Transmexican volcanic belt, the Appalachian belts (Peters 2008), predominantly Variscian mountains in Europe (e.g., numerous sulfide mineral deposits of Spain, Portugal, Germany, Austria, Czech Republic, Slovakia), Central Balkan peninsula in Siberia (e.g., Dangic and Dangic 2007), Albania (e.g., Lazo *et al.* 2007), Ghana (e.g., Smedley 1996) and Nigeria (e.g., Gbadebo 2005).

Flood and delta plains of Himalayan rivers comprise most of the Asian alluvial aquifers with high concentrations of dissolved As and predominantly reducing conditions. Examples are the Ganges-Brahmaputra-Meghna plain and delta (India and Bangladesh; e.g., Bhattacharya *et al.* 2002a,b, 2006a, 2007a,b), the Indus plain (Pakistan; e.g., Nickson *et al.* 2005), the Irrawady delta (Myanmar; e.g., WRUD 2001), the Red River delta (Vietnam; e.g., Berg *et al.* 2001) and the Mekong river delta and plain (Cambodia and Laos; e.g., Feldman and Rosenboom 2001).

However, As contamination occurs also in Pre-Tertiary geological provinces with rocks such as shales and granites (e.g., Black Forest, Vosges) as well-known rocks, whose weathering contributes to the genesis of As-rich groundwater.

Arsenic-contaminated aquifers in SE Australia are probably related to As from marine sediments (e.g., Smith *et al.* 2003). This source is also discussed in chapter 6 as one of the hypotheses to explain the high As concentrations in the groundwater of Chianan plain in southwestern Taiwan.

At several sites As-rich geothermal waters, which mix with aquifers used for water supply or which flow into surface water bodies form an important source of geogenic As. Numerous examples are found predominantly but not exclusively along active plate boundaries and other tectonically active zones in young volcanic areas, young orogenes, rifts, ocean-spreading zones, etc. Additionally, other geothermal manifestations such as fumaroles or emissions of volcanoes, contribute to the As influx into ground- and surface-water bodies. The identification of geothermal components mixed with cold groundwater in the underground is often difficult, and this is the reason why until now the geothermal component as As source was identified only in a few aquifers used for drinking water purposes as we will see later in section 2.5.1.

Arsenic sources, processes of release into the environment and mobility controls will be treated in more detail in chapter 2.

1.1.2   *Geochemical arsenic mobility controls*

In an oxidizing environment the principal geochemical As mobilization is due to (1) desorption of As at high pH values (pH $\geq$ 8) and (2) sulfide oxidation at low pH values (pH $\ll$ 7). Under reducing conditions, As can be mobilized due to reductive dissolution at circum-neutral pH. However, the released As may be again immobilized if it reaches specific geochemical conditions (e.g., of pH and Eh), and may coprecipitate during the formation of (oxy)hydroxides of iron and other metals, or may adsorb on the surfaces of these or other minerals such as clay minerals. From these secondary As sources, which are very common, the As can be remobilized by different processes, which depend on the local geochemical and other conditions. So in oxidizing environment it can be desorbed under high pH conditions, whereas under reducing condition the reductive dissolution takes place. These antagonists of immobilization and remobilization, which are controlled by the local redox, climate, geomorpological and other conditions, are responsible for the concentration of As, which is bioavailable in the environment, e.g., dissolved in the groundwater. This immobilization may either occur *in situ* or in places far away from the original source e.g., if redox conditions change from oxidizing to reducing. This phenomenon explains many of the examples from the alluvial flood and delta plains of rivers originating from the Himalayas (e.g., Ganges-Brahmaputra-Meghna plain and delta, Indus plain, Irrawady delta, Red River delta, Mekong river delta), where As originated from release of rocks and from where the As is transported by the rivers either in dissolved form or as sediment load to the sites where it is found today, predominantly within coatings of Fe (oxy)hydroxides on sediment particles or dissolved. In these alluvial aquifers with predominantly reducing conditions, As is released due to reductive dissolution of the (oxy)hydroxides.

The second mechanism of As release from metal-(oxy)hydroxides is the one arising under oxidizing aquifer conditions and at pH ≥ 8. This is the principal mechanism that explains the high concentrations of As in the groundwater of extended areas of the Chaco-Pampean plain in Argentine and its continuation into the adjacent plains of Uruguay, Paraguay and Bolivia.

1.1.3   *Other arsenic mobility controls*

The As mobilization is further controlled by the geomorphological, geological and hydrogeological conditions as well as climate, land use pattern, groundwater exploitation as we will discuss in more detail in chapter 2 (e.g., Hasan *et al.* 2007, Mukherjee *et al.* 2007). Arid and semiarid climate can further contribute to the genesis of As-rich groundwater due to evaporative increase of As concentration in water. The group of climate-influenced or climate-controlled examples of aquifers with high concentrations of As in groundwater comprise the Atacama desert (northern Chile; e.g., Borgoño and Greiber 1972, Bundschuh 2008a,b,c, 2009a,b), the Chaco-Pampean plain (Argentina; e.g., Bundschuh *et al.* 2000, 2004, 2008a,b,c, 2009b,c, Bhattacharya *et al.* 2006b), the aquifers of the Carson desert (Nevada; e.g., Fontaine, 1994, Welch and Lico 1998) and the southern San Joaquin valley (California; e.g., Fujii and Swain 1995, Swartz 1995, Swartz *et al.* 1996).

1.1.4   *Remediation of arsenic-contaminated sites*

For many countries the simplest and cheapest solution to the problem of As in drinking water supplies is to switch to another source of supply. For example, this has been the solution in Taiwan once the seriousness of the problem was recognized. But in other situations "remediation" of As-contaminated water can be done and there is no shortage of methods. Their individual benefits, in terms of ease of use and of cost—e.g., in the case of drinking water treatment—depend on numerous factors, such as size of treatment device, As concentration and distribution of species in the raw water, chemical composition and grade of mineralization of the water to be treated, requisite for the remaining As concentration in the purified water, economical constrains, etc. *In-situ* methods comprise, e.g., reactive barriers such as permeable reactive barriers (PRB), colloidal reactive barriers (CRB), and mobile reactive barriers (MRB) whose usefulness strongly depends on the local hydrogeological conditions and the specific remediation tasks. *Ex-situ* methods comprise methods based on membrane technologies (micro-, nano-, pico-membranes and reverse osmosis), adsorption methods which use either natural materials (e.g., limestone, clays, zeolites, biomass) or treated natural materials (e.g., clay or sand particles coated by Fe or Mn oxides or hydroxides) or artificial materials (e.g., activated alumina, ferric (oxy)hydroxides), adsorption/coprecipitation in ferric (oxy)hydroxides e.g., by adding a ferric salt (typically sulfate or chloride), chemical or photochemical or photocatalytic oxidation to oxidize As(III) to As(V), ion exchange (anion or cation exchange), and coagulation methods (e.g., electrocoagulation). For overviews on treatment technologies see e.g., Newcombe and Möller 2006, Pirnie 2000, Bianchelli 2004, Mohana and Pittman 2007, Jekel and Amy 2006. Since many of these technologies comprise very advanced technologies, they are applied in most of the cases at industrial scale to remove As from drinking water. This has been used to solve the As problem in many urban areas and in other areas with access to centralized water supplies. Unfortunately there may still exist urban areas with centralized water supplies, where the drinking water was not tested for the parameter As and therefore no switching of water source or remediation has yet been attempted.

If we look on isolated urban areas, where 'isolated' means that those urban areas do not count on a centralized water supply, and the countryside, the situation is quite different. Especially in countries of transition and in developing countries, many isolated urban areas at the periphery of large cities and in rural areas, comprising smaller cities, towns, villages and dispersed settlements still depend on untreated drinking water containing As in toxic levels. There are different reasons. One of them is that the above-described technologies for As removal from water are not suitable, or not suitably modified for small communities and especially not for dispersed rural settlements, which require simple and low-cost equipment, which the population can handle and maintain by themselves. However, there exist methods that meet the requirements, such as solar oxidation, simplified equipment based on adsorption and membrane technologies. Most of them have been developed in laboratory scale, but only in a limited number of cases these techniques have been tested and proven in the field and been shown to help mitigate As problems. A reason for this lack is that in contrast to urban areas with centralized water supply, few actions have been taken by the authorities and the industry to mitigate the As problem for the isolated urban areas and in the countryside. The rural population, then often depends on As-contaminated water as their only available drinking water resource. This lack of interest has slowed the commercial development and implantation of low-cost remediation methods for small communities or single houses despite that various researchers have developed sustainable solutions for rural areas. Hence, the problem is not a technological one since viable solutions have been already available.

Therefore, the key problem is to convince the responsible authorities and the industry to become interested in the problem. The remediation of the As in groundwater used for drinking for the population of isolated urban areas as well as for rural areas will not be possible if the local and national authorities of the affected countries, along with bilateral or international cooperation agencies, do not recognize that the As contamination of (ground)water in both urban and rural areas creates one of the most important natural health risks of the present century, and that it is their responsibility to solve it.

As an example we may use that from the Chaco-Pampean plain in Argentina (Bundschuh *et al.* 2008a,c, 2009c), where about 12% of the population is living in dispersed settlements consisting of less than 50 inhabitants, which belong mostly to the poorest members of the regional population. Despite the information and the possible solutions provided by numerous national and international scientists (e.g., Luján and Graieb 1994, 1995, Luján 2001, Bhattacharya *et al.* 2002c, Claesson and Fagerberg (2003), Mellano and Ramirez (2004), Storniolo *et al.* (2005), Lindbäck and Sjölin (2006), Morgada *et al.* 2006, 2008, 2009a,b, Bundschuh *et al.* 2007b, 2009c), the local, provincial and national governments, have never seriously tried to solve the problem and to provide the population of rural areas a long-lasting solution.

Another example is those of isolated urban areas close to Buenos Aires city, where in the Matanza-Riachuelo basin at least 200,000 persons depend on drinking water with As concentrations exceeding the national regulatory limit of $10 \mu g L^{-1}$. These isolated urban areas not connected to water distribution network have been recently proven to be an important target to mitigate the problem of As because of the huge population living there, close to industrial zones.

1.1.5 *Human exposure to arsenic and related health effects*

Arsenic affects water resources in several different ways and degrades the suitability of the resource for human consumption, livestock production and irrigation. Further it may contaminate sediments and soils used for cultivation of edible plants, whose As uptake through the food chain exposes humans to As. Besides direct ingestion as drinking water and by water used for food preparation, humans can be exposed to As through the food chain. Consuming edible plants grown on As-contaminated soil or irrigated with As-rich water, shellfish and fish cultivated in ponds fed by water with As high concentrations are important sources for human uptake. The genesis and occurrence of As in drinking ground- and surface waters, soils, plants and air, its bioavailability and propagation in the food chain, its chronic toxicological effects from ingestion and inhalation by animals and humans, and the related public health risk assessment and social and economic consequences have been described from many parts of the world, and every year the As problem is discovered in new countries or regions as we will see in more detail in sections 1.2–1.4.

Since As accumulates in the body, a long-term (chronic) exposure and certain concentrations, may result in carcinogenic and non-carcinogenic health effects such as skin lesions (hyperpigmentation and hyperkeratosis of the palms of the hands and soles of the feet), affections to the nervous system, irritation of respiratory organs and gastrointestinal tract, anemia, liver disorders, vascular illnesses and even diabetes mellitus, skin, lung and bladder cancer (Del Razo *et al.* 2000, Albores *et al.* 2001, Endo *et al.* 2003, Kirk and Sarfaraz 2003, Rossman 2003, Del Razo *et al.* 2005). Further on, chronic As exposure affects the intellectual development of children (e.g., Borja *et al.* 2001, Wasserman *et al.* 2004). In the particular case of Taiwan, the endemic blackfoot disease (BFD) is a consequence of As uptake through drinking water (e.g., Chen *et al.* 1985) as we will discuss in detail in chapters 3 and 4. Estimates of the number of exposed people will be discussed jointly with the discussion of regulatory limits in section 1.5.

1.2 ARSENIC: FROM HISTORY TO TAIWAN

Arsenic has been known as a toxin since ancient time, but it was generally considered to be toxic only at high concentrations until recently. It is, and was, obvious when it was first realized that As could be an acute poison, and when it was isolated from geological deposits, that small concentrations of As would be found. But small amounts of, and exposures to, As, whether due to artificial causes such as secondary effects of mining, or natural contaminants were not considered to have adverse medical effects. Indeed medicinal uses of As go back more than 2000 years (Cullen 2008). In 1792 Dr. Fowler of Edinburgh proposed a 1% solution of As to cure stomach upsets. Fowler's solution still remains in the British Pharmacopeia. That As was an important contaminant of groundwater was shown by Hitchcock (1878) who described the first cases of widespread As poisoning due to consumption of groundwater from the northern Appalachian mountains. This was followed on the medical side by the publication by Hutchinson (1887, 1888) who showed that continued use of Fowler's solution led to skin lesions—dispigmentation, keratosis and cancer. On the geological side, As was described in small areas in Germany (1885) and Poland (1898) (Ravenscroft 2007). However, in the pre-Taiwan period (before the beginning of the 1960s), the most important occurrences of As-rich groundwater and the scientifically best studied case are those of Argentina (from 1913 on) whereas small discoveries of the problem in this period such as those from Canada and New Zealand in the 1930s received only marginal attention. The discoveries of As-contaminated groundwater and surface water in Chile and Mexico, fall about together with those of Taiwan at the beginning of the 1960s. Hence pre-Taiwan studies on As in drinking water resources are few, and only the example of Argentina has derived much scientific attention. Using the circumstance that the discoveries of As in groundwater of Mexico, Chile and Taiwan fall in the same time, it is an interesting approach and task to compare—under consideration of the large differences of the cases, e.g., regarding exposure time to As, social conditions, climate, hydrochemical pattern of the water, etc.—how different these three countries managed their As problem. However these discoveries seemed only of academic concern and the possible adverse medical outcomes were almost universally ignored, until the work in Taiwan. In retrospect this delay of a century in recognition of the seriousness of the problem remains a blot on all concerned.

1.2.1 *Arsenic discoveries in groundwater of Argentina*

The first scientific description of the occurrence of As in groundwater is from Argentina, where the toxic effects on public health were described from the locality of Bell Ville, near Córdoba city and these effects were denominated a "new disease" called "Bell Ville disease" in several publications in the years 1913–1917. In the year 1913, the physician Dr. Mario Goyenechea described for the first time skin lesions in his work "About a new disease detected in Bell Ville" (published in Spanish with the original title "*Sobre una nueva enfermedad descubierta en Bell Ville*") (Goyenechea 1917, Círculo Médico del Rosario 1917) and related it to the uptake of As from groundwater tapped from aquifers of Tertiary and younger eolian loess-type deposits, used for drinking and food preparation. In 1917/18, Dr. Abel Ayerza described exhaustively the cardiovascular and cutaneous manifestations of this disease and named it "Chronic Endemic Regional Arsenicism" in Spanish: "*Arsenicismo Regional Endémico*" (Ayerza, 1917a,b, 1918). In 1921, Reichert and Trelles (1921) described As from many sites of arable lands in Argentina, showing highest contents in the basin of the Río Tercero river in the Pampean plain in the province of Córdoba, and in the north of Santa Fe province. Ayerza (1917) had stated, that from a total population of 8534 inhabitants, 1300 (15%) suffered from the Bell Ville disease. More recently, post-Taiwan, Trelles *et al.* (1970) transcribed the data statistics of the period 1934–1944 registered by the regional hospital of Bell Ville where 511 persons affected by arsenicism were registered. However, much before, the discovery of aquifers with high concentrations of As in the groundwater near to the town of Bell Ville was followed soon (1921) by detections of the problem in other areas of the Chaco-Pampean plain, as in the Río Tercero river basin in the Córdoba province, and in the north of the Santa Fe province (Reichert and Trelles 1921). In 1951, the physician Prof. Enrique E. Tello, in its work entitled "Chronic Endemic Hydroarsenicism (CEHA), its Clinical Manifestations" published in Spanish with the original title "*Hidroarsenisismo Crónico Regional Endémico (HACRE), Sus Manifestaciones Clínicas*" Tello (1951) introduced a new name for the disease which relates it to its source water. In following up investigations, the same author registered 339 patients originating from the provinces of Buenos Aires, Córdoba, Chaco, Santa Fe and Salta (Tello 1986, 1988). Biagini *et al.* (1995) found in the period 1972–1993 other 87 patients originating, the majority of them, from the provinces of Santiago of the Estero, Chaco and Salta. In spite of these excellent investigations, the problem did not achieve the necessary attention by the authorities, and still today the problem remains unresolved in many rural and some urban areas. Other countries did not learn either from the Argentine experiences.

1.2.2 *Arsenic discoveries in groundwater of Mexico*

In Mexico, the chronic exposure to groundwater As, tapped from sedimentary aquifers was described as endemic problem for the first time in the year 1958, in the Lagunera region, which extends over large parts of Durango and Coahuila states (Cebrián *et al.* 1994, Armienta and Segovia 2008). In 1962, 40 serious cases and one death were reported in the urban area of Torreón (state of Coahuila; Castro de Esparza 2009). Later on the presence of As in drinking water supplied from groundwater resources was found to be a problem in many other states comprising Durango, Coahuila, Zacatecas, Morelos, Aguas Calientes, Chihuahua, Puebla,

Nuevo León, Guanajuato, San Luis Potosí where As concentrations have been found that exceed the national regulatory values for drinking water. Based on the old national regulatory limit (50 µg L^{-1}) an estimation of exposed people amounts to about 450,000 (Castro de Esparza 2009, Armienta *et al.* 2008). If we consider the present-day national regulatory limit (25 µg L^{-1}) this number may be significantly higher.

1.2.3   *Arsenic discoveries in ground- and surface-water of Chile*

In northern Chile, the first cases of chronic As-related health effects were reported in 1962 from the city of Antofagasta, as consequence of the 4 years before introducing new drinking water supply, which taps water from the Loa river and its tributaries, which has average As concentration of about 800 µg L^{-1} of As (Bundschuh *et al.* 2009b). The As is predominantly released by weathering/dissolution of volcanic rocks and sulfide ore deposits at the Andean volcanic chain, where As is mobilized following the snow melt and rain, which adds As to the overland flow and infiltrating water and transport it to the rivers and springs, respectively, which originate at the flanks of the Andean mountains and which form the only freshwater resource in northern Chile.

At the beginning of the 1960s, the first skin lesions were detected, especially in children, and later on other As-related health effects were discovered. In the entire NW Chile, a total of 500,000 people were exposed to As through drinking water (Castro de Esparza 2009). In Antofagasta, 130,000 persons were exposed for 12 years to As-containing drinking water until, in 1970, an As removal plant was installed. In 1978, in Kalama, the second largest city in the As-affected area, a removal plant was installed. Both treatment plants solved the problem in these two cities and in some other larger urban areas (Rivara *et al.* 1997). However, in many smaller towns, villages, and isolated houses, the As-contaminated water is still used for drinking water and irrigation purposes (e.g., Chiu Chiu; Smith *et al.* 2000). The Chilean As showcase had a large benefit for health studies since the As exposure time was well defined and, in consequence, the Chilean case was studied intensively through the following decades (Borgoño and Greiber 1972, Pizarro and Balabanoff 1973, Puga *et al.* 1973, Klohn 1974, Borgoño *et al.* 1977, Smith *et al.* 1998, Karcher *et al.* 1999, Ferreccio *et al.* 2000, Sancha and Frenz 2000, Sancha *et al.* 2000, Hopenhayn-Rich *et al.* 2000, Bates *et al.* 2004, Christian and Hopenhayn 2004, Cáceres *et al.* 2005, Bundschuh *et al.* 2009b).

1.2.4   *Arsenic discoveries in groundwater of Taiwan*

In Taiwan, mass poisoning of the population by As from groundwater resources used for drinking water supply occurred since the beginning of the 20th century in southwestern Taiwan (Chianan plain) due to still unknown reasons as we will discuss in chapter 6, before the drinking water supply of most of the affected households was changed from well to tapwater using As-safe water. A unique peripheral vascular disease, blackfoot disease (BFD) and other As-related diseases were the consequence. The disease ("spontaneous gangrene") was first reported in 1954 (Kao and Kao 1954). Awareness of the As problem started during the 1960s when Tseng *et al.* (1968) found skin cancers, 90% benign, in waters NE of Tainan city. These were analyzed widely, by the US EPA for example, and appeared to have a threshold below which they did not occur. But it was not until Chen *et al.* (1985) looked at internal cancers, that the magnitude of the problem became apparent. The fatality rate was 100 times larger than the fatal skin cancers and seemed to have no threshold below which they did not occur. Although the first studies found high levels of As (350–1100 µg L^{-1}) in drinking water from wells (e.g., Chen and Wu 1962) they also found the presence of algae (e.g., Chen *et al.* 1962). However, the high As concentration in the groundwater was suspected to be the most likely causal factor and not the algae (Chen *et al.* 1962). A dermatological survey performed in 1965 in 37 villages of the BFD area (Chianan plain) found a close correlation between the occurrence of health effects such as BFD, hyper-pigmentation, keratosis, and skin cancer and confirmed the exposure to As as trigger (Tseng *et al.* 1968). A detailed description of the BFD history will be presented in chapter 3. Nonetheless public health authorities in other parts of the world were skeptical. Committees of the US EPA did not mention these data for another 4 years! But the discovery of As-related lesions in other countries without the algae present in Taiwan convinced the waverers.

1.3   ARSENIC: FROM TAIWAN TO THE END OF THE 20TH CENTURY

In the post-Taiwan period, the As problem was discovered in an increasing number of regions, where the problem had not been described so far. So the problem was described in Peru in the beginning of the 1990s from the vicinity of Ilo valley (Castro de Esparza 2009, Bundschuh *et al.* 2008c, 2009b). In Argentina, new regions with high concentrations of As in groundwater were detected predominantly in the Chaco-Pampean plain, but also in the Andean highlands (Astolfi *et al.* 1981, 1982, Nicolli *et al.* 1989, 2001, 2009, Smedley *et al.* 2005, Bundschuh *et al.* 2004, 2008a,b,c, 2009b). Until the end of the 20th century, As-rich groundwater was confirmed in the provinces of Córdoba, Río Negro, Tucumán, Santa Fe, La Pampa, Santiago del Estero, Salta and Chaco (Ministerio de Salud 2001). Different epidemiological studies prove severe health effects due to As uptake through groundwater used for drinking purposes and food preparation not only in the Chaco-Pampean plain in the Córdoba province (Hopenhayn-Rich *et al.* 1996, 1998) and in the Chaco province (Concha *et al.* 1998a,b), but also in the Andean highlands, e.g., in San Antonio de Los Cobres (Salta province; Concha *et al.* 1998a,b).

However, it was not before half a century after the description in Argentina, and two decades after the Taiwan As contamination was detected that the As problem started in the Bengal delta, where finally it achieved world interest. Although the Bengal delta As history starts much earlier when UNICEF launched in the 1970s a big program to provide the rural population of Bangladesh with microbe-free drinking water. In this program, it was attempted to change the drinking water supply coming from surface- and rainwater, used until then, to groundwater exploited from shallow aquifers by using hand pumps. However, the quality of this groundwater had not been tested for As and caused, 10 years later, toxicological effects in large parts of the population. Although the first cases of As intoxication were reported in 1983 by Krishna Chandra Saha in the Indian part of Bengal delta (West Bengal) and in 1987 in the Bangladeshi part (Saha 1995, Mandal *et al.* 1996), the problem was not considered to be serious and did not achieve attention until 1995 when Dipankar Chakraborti, an Indian scientist, attributed the severeness of the mass poisoning by As from groundwater consumption, and dramatically brought, and continues to bring, the topic of the risks of drinking groundwater containing As to the attention of the whole world and excited wide scientific and public interest. It was Chakraborti together with Quamruzzaman in Dhaka Community Hospital in Bangladesh, who organized in 1998 the first of many yearly conferences on the problem which attracted and inspired scientists and clinicians worldwide. From 1998 on, the international community, which had urged a use of groundwater instead of the traditional surface water, accepted some responsibility for the disaster, and extended international aid efforts to mitigate the As problem in Bangladesh and adjacent West-Bengal, India, where many treatment methods have been developed and tested, and brought the groundwater As topic to wide scientific and public interest. About 30% of the 10 million groundwater wells in Bangladesh have been contaminated with As and one third to one half of this huge number of As-contaminated wells are still in use in 2010, and most are still used for drinking. Most of the 50 million people in risk in Bangladesh, which corresponds to nearly half of the population, are still drinking As-contaminated water. As a result it has been widely said that the As problem in Bangladesh is the largest mass poisoning in world history, and the effect on health exceeds the effects of the Chernobyl accident by one or two orders of magnitude (Wilson 2006).

Even as late as 1994 international bodies were still urging, and funding, 'tube wells" pumping groundwater without testing for As. The medical problem of As in drinking water supplied from groundwater and to a minor extent from surface water resources was considered to be a problem in limited areas and of the aforementioned few countries. Hence, in many countries, especially in the developing world, As was not included as a parameter of standard drinking water analysis. As a consequence, there was no worldwide search and aquifers with high concentrations of dissolved As were often detected by accident, either as a consequence of scientific investigations performed originally for other purposes or by observation of a significant increase of As-related lesions. In the last two decades of the 20th century several sites with As in groundwater were described as in northwest China. Wang (1984) reported on concentrations in the Xinjiang Uyghur Autonomous Region (NW China). Dr. Zheng-Dong Luo of the Huhhot Sanitation and Anti-epidemic station had been troubled for many years by unexplained lesions in villages in the Huhhot plain in the Mongol Autonomous Region (Inner Mongolia). When he saw the data reported (in Chinese) from Taiwan he searched for, and found, As in the groundwater. The data are published in a series of papers in Chinese (Luo *et al.* 1993a,b, Zhang *et al.* 1994). The data have since been more extensively studied with American collaborators (Luo *et al.* 1997, Lamm *et al.* 2007). Later As was found in the neighboring province of Shanxi (northern China; e.g., Wang 1984, Wang and Huang 1994, Niu *et al.* 1997, Smedley *et al.* 2001). From then onward, a study of possible adverse effects on health and a geological study of the presence of As have proceeded hand in hand. While the search for As in groundwater is more sensitive than the search for skin lesions and cancers, it is evidently the concern for the latter that is now the main driving force for the former.

In the 1990s, As concentrations of As above 50 µg L^{-1} have been detected in groundwater from alluvial aquifers in the southern part of the Great Hungarian Plain (Pannonian basin) (Varsányi *et al.* 1991). Also in the decade of the 1990s, elevated As concentrations have been identified at different sites of the USA; they are summarized by Welch *et al.* (1988, 2000) (see also chapter 2). Arsenic also occurs in groundwater from the principal Cenozoic sedimentary basins of Spain Duero, Ebro, and the Madrid Tertiary detrital aquifer (for summaries see Cama *et al.* 2008) and at different sites of the USA (for summaries see Welch *et al.* 1988, 2000).

In the period 1970s to 1990s, several sites where aquifers with high concentrations of As in groundwater are related to predominantly sulfidic ore deposits and/or mining activities were described. In southern Thailand in 1987 a severe case of As poisoning related to mining activity was detected in the Ron Phibun district (Nakhon Si Thammarat province). There about 1000 people were affected by As-related skin disorders (Williams 1996, Choprapawon and Rodcline 1997). The occurrence of As in groundwater in Ghana (Smedley 1996), Burkina Faso (De Jong and Kikietta 1980), Zimbabwe and South Africa were related to mining activities and/or the presence of sulfidic ore bodies, which also explain groundwater As concentrations at many other sites of mostly very limited size and mostly no regional importance as can be found many examples in classical mining areas in sulfidic ore deposits such as in Mexico (Armienta *et al.* 2008), the Iberian peninsula (Cama *et al.* 2008).

1.4   ARSENIC IN THE 21ST CENTURY—RECOGNIZING GROUNDWATER ARSENIC AS A GLOBAL PROBLEM

In half of the 70 countries where the problem of aquifers with high As in groundwater is now known, the problem was only detected within the last ten years (Ravenscroft 2007, Ravenscroft *et al.* 2009). These new discoveries of areas in natural hydrologic systems with elevated concentrations of As of predominantly geogenic origin show that at least 230 sites distributed over all the habited continents are affected worldwide in a significant manner by the problem (Ravenscroft 2007, Ravenscroft *et al.* 2009, Nriagu *et al.* 2007) (Fig. 1.1);

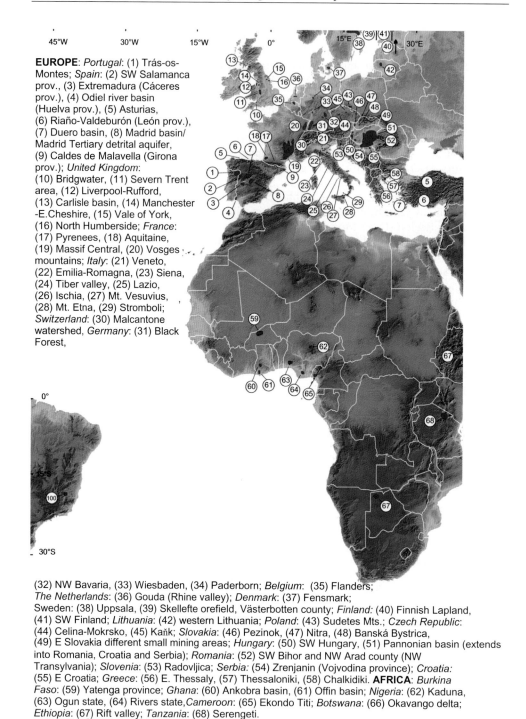

EUROPE: *Portugal*: (1) Trás-os-Montes; *Spain*: (2) SW Salamanca prov., (3) Extremadura (Cáceres prov.), (4) Odiel river basin (Huelva prov.), (5) Asturias, (6) Riaño-Valdeburón (León prov.), (7) Duero basin, (8) Madrid basin/ Madrid Tertiary detrital aquifer, (9) Caldes de Malavella (Girona prov.); *United Kingdom*: (10) Bridgwater, (11) Severn Trent area, (12) Liverpool-Rufford, (13) Carlisle basin, (14) Manchester -E.Cheshire, (15) Vale of York, (16) North Humberside; *France*: (17) Pyrenees, (18) Aquitaine, (19) Massif Central, (20) Vosges mountains; *Italy*: (21) Veneto, (22) Emilia-Romagna, (23) Siena, (24) Tiber valley, (25) Lazio, (26) Ischia, (27) Mt. Vesuvius, (28) Mt. Etna, (29) Stromboli; *Switzerland*: (30) Malcantone watershed, *Germany*: (31) Black Forest,

(32) NW Bavaria, (33) Wiesbaden, (34) Paderborn; *Belgium*: (35) Flanders; *The Netherlands*: (36) Gouda (Rhine valley); *Denmark*: (37) Fensmark; Sweden: (38) Uppsala, (39) Skellefte orefield, Västerbotten county; *Finland:* (40) Finnish Lapland, (41) SW Finland; *Lithuania*: (42) western Lithuania; *Poland*: (43) Sudetes Mts.; *Czech Republic*: (44) Celina-Mokrsko, (45) Kaňk; *Slovakia*: (46) Pezinok, (47) Nitra, (48) Banská Bystrica, (49) E Slovakia different small mining areas; *Hungary*: (50) SW Hungary, (51) Pannonian basin (extends into Romania, Croatia and Serbia); *Romania*: (52) SW Bihor and NW Arad county (NW Transylvania); *Slovenia*: (53) Radovljica; *Serbia*: (54) Zrenjanin (Vojvodina province); *Croatia:* (55) E Croatia; *Greece*: (56) E. Thessaly, (57) Thessaloniki, (58) Chalkidiki. **AFRICA**: *Burkina Faso*: (59) Yatenga province; *Ghana*: (60) Ankobra basin, (61) Offin basin; *Nigeria*: (62) Kaduna, (63) Ogun state, (64) Rivers state,*Cameroon*: (65) Ekondo Titi; *Botswana*: (66) Okavango delta; *Ethiopia*: (67) Rift valley; *Tanzania*: (68) Serengeti.

Figure 1.1.   World distribution of aquifers and surface water bodies with high concentrations of dissolved As as known today. Adapted from Bundschuh and Litter (2010) based on

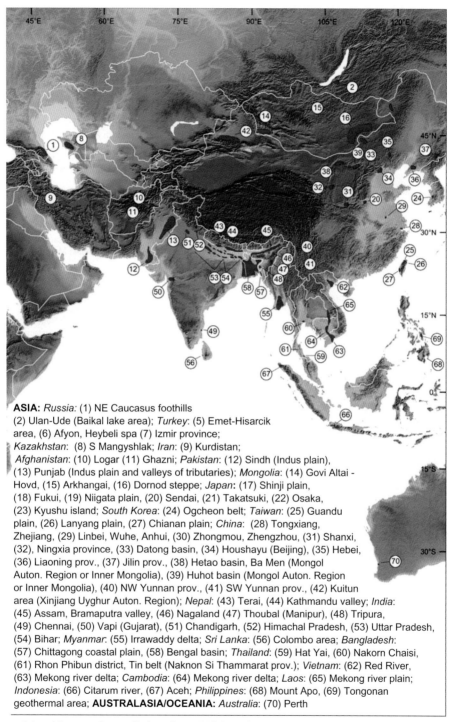

ASIA: *Russia:* (1) NE Caucasus foothills
(2) Ulan-Ude (Baikal lake area); *Turkey:* (5) Emet-Hisarcik
area, (6) Afyon, Heybeli spa (7) Izmir province;
Kazakhstan: (8) S Mangyshlak; *Iran:* (9) Kurdistan;
Afghanistan: (10) Logar (11) Ghazni; *Pakistan:* (12) Sindh (Indus plain),
(13) Punjab (Indus plain and valleys of tributaries); *Mongolia:* (14) Govi Altai -
Hovd, (15) Arkhangai, (16) Dornod steppe; *Japan:* (17) Shinji plain,
(18) Fukui, (19) Niigata plain, (20) Sendai, (21) Takatsuki, (22) Osaka,
(23) Kyushu island; *South Korea:* (24) Ogcheon belt; *Taiwan:* (25) Guandu
plain, (26) Lanyang plain, (27) Chianan plain; *China:* (28) Tongxiang,
Zhejiang, (29) Linbei, Wuhe, Anhui, (30) Zhongmou, Zhengzhou, (31) Shanxi,
(32), Ningxia province, (33) Datong basin, (34) Houshayu (Beijing), (35) Hebei,
(36) Liaoning prov., (37) Jilin prov., (38) Hetao basin, Ba Men (Mongol
Auton. Region or Inner Mongolia), (39) Huhot basin (Mongol Auton. Region
or Inner Mongolia), (40) NW Yunnan prov., (41) SW Yunnan prov., (42) Kuitun
area (Xinjiang Uyghur Auton. Region); *Nepal:* (43) Terai, (44) Kathmandu valley; *India:*
(45) Assam, Bramaputra valley, (46) Nagaland (47) Thoubal (Manipur), (48) Tripura,
(49) Chennai, (50) Vapi (Gujarat), (51) Chandigarh, (52) Himachal Pradesh, (53) Uttar Pradesh,
(54) Bihar; *Myanmar:* (55) Irrawaddy delta; *Sri Lanka:* (56) Colombo area; *Bangladesh:*
(57) Chittagong coastal plain, (58) Bengal basin; *Thailand:* (59) Hat Yai, (60) Nakorn Chaisi,
(61) Rhon Phibun district, Tin belt (Naknon Si Thammarat prov.); *Vietnam:* (62) Red River,
(63) Mekong river delta; *Cambodia:* (64) Mekong river delta; *Laos:* (65) Mekong river plain;
Indonesia: (66) Citarum river, (67) Aceh; *Philippines:* (68) Mount Apo, (69) Tongonan
geothermal area; **AUSTRALASIA/OCEANIA:** *Australia:* (70) Perth

Figure 1.1.   (*Continued*) compilations from: Welch *et al.* 2000, Smedley 2006, Ravenscroft 2007, Chandrasekharam and Bundschuh (2008), Bundschuh *et al.* 2008a, 2009a, from the references

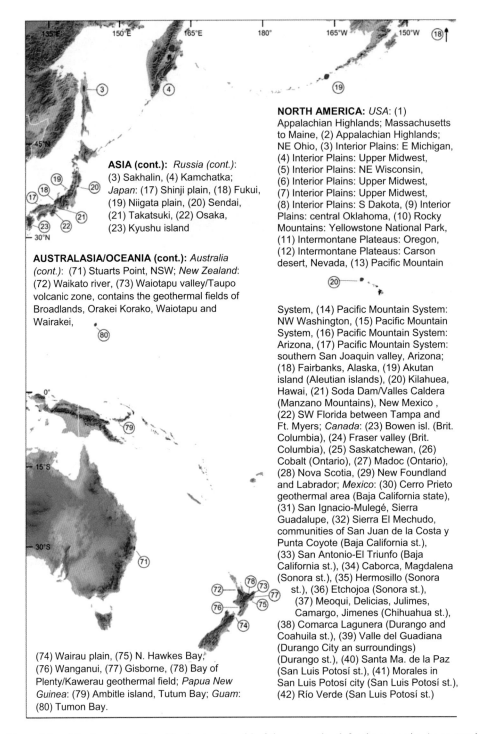

ASIA (cont.): *Russia (cont.)*:
(3) Sakhalin, (4) Kamchatka; *Japan*: (17) Shinji plain, (18) Fukui, (19) Niigata plain, (20) Sendai, (21) Takatsuki, (22) Osaka, (23) Kyushu island

AUSTRALASIA/OCEANIA (cont.): *Australia (cont.)*: (71) Stuarts Point, NSW; *New Zealand*: (72) Waikato river, (73) Waiotapu valley/Taupo volcanic zone, contains the geothermal fields of Broadlands, Orakei Korako, Waiotapu and Wairakei,

(74) Wairau plain, (75) N. Hawkes Bay, (76) Wanganui, (77) Gisborne, (78) Bay of Plenty/Kawerau geothermal field; *Papua New Guinea*: (79) Ambitle island, Tutum Bay; *Guam*: (80) Tumon Bay.

NORTH AMERICA: *USA*: (1) Appalachian Highlands; Massachusetts to Maine, (2) Appalachian Highlands; NE Ohio, (3) Interior Plains: E Michigan, (4) Interior Plains: Upper Midwest, (5) Interior Plains: NE Wisconsin, (6) Interior Plains: Upper Midwest, (7) Interior Plains: Upper Midwest, (8) Interior Plains: S Dakota, (9) Interior Plains: central Oklahoma, (10) Rocky Mountains: Yellowstone National Park, (11) Intermontane Plateaus: Oregon, (12) Intermontane Plateaus: Carson desert, Nevada, (13) Pacific Mountain System, (14) Pacific Mountain System: NW Washington, (15) Pacific Mountain System, (16) Pacific Mountain System: Arizona, (17) Pacific Mountain System: southern San Joaquin valley, Arizona; (18) Fairbanks, Alaska, (19) Akutan island (Aleutian islands), (20) Kilahuea, Hawai, (21) Soda Dam/Valles Caldera (Manzano Mountains), New Mexico , (22) SW Florida between Tampa and Ft. Myers; *Canada*: (23) Bowen isl. (Brit. Columbia), (24) Fraser valley (Brit. Columbia), (25) Saskatchewan, (26) Cobalt (Ontario), (27) Madoc (Ontario), (28) Nova Scotia, (29) New Foundland and Labrador; *Mexico*: (30) Cerro Prieto geothermal area (Baja California state), (31) San Ignacio-Mulegé, Sierra Guadalupe, (32) Sierra El Mechudo, communities of San Juan de la Costa y Punta Coyote (Baja California st.), (33) San Antonio-El Triunfo (Baja California st.), (34) Caborca, Magdalena (Sonora st.), (35) Hermosillo (Sonora st.), (36) Etchojoa (Sonora st.), (37) Meoqui, Delicias, Julimes, Camargo, Jimenes (Chihuahua st.), (38) Comarca Lagunera (Durango and Coahuila st.), (39) Valle del Guadiana (Durango City an surroundings) (Durango st.), (40) Santa Ma. de la Paz (San Luis Potosí st.), (41) Morales in San Luis Potosí city (San Luis Potosí st.), (42) Río Verde (San Luis Potosí st.)

Figure 1.1.   (*Continued*) mentioned in chapters 1 and 2 of the present book for the respective As-contaminated sites and from unpublished information of the authors. Abbreviations: st.: state, prov.:

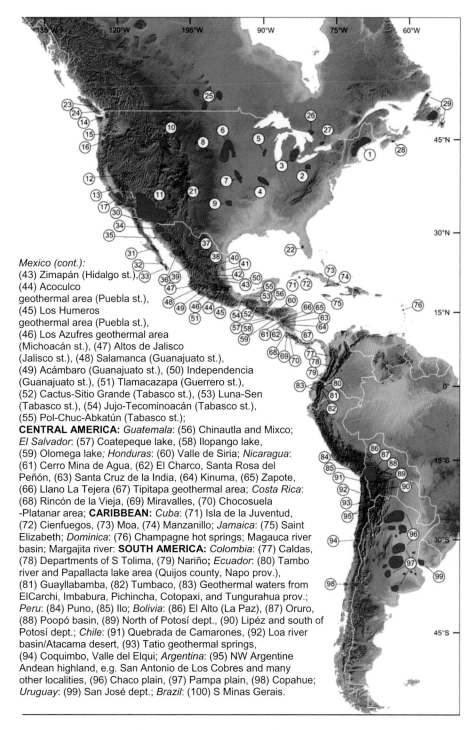

Mexico (cont.):
(43) Zimapán (Hidalgo st.), (44) Acoculco geothermal area (Puebla st.), (45) Los Humeros geothermal area (Puebla st.), (46) Los Azufres geothermal area (Michoacán st.), (47) Altos de Jalisco (Jalisco st.), (48) Salamanca (Guanajuato st.), (49) Acámbaro (Guanajuato st.), (50) Independencia (Guanajuato st.), (51) Tlamacazapa (Guerrero st.), (52) Cactus-Sitio Grande (Tabasco st.), (53) Luna-Sen (Tabasco st.), (54) Jujo-Tecominoacán (Tabasco st.), (55) Pol-Chuc-Abkatún (Tabasco st.);
CENTRAL AMERICA: *Guatemala:* (56) Chinautla and Mixco; *El Salvador:* (57) Coatepeque lake, (58) Ilopango lake, (59) Olomega lake; *Honduras:* (60) Valle de Siria; *Nicaragua:* (61) Cerro Mina de Agua, (62) El Charco, Santa Rosa del Peñón, (63) Santa Cruz de la India, (64) Kinuma, (65) Zapote, (66) Llano La Tejera (67) Tipitapa geothermal area; *Costa Rica:* (68) Rincón de la Vieja, (69) Miravalles, (70) Chocosuela -Platanar area; **CARIBBEAN:** *Cuba:* (71) Isla de la Juventud, (72) Cienfuegos, (73) Moa, (74) Manzanillo; *Jamaica:* (75) Saint Elizabeth; *Dominica:* (76) Champagne hot springs; Magauca river basin; Margajita river: **SOUTH AMERICA:** *Colombia:* (77) Caldas, (78) Departments of S Tolima, (79) Nariño; *Ecuador:* (80) Tambo river and Papallacta lake area (Quijos county, Napo prov.), (81) Guayllabamba, (82) Tumbaco, (83) Geothermal waters from ElCarchi, Imbabura, Pichincha, Cotopaxi, and Tungurahua prov.; *Peru:* (84) Puno, (85) Ilo; *Bolivia:* (86) El Alto (La Paz), (87) Oruro, (88) Poopó basin, (89) North of Potosí dept., (90) Lipéz and south of Potosí dept.; *Chile:* (91) Quebrada de Camarones, (92) Loa river basin/Atacama desert, (93) Tatio geothermal springs, (94) Coquimbo, Valle del Elqui; *Argentina:* (95) NW Argentine Andean highland, e.g. San Antonio de Los Cobres and many other localities, (96) Chaco plain, (97) Pampa plain, (98) Copahue; *Uruguay:* (99) San José dept.; *Brazil:* (100) S Minas Gerais.

Figure 1.1.   (*Continued*) province, dept.: department. The digital elevation model of the world PIA3388 is courtesy of NASA/JPL-Caltech (http://photojournal.jpl.nasa.gov).

this leads to the expectation that there will be a problem at many other sites. These massive findings during the first decade of the 21st century changed finally the outdated opinion of very limited occurrence of the As problem (such as in Bangladesh, Argentina, Mexico, Chile and Taiwan) and groundwater As finally was recognized to be a global problem. Some examples: in Latin America, elevated concentrations of As were discovered in groundwater of Nicaragua (1996/2000), Bolivia (2001), El Salvador (2005), Ecuador (2005), Honduras (2006) and Baja California State (Mexico) (2006), and recently in Uruguay, Colombia, Guatemala, Costa Rica, Cuba, and Venezuela (Bundschuh *et al.* 2008a, 2009a). In 2008, groundwater with high As concentrations used for drinking water supply was reported in Turkey (Izmir). In Asia, in the years 2000–2002, high As concentrations have been detected in groundwater of six more nations: Cambodia and Laos (e.g., Feldman and Rosenboom 2001), Pakistan (e.g., Nickson *et al.* 2005), Myanmar (e.g., WRUD 2001), Vietnam (e.g., Berg *et al.* 2001) and Nepal (Tandukar *et al.* 2001, 2005, 2006, Shrestha *et al.* 2003). In India, new As-affected areas were delimited: Bihar (year 2002; Chakraborti *et al.* 2003), Uttar Pradesh (year 2003; Ahamed *et al.* 2006), Jharkhand (year 2004; Bhattacharjee *et al.* 2005), and Assam (year 2004; Chakraborti *et al.* 2004).

1.5   REGULATION OF ARSENIC CONTENTS IN DRINKING WATER AND ITS IMPACT ON THE EXPOSED POPULATION

The global spreading of As-contaminated sites, together with the fact that every year new regions affected by As are discovered, highlight the global importance of the problem. Consequently, studies of chronic toxicological effects of As in drinking water were strengthened and as result of these investigations the regulatory bodies became aware of the severeness of the problem and started to introduce or to strengthen permissible limits of As for safe drinking water consumption. The World Health Organization (WHO) provisionally reduced the As guideline value for drinking water from 50 to 10 µg L^{-1} in 1993 (WHO 1993), which became Maximum Contaminant Level (MCL) in 2001 (WHO 2001). For some 90 years clinicians, toxicologists and academic scientists have wrestled with the effects of poisonous substances, and adverse exposures at low doses. Starting with radiation in 1927 when ICRP recommended that no one be exposed to radiation without expectation of some benefit, this has expanded to "genotoxic" carcinogenic chemicals whose primary action is on the cell. But Crump *et al.* (1976) pointed out that the same argument applies to any substance which produces an adverse outcome indistinguishable from one which occurs naturally. The full implications of this have been slow to be realized but if applied to As, the level of As in drinking water would have to be lower than 1.5 ng L^{-1} to meet the low level of risk the US EPA and many international bodies have used for regulation of other toxic chemicals such as trichloroethylene (A. Fan 1992, *pers. commun.*). Practically this is unattainable and there is no pressure to further reduce the limits for As in drinking water the required factor of 7000 below the new 10 µg L^{-1}. The reduction of the dose limit to 10 µg L^{-1} already increases the number of population exposed to As above the regulatory level by a several fold and classifies several countries that until now had "safe" As levels as having unsafe concentrations.

This new As limit for drinking water became law in Jordan (1991), Japan (1993), Namibia, (1994), Syria (1994), Nicaragua (1994), Honduras (1995), Costa Rica (1997), El Salvador (1997), European Union (1998), Mongolia (1998), Colombia (1998), Guatemala (1998), Panama (1999), Peru (1999), Laos (1999), Taiwan (2000), USA (2001), Chile (2008), Argentina (2008), Vietnam (2008) and other countries will follow. Most of them still maintain the 50 µg L^{-1} limit, some others have regulatory limits in between the 10 and 50 µg L^{-1}: Canada 0.025 µg L^{-1} (1999), Mexico 0.025 µg L^{-1} (2005), only Australia has a limit lower than 10 µg L^{-1} (7 µg L^{-1} since 1996). Further on, even if the 10 µg L^{-1} limit for water and other As limit values are not adopted by law, it may be essential for many industries to adopt this value, in order to maintain their exports of goods (e.g., food industry and agricultural products) as well as not to risk their tourism industry.

The delay in the USA in accepting the lower standard is instructive. Strangely, advisory committees of the US EPA ignored the paper of Chen *et al.* (1985) until 1992. A simple plot of the internal cancer rates versus As exposure in Taiwan showed a clear linear dose-response relationship (Byrd *et al.* 1996), albeit in an ecological study. Many scientists from 1990 on calculated the correct standard using criteria used by the US EPA for other carcinogens much less dangerous, and as noted above suggested an extraordinarily low number for a standard of 1.5 ng L^{-1}. A group filed a lawsuit which resulted in a court order, requiring the US EPA to promulgate a new standard, a task which took 10 years until 2001. Meanwhile it was discussed at many meetings such as a series organized in San Diego California by Chappell *et al.* (1994, 2001). In this and subsequent meetings a surprisingly large group of toxicologists was insisting that the adverse medical effects were only present at high concentrations, and implicitly rejecting a need for a large safety factor. The simple calculation would have entailed a reduction of a factor of 30,000 from the old standard and 7000 below the new one. This reluctance was in spite of the fact that the USA, a wealthy country, can easily reduce the As levels without excessive financial burden.

The consequences of neglect become evident when we look e.g., on the example of Nicaragua. Here, the national water company drilled in 1994 a well in El Zapote in order to supply the population with clean drinking water. However, as in Bangladesh, some decades before, the exploited water was not analyzed for As, resulting in a poisoning of the population for 2 years with As-rich (1320 µg L^{-1}) drinking water (Barragne 2004, Bundschuh *et al.* 2007a, Altamirano Espinoza and Bundschuh 2009). Similar negligence can be observed in many other countries.

Although toxic manifestations and health impacts due to As exposure have been reported from many sites around the globe, at present it is not possible to precisely assess the number of persons potentially exposed worldwide. However, there exist some regional or countrywide estimations. The largest population exposed to groundwater As is in the area of Bengal delta (India and Bangladesh) (Pearce 2003). Bhattacharjee (2007) has described this situation as 'humanity's biggest mass poisoning' since in Bangladesh about 50 million people are at risk of drinking water with toxic As levels, which corresponds to nearly half of the countries population (Mandal *et al.* 1996, Pearce 2003). Wilson (2006) has noted that the effects on the population exceed the effects of the Chernobyl accident by one or two orders of magnitude, without the concomitant public attention. No person has yet questioned this surprising comparison. In Latin America, at least 4.5 million people are exposed to drinking water with more than 50 µg L^{-1} of As (Castro de Esparza 2009) which is still in many countries the regulatory limit. If the 10 µg L^{-1} limit is considered, value increasingly adapted by the countries of the region (see section 1.1.6), then the number of exposed people increases to around 14 million. This means that the number of people exposed to As with more than 50 µg L^{-1} of As are by several fold higher if we apply the 10 µg L^{-1} limit. This becomes evident if we look on the example of Argentina. In this country between 1.2 and 2 million people are exposed to the As problem if we consider the 50 µg L^{-1} limit for As and about 3 to 8 million people if we consider the 10 µg L^{-1} national limit, which was introduced in 2007 (CAA 2007, Bundschuh *et al.* 2008c, Bundschuh 2009b, Morgada *et al.* 2009b).

This sharply increasing awareness of the As problem, which could be especially observed during the present decade, makes it necessary to reassess the Taiwan As case, which was thought at its time of reconnaissance to be only a local problem. Now, knowing, that the environmental contamination by As from predominantly geogenic sources, is a principal problem and challenge for humanity, we need to re-elaborate the findings of the Taiwan As case from its beginning to present, all in all to contribute to improve our understanding and to evaluate which findings and experiences can be transferred to other affected sites around the world and what can be learned from the positive and negative experience and mistakes made by this country and the international aid agencies, which in that time did not recognize As as a worldwide problem, hence underestimating it.

1.6   WHY WAS THE "TAIWAN SIGNAL" NOT IMMEDIATELY RECOGNIZED WORLDWIDE?

As with all other discoveries of large sites affected by aquifers with concentrations of dissolved As at toxic levels, such as Argentina at the beginning of the 20th century, the Taiwan case, was considered as a unique situation. This was despite that here the discovery of the human As exposure due to uptake from drinking water was made at about the same time as in Chile and Mexico, which should have inspired already in the 1960s the scientific community that groundwater As may be a global problem. Also the example of the Bengal delta case in the 1990s did not contribute to identify the groundwater As problem as a global issue. So, none of the cases were recognized as a signal for the need to extend the knowledge (by testing drinking water) of the As problem on a global scale. Among the reasons were missing global cooperation, missing information and still the outdated thinking that the As occurrences are only limited to few sites. This missing awareness delayed the recognition of the As problem as a widespread issue and hence a global problem, which deserves global solutions. However this should have been assumed because the specific geological conditions where As problems are generally found should have implied to think on a widespread distribution of As-affected aquifers. So, the occurrence of As in the province of Córdoba (Argentina) about 100 years ago, should have implied that the problem also exists in other areas with similar geological conditions, which are either allocated in the Andean area and the foreland areas affected by weathering products of volcanic rocks, as later on the discoveries of high concentrations of As in groundwater in different Latin American countries have proven. The same is valid for the SE Asian countries, where the weathering products of the Himalayas and the SE continuation as As source areas have transported As by rivers to the Bengal delta. When investigating the occurrence of high concentrations of As in these alluvial sediments during decades, it took long time that the idea rose, that As may be also present in the alluvial plains of other rivers originating from the same mountain range, such as in the delta and flood plains of the Indus, Irrawady, Red river, and Mekong, which comprise most of the Asian aquifers with high concentrations of dissolved As as the discoveries during the last 10 years have proven.

1.7   WHY DOES ARSENIC CONTINUE TO AFFECT PEOPLE WORLDWIDE?

Despite the fact that the United Nations have declared the access of the population from all around the world to safe drinking water as a human right, a significant percentage of population, especially those from isolated urban areas, and those from rural areas, consumes water with toxic concentrations of As and other chemical or microbiological contaminants. This presence of As in water resources used for drinking (humans and livestock) and irrigation is therefore a principal environmental problem, whose mitigation is prerequisite for any social and economic development of the affected regions.

Reasons, why As continues to affect the populations in many regions of the world, are manifold:

- Some people are still convinced that As problems are only local problems.
- A belief among many toxicologists that there is a definite threshold below which chronic As poisoning will not occur. This belief is accompanied by a refusal to allow the adequate margin of safety that is used for almost all other toxins.
- As a consequence, As is still not a standard obligatory parameter for drinking water analysis.

- The failure by the national regulatory bodies to adopt the new 10 µg L^{-1} limit for As in drinking water, which was made in 2001 Maximum Contaminant Level (MCL) by the World Health Organization (WHO 2001).
- Failure to mitigate the As pollution, or implementation of remediation methods, which are technically or economically not suitable, or which does population not accept.
- Isolated urban areas (without centralized water supply) and rural areas are neglected by the authorities and development is hindered due to missing political interest, making these areas the most affected areas to the As problem.
- Although that human uptake of As through drinking water is now widely studied, much less studies of As intake from food, e.g., vegetables and cereals irrigated by As-rich water or fish and shellfish exposed to As-rich water can be found in different places of the world including Taiwan, where commercial fish and shellfish farming is an important economic activity which is partly practiced in the areas where groundwater used for these purposes has high As concentrations.
- Lacking knowledge about As sources, but especially about mobilization processes and their triggering factors including the kinetics of these mobilization processes.
- The public pressure from victims is limited, especially in the USA. Since no one is to blame for putting the As there, there is no one to sue!

1.8 DEMANDS FOR INTERNATIONAL COOPERATION AND NETWORKING

In consequence of the deficiencies listed in section 1.7 and the there from resulting demands, more detailed and complete knowledge and integral understanding on the As issue is required. A first simple, but urgent need is to include worldwide As as a parameter for standard drinking water quality analysis, and the adaption of the 10 µg L^{-1} limit for As in drinking water, neither of which demands are fulfilled by the regulations of many countries. Further on, it calls for common global approaches and exchange of information/experiences from regions where the problem and its mitigation have previously been studied. Thereby measures for problem mitigation comprise two groups: (1) tapping alternative water sources which includes the approach of targeting As-safe aquifers (or aquifer sections) within As-contaminated areas and (2) treatment of As-contaminated water (or combinations of both). Both measures require a detailed understanding of the As source and its release into the groundwater as well as a conceptual flow and solute transport model of the aquifer system to understand the present As occurrence in the well water and to forecast future changes of the behavior, e.g., caused by increasing exploitation of the aquifer, which may change the groundwater flow field and the hydrochemical conditions and may cause changes of As concentration within the same well.

As we have already mentioned and as we will discuss in more detail in chapter 2, the occurrences aquifers with high As concentrations in groundwater can be attributed to particular geological settings and climate conditions. That means that if we detect (or not detect) As-rich groundwater in a specific environment, we can also expect to find (or not to find) it in another country or region with similar geological and climate conditions. This likelihood approach of finding or not finding was also used by Ravenscroft (2007) on a global scale to determine probability distributions of As occurrence. As example we may use the case of Central America and those of the large plains the southern part of South America. We have already mentioned that in western Nicaragua some few spots with high concentrations of groundwater As were discovered in shallow aquifers, whose sediments correspond to weathering products from volcanic Cenozoic rocks containing different hydrothermal deposits. However the similar geological conditions found in large parts of Nicaragua as well in neighboring countries would make discovery in these vast regions probable. The same is valid if the Argentine Chaco-Pampean plain and its extension into the neighboring countries Bolivia, Paraguay and Uruguay, where sediments of the same origin as in the Chaco-Pampean plain occur are considered. The finding of numerous sites in at present 16 Argentine provinces (Buenos Aires, Catamarca, Chaco, Chubut, Córdoba, Jujuy, La Pampa, La Rioja, Mendoza, Salta, Santa Fe, Santiago del Estero, San Juan, San Luis, Río Negro, Tucumán; Bundschuh *et al.* 2008c), together with the similar geological conditions makes probable the discovery at other sites where the problem was not yet described, such as in the Argentine provinces of Formosa, Entre Ríos and Misiones, as well as in the adjacent plains in Bolivia, Paraguay and Uruguay (probably due to the absence of sampling sites or As analysis data). Therefore it is not surprising that in 2006—nearly 100 years after the detection of the problem in the same regional sediment basin (near to the locality of Bell Ville in Cordoba, Argentina)—As was found in the aquifers of SW Uruguay (Guérèquiz *et al.* 2006, Bundschuh *et al.* 2008c). A third example is those from SE Asia. Considering the alluvial aquifer of Bengal delta, whose sediments derived from the Himalayas, the similar geological and geochemical conditions of other flood and delta plains deposited by other Himalayan rivers such as Indus plain, Irrawady delta, Red River delta, and Mekong river delta would have suggested the probable occurrence of As in the groundwater of these regions many years ago (instead, the discoveries in these areas were not made until the first decade of the 20th century, exposing many people unnecessarily to As).

In order to fulfill all these tasks, we need an international, multi- and interdisciplinary cooperation including the establishment of solid sources and on-line databases (which be periodically updated) with information on worldwide occurrences of As in water resources and specific characteristic information (e.g., geological, geochemical, hydrogeological, geomorphological and climatic conditions, as well as information on land use and aquifer exploitation). This would allow an easy access to information, which otherwise would not be or only be hardly possible to access. We need a dynamic global network and forum for As research oriented discussions to the direct solution of problems with considerable social impact and relevance rather than only focusing on cutting-edge and breakthrough research in physical, chemical, toxicological, medical and other specific issues on As on a broader environmental realm. It is not sufficient that these activities and efforts remain in the hands of some scientists, working extra hours on them. These activities need to be financed to allow the establishment of powerful tools to combat the global As problem. Only such a global, multi- and interdisciplinary approach will allow us to present an integrated approach from the occurrence of As in rocks and their mobilization into the ground- and surface water, soil and air, its transport therein, the pathways of As and their introduction into the food chain and finally the uptake by humans. Human As exposure, bioavailability, metabolism and toxicology need to be treated together with related public health effects and risk assessments in order to better manage the As-bearing terrestrial and aquatic environments to reduce human As exposure. Arsenic removal technologies and other methodologies to mitigate the As problem must be addressed not only from the technological, but also from economic and social point of view considering legislative and political issues and international cooperation, as e.g., international agreements or programs for mitigating the As problem. Only such inter- and multidisciplinary approaches would allow case-specific selection of optimal mitigation measures to provide As-safe drinking water, food, and air. Assembly of human resources is essential and we need to increase awareness and knowledge among administrators, policy makers and company executives, on the problem and to improve the international and bilateral cooperation on geogenic As hazard and its effects globally.

CHAPTER 2

Geological controls of arsenic concentrations in ground- and surface-waters—An overview of our worldwide state-of-the-art knowledge

Arsenic (As) release into ground- and to some less extent to surface-water bodies occurs predominantly from natural geogenic As sources (for overviews see e.g., World: Smedley 2002, Nordstrom 2002, Nriagu *et al.* 2007, Ravenscroft *et al.* 2009; North America: Welch *et al.* 2000; Latin America: Bundschuh *et al.* 2008a,b, 2009; Iberoamerica and Iberian peninsula: Bundschuh *et al.* 2008c; Asia: Smedley 2005). As we saw in chapter 1, it is a very common process found in many regions all around the world. Thereby the concentrations of arsenic obtained from these predominantly geogenic sources in groundwater vary regionally depending on the geological and geomorphological setting, hydrogeological pattern and climate. Arsenic release into groundwater further depends on geochemical and biogeochemical conditions, and available As sources: primary and secondary arsenic-bearing rocks and mineral phases (e.g., iron oxyhydroxides, organic carbon, mineralized areas, or coal deposits where the release of As into the environment is further increased by exploitation of ores, coal and hydrocarbons) and fluids and gases from volcanic and geothermal activities. In Taiwan, all of these potential As sources and mobility controls coexist.

In this chapter, processes that lead to similarities and differences between arsenic contamination in Taiwan and other regions of the world will be reviewed, especially geochemical and microbiological controls that result in high arsenic concentrations in groundwaters. However, it must remain clear that this summary is incomplete and for more details we refer to the specific references. For the locations mentioned in this chapter we refer to Figure 1.1 in chapter 1.

2.1 ARSENIC IN THE EARTH'S ENVIRONMENTS AND INTRODUCTION INTO GROUND- AND SURFACE-WATER RESOURCES

Arsenic (As; atomic number 33, atomic weight 74.922) is widely distributed in all the earth's different environments: atmosphere, lithosphere, hydrosphere, and biosphere. Its primary origin is natural and its release is predominantly related to geogenic processes, such as the weathering of rocks and their minerals (including those of mine-wastes), volcanic activity, geothermal activity, and far less importantly natural forest fires and marine spray (Nriagu and Pacyna 1988). Impacts through anthropogenic activities (smelting and ore processing, pesticides, fertilizers, chemical industries, wood preservation agents, etc.) are locally of importance.

Arsenic may be introduced into ground- and surface waters by two principal methods: (1) rock weathering, erosion, transport, deposition, and diagensis, and (2) through atmospheric deposition. However, the content of As in the atmosphere is generally low. The mean annual global emission of As into the atmosphere is estimated to be about 42,600 t year^{-1} (Matschullat 2000). Principal sources for atmospheric As correspond to limited areas with: (1) geothermal activity (hot springs, fumaroles, solfatares), (2) volcanic eruptions accompanied by the release of gases, fluids, and particles to the atmosphere, (3) wind-blown dust from weathered continental crust, (4) natural forest fires, (5) sea spray, and (6) anthropogenic activities. The main sources are anthropogenic, contributing 25,450 t year^{-1}, and volcanic emissions which contribute 17,150 t year^{-1} to the total atmospheric As emission (Matschullat 2000).

Particles, containing As, can be transported, depending on particle properties (size, shape, specific weight) and atmospheric conditions over large distances before being deposited in a dry or wet form on the earth's surface. The aeolic transport of weathering products is important in some regions. This process deposited loess-type sediments (originating from the Andes) containing highly water-soluble As-bearing volcanic ash over an area of one million square kilometers covering the Chaco-Pampean plain in Argentina, which is one of the world's largest areas with high groundwater As concentrations (Smedley *et al.* 2005, 2008, Bundschuh *et al.* 2004, Bhattacharya *et al.* 2006b) (see example 2.6.2.3).

Atmospheric deposition generally contributes little to the arsenic content of soil, surface-water, and groundwater environments because uncontaminated rainwater and surface water have very low concentrations of dissolved As (<1 µg L^{-1}; e.g., Andreae 1980, Smedley and Kinninburgh 2002). Also, seawater contains only 1.5 µg L^{-1} of As on average (Adriano 2001, Navarro *et al.* 1993). Exceptions are geothermal water discharges into surface water (e.g., in the Andean range of Ecuador and in Yellowstone National Park; see section 2.5.2 for details), or where sediments and soils with easily leachable As are exposed to surface runoff (e.g., Chaco-Pampean plain; see section 2.6.2.3).

In consequence, in the following sections, we will only consider As from geogenic sources as responsible source for high concentrations of As in groundwater, and, to a limited extent, in surface waters used for drinking or irrigation purposes resulting from weathering of rocks/minerals, geothermal activity, and volcanic activity.

2.2   GEOGENIC ARSENIC: OCCURENCE AND SOURCES

In the scale of abundance of the naturally occurring elements, As ranks 41st of the 88, and ranks 20 when considering the elements which compose the rocks of the earth's crust (Cullen and Reimer 1989).

The chemical characteristics of As facilitates sorption and desorption on solid particles. The redox state can easily change due to reactions with oxidants, such as oxygen and organic matter, or by microbes present in air, soil, surface- and groundwater environments. These reactions determine the mobility and bioavailability of arsenic in these environments.

2.2.1   *Arsenic in minerals and amorphous phases*

In the lithosphere, As is found as native arsenic, arsenate, sulfide, oxide, sulfate, sulfosalts, and arsenite minerals. The arsenates constitute the most frequent form (60%) of total As, sulfates and sulfosalts contribute 20%, and arsenites, oxides, silicates mixtures and polymorphs of the elemental As contribute the rest (20%) (Plant *et al.* 2004, Baur and Onishi 1969). Arsenic can be found in the crystal structure of over 200 minerals, however, only 10% of them are generally of importance. The principal As minerals are shown together with the respective geological environments where they are typically found in Table 2.1. Table 2.2 shows the typical ranges of As concentrations in rock-forming minerals.

In many sediments, which comprise most principal arseniferous aquifers, As is found in high concentrations in metal oxyhydroxides, especially those of Fe, Mn, and Al. Fe oxyhydroxides may contain up to 76,000 mg kg^{-1} of As since these minerals can include As in their crystal structure or adsorb it as oxyanions on the particle surface.

Because As chemistry is similar to that of sulfur, As can often be found within the crystal structure of sulfides, with the highest concentrations in minerals such as pyrite, chalcopyrite, galena, and marcasite, where As concentrations cover a wide range from 5 to126,000 mg kg^{-1} (Smedley and Kinniburgh 2002). Arsenian pyrite and arsenopyrite seem to be the most abundant As sulfide minerals. The presence of sulfides is typical in geological settings, such as sulfidic ore deposits (hydrothermal deposits), and can release large amounts of As as we will

Table 2.1.   Principal arsenic minerals and typical geological environments (modified from compilations of Smedley and Kinniburgh 2002, Lillo 2003, Bundschuh *et al.* 2008c).

Mineral	Composition	Occurrence
Native arsenic	As	Hydrothermal veins
Niccolite	NiAs	Vein deposits and norites
Realgar	AsS	Vein deposits, in many cases found together with orpiment, clays and carbonates; frequently in geothermal spring deposits
Orpiment	As_2S_3	Hydrothermal veins, geothermal spring deposits and sublimation deposits of volcanic exhalations
Cobaltite	CoAsS	Metamorphic rocks and hydrothermal veins
Arsenopyrite	FeAsS	Mineral veins, sulfidic mineral deposits, geothermal deposits
Arsenian pyrite	$Fe(S, As)_2$	Mineral veins, sulfidic mineral deposits
Tennantite	$(Cu, Fe)_{12} As_4S_{13}$	Hydrothermal veins
Enargite	$Cu_3 AsS_4$	Hydrothermal veins
Arsenolite	As_2O_3	Secondary As mineral, formed by oxidation of As minerals such as arsenopyrite and native As; also forms from realgar and scorodite
Claudetite	As_2O_3	Secondary As mineral, formed by oxidation of As minerals such realgar and arsenopyrite; also forms from realgar and scorodite
Scorodite	$FeAsO_4 \cdot 2H_2O$	Secondary As mineral, also found in hydrothermal vein deposits
Annabergite	$(Ni, Co)_3(AsO_4)_2 \cdot 8H_2O$	Secondary As mineral
Hoernesite	$Mg_3(AsO_4)_2 \cdot 8H_2O$	Secondary As mineral, frequent in waste products of mineral ore processing
Hematolite	$(Mn, Mg)_4 Al(AsO_4)(OH)_8$	Secondary As mineral
Conicalcite	$CaCu(AsO_4)(OH)$	Secondary As mineral
Pharmacosiderite	$Fe_3(AsO_4)_2(OH)_3 \cdot 5H_2O$	Secondary As mineral, formed by oxidation of As minerals such arsenopyrite

Table 2.2.   Arsenic concentrations in the principal rock-forming minerals (compiled from: Baur and Onishi 1962, Boyle and Jonasson 1973, Pichler *et al.* 1999, Smedley and Kinniburgh 2002).

Mineral	As (mg kg^{-1})	Mineral	As (mg kg^{-1})
Sulfides		Silicates	
pyrite	<1–100000	amphiboles	1.1–2.3
marcasite	20–126000	biotite	1.4
galena	5–10000	feldspars	<0.1–2.1
esfalerite	5–17000	olivine	0.08–0.17
chalcopyrite	10–5000	pyroxenes	0.05–0.8
Oxides		quartz	0.4–1.3
hematite	≤160	Carbonates	
Fe oxides	≤2000	calcite	1–8
Fe(III) oxyhydroxides	≤76000	dolomite	<3
magnetite	2.7–41	siderite	<3
ilmenite	<1	Sulfates	
Phosphates		barite	<1–12
apatite	<1–1000	gypsum/anhydrite	<1–6
		jarosite	34–1000

see in section 2.3.2. Sulfides can form diagenetically under anaerobic, strongly reducing conditions such as black shales, coal, peat deposits, and phosphorites (see section 2.2.2).

Arsenic is commonly found as sulfides in hydrothermal ore deposits (e.g., Au) especially in epithermal quartz veins, and epithermal quartz stockworks, which constitute fossil geothermal systems (see also section 2.5 on geothermal arsenic). Arsenic sulfides are also common in W and/or Sn ore deposits related to granites where As is generally found as arsenopyrite. It is also found in skarn and porphyry deposits, as well as in pegmatites. Arsenic can be common in altered zones of mineralized faults and hydrothermal conduits (e.g., feldspatic, argillic alteration, propylitic alteration).

Arsenic is present in variable amounts in all rock-forming primary and secondary minerals. Relatively high As contents are also found in phosphates (e.g., apatite: 1–100 mg kg^{-1}). The arsenic content of sulfates is generally low (<1–12 mg kg^{-1}), however, concentrations in jarosite are high (34–1000 mg kg^{-1}). The concentration of As in calcite ranges from 1–8 mg kg^{-1}, whereas the concentration in dolomite and siderite is <3 mg kg^{-1}.

Silicates contain generally <2 mg kg^{-1}, however the weathering of biotite can be an important source for As release and can explain the positive correlation between concentrations of dissolved Mg and As, which has been observed in different regions (Appelo and Postma 2006). The relatively high As concentrations of magnetite (2.7–41 mg kg^{-1}) can be explained by substitutions in the crystal structure between As and Si^{4+}, Al^{3+}, Fe^{3+} and Ti^{4+}, due to their similar ionic radii (Baur and Onishi 1969). Volcanic glass, which is a principal component in many volcanic ashes, deserves a special consideration. Even though the As concentration is relatively low (<20 mg kg^{-1}), the high solubility of amorphous glass makes it an important source for dissolved As (example Chaco-Pampean plain, Argentina; see example 2.6.2.3; Smedley *et al.* 2005, 2008, Bundschuh *et al.* 2004, Bhattacharya *et al.* 2006b). Volcanic ash was also described as an As source from the Interior Plains of southern Dakota (Carter *et al.* 1998).

2.2.2   *Arsenic in rocks*

The ranges of As concentrations found in different rocks are compiled in Table 2.3. Igneous rocks have generally uniform As contents, which correspond to the order of the average crust (<5 mg kg^{-1}; Goldschmidt 1937, Jacks and Bhattacharya 1998). The average As concentration in igneous rocks is only 1.5 mg kg^{-1} (Baur and Onishi 1969, Ure and Berrow 1982), however higher values of up to 113 mg kg^{-1} are reported as an upper limit for basic igneous rocks (Smedley and Kinniburgh 2002).

In metamorphic rocks, the As concentration is controlled by that of the original host rocks. However, we must consider that during the process of metamorphism, the rocks may be enriched or depleted in As. Most metamorphic rocks contain <5 mg kg^{-1} of As, with the highest values in schists and phyllites (up to 143 mg kg^{-1}; the average is 18 mg kg^{-1}; Smedley and Kinniburgh 2002) because these rocks are derived from claystones and mudstones and their unconsolidated equivalents, which have elevated As.

The As concentration in sediments is variable and depends on many factors including: (1) original rock type, (2) type of weathering, (3) mechanism of transport from weathering to deposition area, including the prevailing geochemical, mechanical and sedimentological processes, and (4) formation of secondary As minerals. Especially in unconsolidated sedimentary rocks, additionally leaching and flushing by groundwater may affect the composition of the solid phase. Depending upon geochemical conditions, these processes may result in a net release of As from the sediment.

Excluding coal, the As concentration in sedimentary rocks is <1–490 mg kg^{-1} (Smedley and Kinniburgh 2002); however, most sedimentary rocks have As concentrations of 5–10 mg kg^{-1} (Webster 1999). Carbonate rocks have As concentrations of 0.1–20.1 mg kg^{-1} (Smedley and Kinniburgh 2002), though they generally have levels <5 mg kg^{-1} (Baur and Onishi 1969), even though carbonate minerals can integrate As in their crystal lattice by substituting AsO$_3^{3-}$ for CO$_3^{2-}$ (Di Benedetto *et al.* 2006) (see sections 2.2.1 and 2.8).

Table 2.3.   Typical contents of arsenic in consolidated and hard rocks (compiled from Smedley and Kinniburgh 2002).

Rock	Arsenic (mg kg^{-1}) Mean (range)	Number of samples
Igneous rocks		
ultramafic	1.5 (0.03–15.8)	40
mafic	1.8 (0.06–113)	190
intermediate	1.7 (0.09–13.4)	69
felsic	1.4 (0.2–15)	118
Metamorphic rocks		
quartzites	5.5 (2.2–7.6)	4
hornfels	5.5 (0.7–11)	5
phyllites and slates	18 (0.5–143)	75
schist/gneiss	1.1 (<0.1–18.5)	16
amphibolites and greenstone	6.3 (0.4–45)	45
Sedimentary rocks		
marine shales and mudstones	3–15 (up to 490)	–
non-marine shales	3–12	–
sandstones	4.1 (0.6–120)	15
limestones, dolomites	2.6 (0.1–20.1)	40
phosphorite	21 (0.4–188)	205
iron formations and Fe-rich sediments	1–2900	45
evaporites (gypsum, anhydrite)	3.5 (0.1–10)	5
coal	0.3–35.000	–
Sediments, unconsolidated		
alluvial sands (Bangladesh)	2.9 (1.0–6.2)	13
alluvial mud and clay (Bangladesh)	6.5 (2.7–14.7)	23
lake sediments (British Columbia)	5.5 (0.9–44)	119
glacial tills (British Columbia)	9.2 (1.9–170)	–
river sediments (world average)	5	–
stream and lake silts (Canada)	6 (<1–72)	310
loess-type sediments (Argentina)	5.4–18	–

Sandstones and sand have the lowest As concentrations, because the As concentrations of their primary minerals, quartz and feldspar, are low (average As concentration: 1–4 mg kg^{-1}; Ure and Berrow 1982). Sands from alluvial deposits, tills and lacustrine sediments contain 1–15 mg kg^{-1} of As (Plant *et al.* 2004). However, some sandstones and sands contain significantly higher As (up to 120 mg kg^{-1}, Smedley and Kinniburgh 2002) and the concentration can be even higher if large amounts of Fe are present.

Compared to sandstone and sand, silt- and claystones (and their unconsolidated equivalents: silt and clay) have higher average As concentrations of 13 mg kg^{-1} (Ure and Berrow 1982, Smedley and Kinniburgh 2001) though the concentration range varies between 1 and 490 mg kg^{-1}. If there is a high Fe content, the As concentration can be even higher (Smedley and Kinniburgh 2002), which can be explained by the presence of more minerals and phases with elevated As contents, such as sulfides, oxides, clay minerals, and organic matter present in these rocks, and with the fine grain size.

Unconsolidated and consolidated sediments of silt to sand fraction constitute most of the arseniferous aquifers around the world used to supply drinking-water. Because these sediments are an erosion product of all kinds of rocks and geological features, such as mineralized zones and hydrothermal deposits, they may contain As in the original mineral phase, but in most cases the As has undergone a dissolution processes, often followed by formation of secondary As-containing minerals in the sediment, which are available for remobilization to the aqueous phase. Thus, we need to consider, that the As contents in sediments of arseniferous aquifers are not necessarily higher than those of average sediments. Key issues are the geochemically favorable conditions, which result in the dissolution or desorption of As either from secondary or primary As-containing minerals into the groundwater. The As may be derived from many sources: (1) the presence of diagenetically formed authigenic sulfides, especially pyrite (Keuper aquifer Germany; Ordovician sandstone aquifers from eastern Wisconsin, USA; Bengal Delta, Bangladesh), (2) intercalated lentils of clay, black shale, coal, organic material, and (3) As adsorbed on metal oxyhydroxide coatings on sediment particles (examples are the flood and delta plains with arseniferous alluvial aquifers, such as the Ganges-Brahmaputra-Meghna plain and delta, Indus plain, Irrawady delta, Red river delta, Mekong river delta; the Molasse basin in southern Germany; loess-type aquifers of the Chaco Pampean region, Argentina) (see examples in sections 2.6.2.3 and 2.6.2.4). However, primary sources which lead to metal oxide/oxyhydoxide coatings on sediment particles and adsorbed (or coprecipitated) As may be quite different in the individual sites.

Arsenic is incorporated in sulfides (especially in pyrite), formed diagenetically in sediments rich in organic matter, such as black shales, phosphorites, and coal and peat deposits where As contents are generally in the 20–200 mg kg^{-1} range (Minkkinen and Yliruokanen 1978, Finkelman 1980, Nordstrom 2000, Smedley and Kinniburgh 2002). However, values as high as 35,000 mg kg^{-1} are reported by Smedley and Kinniburgh (2002) as an upper limit for the As content in coal. Depending predominantly on its sulfide and organic matter, the As contents of coals and peat deposits are highly variable, but generally high. The typical As range is 0.5–80 mg kg^{-1} and the world average was established as 10 mg kg^{-1} (Clarke and Sloss 1992, Yudovich and Ketris 2005). However, locally much higher As contents are reported. In the Czech Republic, 3245 mg kg^{-1} was reported in lignites and in SW China, and As content of up to 8300 mg kg^{-1} was reported for bituminous coal deposits associated with gold ore mineralization (province Guanzhou; Zheng *et al.* 1999, Finkelman 2004).

Hydrothermally altered rocks and hydrothermal gold mineral deposits which have the highest arsenic concentration of any rocks are locally of importance for As release into ground- and surface-water. Arsenic is concentrated in hydrothermal veins or rocks affected by hydrothermal fluids (Moore *et al.* 1988, Brannon and Patrick 1987). There are different types of hydrothermal fluids: those which remained in solution until the late stages of magma crystallization (Tarbuck and Lutgens 2002), and geothermal fluids. The fluid-rock interactions result in a removal of metal and metalloid ions, including As, e.g., from intrusive igneous rock, and transports them upwards where they can be precipitated (Tarbuck and Lutgens 2002). In mildly alkaline reducing conditions As easily precipitates as carbonates, whereas in strongly reducing conditions it precipitates as sulfides (Brennan and Lindsay 1996, Smedley and Kinniburg 2003).

2.3   MECHANISMS OF ARSENIC MOBILIZATION INTO AQUEOUS ENVIRONMENTS: AN OVERVIEW

The concentration of As dissolved in ground- and surface-water bodies is a function of space and time and depends on the available primary and secondary As sources, the fluid properties (pH, redox potential, ions presence, ionic strength, organic matter content) and interfacial processes between the solid and fluid phases. These factors determine which As species are present in the fluid and control the As release or sequestration by the solid phase, and its transport in the aqueous phase. In this overview chapter, the As species available in natural waters are summarized (2.3.1), followed by a listing of the principal solid-fluid interface processes which control As release and mobility (2.3.2) and a short discussion of other factors and processes, which may influence the concentrations of dissolved As in ground- and surface waters, such as a specific pattern of climate, geomorphology, geological/tectonic setting, groundwater-surface water interactions and groundwater exploitation (2.3.3). A section dealing with some principal issues relevant to As transport in aquifers (2.3.4) concludes this chapter.

2.3.1 *Arsenic species in natural waters and reaction kinetics*

In natural waters, dissolved As is generally present in inorganic forms either as oxyanions or as neutral species. Organic As forms are generally not of importance in natural waters though methylated As compounds may be formed under anoxic conditions by microbiological processes in soils and sediments. However, these compounds are released quickly into the atmosphere and are transformed to inorganic forms (Litter *et al.* 2008). Additionally, organic acids may form complexes with As or may adsorb on metal oxides/oxyhydroxids (see 2.9).

Depending on the redox potential and pH of the water, inorganic As is found as As(V) and As(III) species. As(0) and As(–III) species are much less common. Arsenate is present as H_3AsO_4 and its corresponding anions are $H_2AsO_4^-$, $HAsO_4^{2-}$ and AsO_4^{3-} with dissociation constants (pK_a) of 2.3, 6.8 and 11.6, respectively. Arsenite occurs as H_3AsO_3 and its principal hydrolysis products found in natural waters are $H_2AsO_3^-$, $HAsO_3^{2-}$ and AsO_3^{3-} with dissociation constants of 9.2, 12.7 and 13.4, respectively (Smedley and Kinninburgh 2002, Adriano 2001, Lillo 2003).

Under oxidizing conditions, As(V) predominates and in the pH range of most natural ground- and surface waters is found as the oxyanions $H_2AsO_4^-$ (pH <6.8) and $HAsO_4^{2-}$ (pH >6.8). Under very acid conditions, As(V) is present as neutral H_3AsO_4 (pH <2.3; e.g., in some acid-spring discharges at the earth's surface and some fumaroles that have a high HCl content). In very alkaline waters, AsO_4^{3-} predominates (pH >11.6; e.g., in alkaline lakes, soda lakes).

In water with reducing conditions, As(III) is the predominant oxidation state and, in most reduced natural groundwater—including the water of geothermal reservoirs an uprising geothermal fluids and geothermal spring water—and surface waters the neutral species H_3AsO_3 predominates (pH <9.2) (Yan *et al.* 2000). Charged As(III) species are only dominant in strongly alkaline waters as $H_2AsO_3^-$ (pH >9.2) and as $HAsO_3^{2-}$ (pH > 12.7). These conditions only occur rarely in some geothermal waters (see e.g., Linklater *et al.* 1996).

Under extremely reducing conditions in strongly sulfidic environments that can be found in some geothermal fluids, waters in aquifers rich in carbons, peat, and some mining waters, As can form sulfide complexes. In sulfidic waters arsenic thioanions (arsenic-sulfide species) replace arsenates $(H_nAs(V)O_4^{3-})$ and arsenites $(H_nAs(III)O_3^{n-3})$ in the water (Webster 1990, Eary 1992, Wood *et al.* 2002, Wilkin *et al.* 2003, Bostick *et al.* 2005, Hollibaugh *et al.* 2005, Planer-Friedrich *et al.* 2007, Beak *et al.* 2008). The stoichiometric compositions, oxidation states and thermodynamic stabilities of the thioanions were extensively discussed during the last two decades (Helz *et al.* 2008). For the condition of near saturation with respect to orpiment (As_2S_3), arsenic thioanions, probably with the generic formula $H_xAs(III)_3S_6^{x-3}$ (x = 1–3), (Spycher and Reed 1989, Webster 1990, Eary 1992) can be formed. In strongly reducing waters with a lower sulfide content and great undersaturation with respect to As_2S_3, monomeric thioanions, $H_xAs(III)S_3^{n-3}$ and $H_xAs(III)OS_2^{n-3}$ are likely to form (Helz *et al.* 1995, Clarke and Helz 2000). However recently it was proven that at least six thiolated As species with S/As ratios approximating 0, 1, 2, 3 and 4, exist simultaneously in synthetic and natural sulfidic waters (Wilkin *et al.* 2003, Stauder *et al.* 2005, Hollibaugh *et al.* 2005, Planer-Friedrich *et al.* 2007). The earlier results are increasingly questioned and contradictory views about the nature of As thioanions continue to have advocates (Helz *et al.* 2008). One implication is that all earlier thermodynamic data on the stabilities of As thioanions are uncertain and incomplete. As a consequence, Helz *et al.* (2008) used computational as well as empirical information to construct a provisional model for equilibrium As thioanion distributions in sulfidic waters. In contrast to previous authors who stated that either As(III) or As(V) exist, these researchers found that both are important and can occur simultaneously under commonly encountered pH and ΣS^{-II} conditions.

In most surface waters, As(V) dominates over As(III), though stagnating waters with high contents of organic material are an exception to this rule. In groundwater, depending upon the redox conditions, the presence of oxidants/reductants, such as organic matter and the presence of microbes, which may catalyze redox processes between As species with different oxidation states, As(III) or As(V) may be the prevailing oxidation state.

However, in many natural waters, there is no redox equilibrium between the As(III/V) redox couple and other redox couples (Lindberg and Runnells 1984, Nordstrom and Muñoz, 1994 and references therein). This is the reason why in several aquifers, As(III) species are found in oxidized aquifers. The oxidation of As(III), e.g., by dissolved oxygen, is a very slow process and does not happen in measurable periods of time without microbes or an oxidant (McCleskey *et al.* 2004, Cherry *et al.* 1979). The presence of catalytic compounds, such as Mn oxides, can significantly increase the oxidation rate of As(III). Furthermore, the oxidation rate of As(III), and the reduction rate of As(V) can be strongly increased by the presence of microbes. This lack of redox equilibrium has also been observed for surface water, where As(III) has been described in oxic water and As(V) has been detected in highly reducing sulfidic water (Spliethoff *et al.* 1995).

2.3.2   *Arsenic release and mobility: solid-fluid interfacial processes*

The principal interfacial processes which are responsible for As release and mobility comprise: (1) chemical processes such as dissolution/precipitation, e.g., reductive dissolution of Fe oxides and hydroxides, reduction of sulfate and the precipitation of pyrite, (2) biological transformations, e.g., microbiological oxidation of organic matter, (3) physicochemical processes, e.g., adsorption/desorption and ion exchange.

In many aquifers, the principal mechanisms of As mobilization could be determined by the prevailing aquifer conditions, such as certain dominant geomorphological, geological, hydrogeological and hydrogeochemical conditions. The As mobilization is governed by: (1) the As source (e.g., leaching from As-bearing rocks, dissolution of secondary As minerals, influx/mixing of As-rich geothermal water) and (2) site- or zone-specific hydrogeochemical conditions that result in extremely heterogeneous lateral and vertical distributions of As in groundwater, as is typical for the Bengal delta aquifers (e.g., BGS 2001, Bhattacharya *et al.* 2002, Zheng *et al.* 2004, Hasan *et al.* 2007) and for different parts of Latin America (e.g., Smedley *et al.* 2005, Bundschuh *et al.* 2004, 2008a,b, 2009, Bhattacharya *et al.* 2006b, Altamirano Espinoza and Bundschuh 2009).

The release of As from rocks and minerals by processes such as dissolution and surface processes, such as desorption and ion exchange, affects all rocks and minerals to some extent. However, there some are specific As release/mobilization/transport processes, which are typical for specific geological settings, and which can explain principal processes, mainly dissolution and desorption mechanisms, which take part under specific redox and pH conditions, releasing and mobilizing As from a geogenic solid source into ground- and surface water. This can explain the genesis of many arseniferous aquifers around the world. The relevant mobilization and re-mobilization processes are:

- *Sulfide oxidation in mineralized areas* (hydrothermal deposits, sulfide ore deposits) but also of diagenetic authigenic sulfides, e.g., contained in sediments. Sulfide oxidation may result in the formation of secondary As minerals, such as Fe sulfates, arsenates, oxides and oxyhydroxides (see section 2.4).
- *As dissolution in deep geothermal reservoirs* and discharge of geothermal fluids into cold aquifers, surface water bodies, or at the earth's surface. In these environments, As is present either dissolved or precipitated as secondary As-containing minerals, such as sulfides, but especially metal oxides and oxyhydroxides, which often form the principal As sinks for further mobilization. The influence of geothermal activities, which in many cases is underestimated, will be discussed in section 2.5.
- *Formation of secondary As minerals*: Metal oxyhydroxides as principal As source are formed by different geogenic processes, such as sulfide oxidation, geothermal activities, and generally dissolution/leaching of rocks and minerals followed by precipitation of these

secondary minerals as a function of local hydrochemical conditions (especially of redox, pH) on sediment particles in sedimentary aquifers.

- *Arsenic remobilization from metal oxides and oxyhydroxides*, which can explain the genesis of many arseniferous sedimentary aquifers around the world, where three principal mechanisms can be distinguished:
  - ° Dissolution of metal oxyhydroxides under very acidic conditions (see 2.6.1).
  - ° Reductive dissolution of metal oxyhydroxides under reducing conditions (see 2.6.1).
  - ° Arsenic sorption by metal oxyhydroxides at high pH and oxidizing conditions (see 2.6.2).
- *Arsenic sorption with respect to clay minerals* (see 2.7).
- *Precipitation/dissolution and sorption processes for calcite* (see 2.8).
- *Arsenic sorption by other solid surfaces*: Ti oxides, which have a pH_{pzc} similar to those of Fe- and Al oxides, can adsorb As(V). Under conditions where sulfides are stable (reducing conditions), As may be adsorbed on sulfides, such as galena and sphalerite (Bostick *et al.* 2003, Wilkin *et al.* 2003).
- *Formation of complexes between humic acids and As species* (see 2.9).

The principal geochemical reactions and influencing parameters, which control the arsenic concentrations in groundwater are listed in Table 2.4, together with some examples.

It is obvious that many of the mentioned processes can occur alongside and in many cases it becomes difficult or impossible to balance the contribution of the individual mechanisms to the total As dissolved in the water.

In the following sections 2.4 to 2.9 we will give a brief description of the principal As release and mobilization mechanisms to compare the Taiwan arsenic case with arseniferous aquifers of the rest of the world. For each As mobilization mechanism, we will present typical case studies and examples.

2.3.3 *Additional factors and processes influencing concentrations of dissolved arsenic*

Additional factors that may increase or decrease As concentrations are related to specific settings of climate, geomorphology, aquifer hydraulics, groundwater-surface water interactions, groundwater exploitation, etc.

The principal factors and processes favoring changes in dissolved As are:

- *Climate*: Arid and semiarid climates with low rates of rainfall and high evapotranspiration favor the evaporative increase of dissolved As concentration. Arsenic is not incorporated in most evaporite minerals until a very high salinity is reached (>9 molar unless it is an acid evaporative water). Consequently, during evaporation the As remains in the fluid phase where it accumulates, which can lead to As concentrations of >100,000 $\mu g\ L^{-1}$ (Levy *et al.* 1999). Especially in arid and semiarid areas, evaporative concentration can significantly increase the As concentrations in ground- and surface waters and in closed or partly closed hydrological basins. Examples are the Chaco-Pampean plain (Argentina), many regions of the Andean highlands, parts of Mexico and the western part of the USA. Specific examples include shallow aquifers of the Carson Desert (Intermontane Plateaus, Nevada; Welch and Lico 1998; Table 2.4) and the southern San Joaquin valley (Arizona; Fujii and Swain 1995, Swartz 1995, Swartz *et al.* 1996; Table 2.4), where evaporation has been identified as a contributing factor producing high As concentrations, together with adsorption or coprecipitation of arsenic on iron oxide as an additional control of As mobility (Fujii and Swain 1995, Welch and Lico 1998). Evaporative increases of As concentration in surface- and groundwater are also of importance at many sites in the Atacama desert (N Chile) and in the Andean highlands in NW Argentina, Bolivia and Peru where arid climates prevail (Bundschuh and Garcia 2008, Bundschuh *et al.* 2008a,b, Bundschuh *et al.* 2009). It may be an additional control (together with As desorption at high pH values under oxidizing aquifer conditions) in the Argentine Chaco-Pampean plain whose climate is characterized

Table 2.4.   Principal geochemical reactions and influencing parameters, which control the arsenic concentrations in groundwater (for locations see Fig. 1.1). Modified from Welch *et al.* 2000.

	Controlling mineral phases and principal reactions	Controlling arsenic mobility conditions	Examples and references
Oxidizing conditions	Fe (Mn, Al) oxides/ oxyhydroxides: Adsorption/ desorption of As	pH; As oxidation state and species (redox equilibrium?); presence of ions competing for adsorption sites; ionic strength; oxygen and Fe^{3+}, organic acids concentrations	Argentina: Chaco-Pampean plain (Smedley *et al.* 2005, 2008, Bundschuh *et al.* 2004, Bhattacharya *et al.* 2006b); USA: Appalachian Highlands (NE Ohio, Matisoff *et al.* 1982): Interior Plains (S Dakota. Carter *et al.* 1998); Carson Desert (Nevada, Welch and Lico 1998); Pacific Mountain System (NW Washington, Goldstein 1988, Ficklin *et al.* 1989, Davies *et al.* 1991 and Arizona, Goldblatt *et al.* 1963, Nadakavukaren *et al.* 1984)
	Fe (Mn, Al) oxides/ oxyhydroxides: Precipitation and coprecipitation of As	(as above)	Molasse aquifer, S-Germany; many geothermal springs, e.g., Ecuador (Tambo area, Cumbal *et al.* 2009); Dead Indian Spring, Welch *et al.* 1988
	Sulfide minerals: Sulfide oxidation	pH and microbial activity; oxygen and nitrate contents	USA: Appalachian Highlands (Massachusetts to Maine, Peters *et al.* 1999, 2008, Zuena and Keane 1985, Boudette *et al.* 1985, Marvinney *et al.* 1994, Ayotte *et al.* 1998); Interior Plains (E Michigan, Westjohn *et al.* 1998, Kolker *et al.* 1998) and many mining sites around the world
Reducing conditions (no sulfide presence)	Fe-oxides/ oxyhydroxides: Adsorption/ desorption and precipitation	Oxidation state of As	Dzombak and Morel 1990
	Fe-oxides/ oxyhydroxides: Dissolution (reductive dissolution)	Presence of organic carbon	Ganges-Brahmaputra-Meghna plain and delta (India and Bangladesh, e.g., Bhattacharya *et al.* 2002a,b, 2006a, 2007a,b), Indus plain (Pakistan, Nickson *et al.* 2005); Irrawady delta (Myanmar, WRUD 2001); Red River delta (Vietnam, Berg *et al.* 2001); Mekong river delta (Cambodia and Laos, Feldman and Rosenboom 2001); USA Carson Desert (Nevada, Welch and Lico 1998)
	Sulfide minerals	Presence of organic carbon	
Reducing conditions (sulfide presence)	Sulfide minerals: Precipitation	Sulfide, iron, and As concentrations	Moore *et al.* 1988, McRae 1995, Rittle *et al.* 1995, Huerta-Diaz *et al.* 1998

by semiarid conditions (*ibid*). Consequently, the climate may explain the occurrence of arseniferous aquifers along the Andean mountain chain and its continuation in Middle America, which all have a similar geological setting and related geogenic As sources (e.g., leaching of young volcanic rocks, abundance of volcanic ash, geothermal activities, sulfide mineral deposits, altered hydrothermal zones, etc.). Whereas arseniferous—mostly sedimentary—aquifers used for drinking water supply are frequent in the regions with arid-semiarid-moderate climates (parts of Argentina, Bolivia, Chile, Peru, Paraguay, Uruguay, Ecuador, Mexico), they are much less frequent or absent in those tropical regions in Central America, where rainfall is high and dilution of surface- and groundwater by rainfall is high (e.g., in Panama and Costa Rica). The same applies if we compare the prevailing semi-arid Chaco-Pampean plain and the tropical Amazon and Orinoco basin of Brazil and Venezuela, which both are located to the east of the Andean mountain chain, which can be considered a principal source of geogenic As. In the case of the Chaco-Pampean plain, the aeolic and fluvial sediment transport from the Andes has caused contamination of significant parts of the aquifer with As. However, the humid tropical areas of Brazil, with high rainfalls, did not allow the formation of arseniferous aquifers from geogenic sources.

- *Geomorphology*: The geomorphological setting affects the shallow aquifers and surface water bodies. Since the shallow subsurface geology often reflects the geomorphological forms of the earth's surface —especially in sedimentary plains— geomorphological depressions together with the geological/tectonic setting groundwater may control flow field and zones with slow-flowing or stagnant groundwater may develop. High groundwater residence time allows long solid-liquid contact time and favors water-rock interactions, such as cation exchange, desorption of As, etc. as well as evaporation processes, which can result in an increase of dissolved As in these zones. Hence, our example from the Chaco-Pampean plain, where zones of high groundwater residence time resulted in cation exchange and a pH increase, causing desorption of As from metal oxyhydroxides (see example 2.6.2.3).

- *Groundwater-surface water interactions*: The interactions may significantly modify the concentration of As dissolved in an aquifer, either simply due to mixing between waters with different As concentrations, or by changing geochemical conditions caused by the mixing, which may contribute to the release of As from the solid phases. There are plenty of examples around the world where geothermal waters with high As concentrations mix with groundwater of cold aquifers or surface water, increasing their As content. An example is the Tatio geothermal springs in Chile, whose water discharges into the Loa river, which supplies water for over half a million of people (Cusicanqui *et al.* 1976, for details see section 2.5.2). In Ecuador, geothermal water discharges into the Tambo river and Papalacta lake, whose water is used for human consumption (Cumbal *et al.* 2009; for details see 2.5.2). In Turkey, in Izmir province, geothermal groundwater discharges into alluvial aquifers. In other areas, where groundwater with high As concentrations discharges from aquifers into surface water bodies are found e.g., in northern Chile. Arsenic is mobilized by the weathering of rocks in the Andes Mountains, accumulates in the aquifer, and emerges at the base of the mountains with 200–900 µg L^{-1} of dissolved As, where it is used for local drinking water supplies. This mechanism also explains the As release into ground- and surface-waters in Peru and Bolivia (Bundschuh *et al.* 2006, Bundschuh and García 2008, Bundschuh *et al.* 2008a,b, 2009).

Infiltration of irrigation water into aquifers may introduce compounds such as organic matter, which can contribute to generate reducing conditions in the aquifer, or phosphates (e.g., from fertilizers), thereby causing a release of adsorbed As (Zeng *et al.* 2008, Manning and Goldberg 1996). Especially in arid and semiarid areas, infiltrating highly mineralized irrigation water may significantly increase the mineralization of the groundwater, influencing As sorption processes. In addition, many irrigation waters are alkaline, and its infiltration into the aquifer may result in a pH increase and corresponding release of As into solution (Ravenscroft *et al.* 2009). The rise of the groundwater table due to irrigation may

further affect the As release and mobility in the aquifer, since it may bring the groundwater in contact with part of the unsaturated zone, where further As may be mobilized.
- *Groundwater exploitation*: The exploitation can change the local geochemical conditions in the aquifer (on a local or regional scale) and cause the mobilization of As from the aquifer material. Intensive groundwater exploitation can significantly change the local or regional groundwater flow field. Therefore, groundwater from greater depths, which may correspond to groundwater with high As concentrations due to a long solid-water contact time, may be attracted and exploited by the wells. The same refers to geothermal waters, which may infiltrate into the water pumped by the wells. This mechanism may explain the appearance of As or the increase of arsenic concentrations in aquifers where arsenic had not been a problem. Additionally, the related downwards movement of the groundwater table may influence As release and mobility. It may expose deeper aquifer zones to dissolved oxygen, and the resulting contact of oxygen with aquifer materials may result in changing geochemical conditions, causing As desorption from aquifer minerals, dissolution of As-bearing minerals, sulfide oxidation, and corresponding As release into the water. This process has been described for a sandstone aquifer in Wisconsin, where the lowering of the groundwater table due to groundwater exploitation has caused oxidation of a deep sulfide-containing cementation layer, which before exploitation was not in contact with oxygen, causing a release of As into the groundwater (Schreiber *et al.* 2000; for example details see 2.4.2).

2.3.4   *Arsenic transport in natural water*

Arsenic is transported in surface water either in dissolved form (influenced by river/lake sediment-water interactions along the flow path) or in solid form as part of the sediment load of the river. Arsenic is transported predominantly in dissolved form in aquifers, however, colloidal transport is also reported.

The concentration of dissolved As depends on the groundwater flow field, and the geochemical conditions of fluid and solid (redox potential, pH, organic matter content, additional ions competing for As adsorption sites, available As for dissolution/desorption, competitive ions, available As adsorption sites, etc.), which are due to changes along a groundwater flow path. These conditions, together with the flow velocity, determine the concentration of dissolved As.

Adsorption processes may cause a retardation of As transport (propagation velocity v_{As}) compared with the transport of the water (flow velocity v_w). The relation v_w/v_{As} is the retardation factor and is a measure for the grade of As adsorption. Due to its strong interaction with solid phases of Fe, Al and Mn oxides and oxyhydroxides, clay minerals, and organic matter, As(V) is less mobile than As(III). Hence, transport of As, e.g., in aquifers, is consequently more retarded for As(V) than for As(III). Mobility studies of As(III) and As(V) using sand columns with Fe (0.6%) and Mn (0.01%) and varying Eh and pH values found (Gulens *et al.* 1979) that:

- Under oxidizing conditions and slightly acid (pH 5.7) conditions, As(III) moves 5–6 times faster than As(V); an increase of pH to about neutral (pH 6.9) increases the velocity of As(V), but it remains below that of As(III);
- Under reducing conditions and alkaline (pH 8.3) conditions, both As(III) and As(V) move quickly.

2.4   SULFIDE OXIDATION

2.4.1   *Mechanism and kinetics of arsenic mobilization through sulfide oxidation*

The release of heavy metals and metalloids, including As in sulfide mining areas due to oxidation, is well studied because of its importance for environment and human health. Much less information is available on sulfide oxidation in circum-neutral water in aquifers, whose sediments contain sulfides particles as a result of weathering, or where diagenetically authigenic sulfide minerals, such as pyrite and arsenopyrite, have formed, or where groundwater is in contact with mineralized faults. Such sediments are found in several areas around the world and they contribute to high arsenic concentrations in the groundwater used for drinking water supply.

Sulfide oxidation, especially pyrite and arsenopyrite, forms an important mechanism for As mobilization in all areas with sulfide mineralization (non-authigenic As source), and hence is a very common process in areas with volcanism or plutonism-related mineralization. Mineralized faults, epithermal stockwork and veins, massive sulfides, and hydrothermal conduits, which are often contained in altered rock that weathers more rapidly, encompass these conditions. The oxidation of authigenic As-bearing minerals is also important because it contributes to dissolved As concentrations in several aquifers around the world.

Depending on geological, hydrogeological and climatic conditions, weathering and related oxidation at depth can release substantial amounts of As into the groundwater. These processes may become accelerated if past or current mining has occurred, where residual mining deposits and/or acid mine drainage from actual or former mines comes into contact with the hydrosphere (ground-, soil, and surface waters). Because of environmental and health concerns, there are many studies on sulfide mineral oxidation mechanisms (especially of pyrite and arsenopyrite) and the mining-related release of As and other contaminants (see summaries of e.g., Moses *et al.* 1987, Nicholson *et al.* 1988, Nicholson 1994, Rimstidt *et al.* 1994 and references therein).

Oxidation of sulfides, especially of pyrite and arsenopyrite, takes place if they are exposed to atmospheric oxygen or in the presence of Fe(III), which oxidizes them faster than atmospheric oxygen and releases As and Fe into solution (e.g., surface- or groundwater; soil water) or into other minerals (solid solution). Due to the high As contents of pyrite (generally 0.02–0.5%; but exceptionally up to 6.5%; Kolker *et al.* 1998), oxidative pyrite dissolution can contribute important amounts of dissolved As to the fluid. These oxidation processes are often catalyzed by abiotic microbe activity, which increases the reaction rate. The bacteria *Acidithiobacillus thiooxidans*, *Acidithiobacillus ferrooxidans*, and *Leptospirillum ferrooxidans* can increase the reaction rates by about five orders of magnitude (Singer and Stumm 1970, Nordstrom 1982, Nordstrom and Southam 1997, Nordstrom and Alpers 1999). The first two bacteria catalyze the oxidation of Fe(II) by oxygen to form Fe(III), which then oxidizes pyrite much more rapidly than oxygen. Hence, in most natural groundwaters with circumneutral pH, sulfides undergo the much slower process of abiotic oxidation through dissolved oxygen as observed in the field studies of Postma *et al.* (1991), Kinniburgh *et al.* (1994), and Schreiber *et al.* (2000). However, Arkesteyn (1980) found that bacterial pyrite oxidation involving oxygen occurs only at pH <4. An exception are waters with high NO_3^- concentrations. Nitrate in the absence of oxygen can oxidize pyrite at pH >5 (Appelo and Postma 1993) and bacterial oxidation of Fe(II) by *Gallionella ferruginea* and S in pyrite by *T. denitrificans* promotes this reaction.

The Fe released by the oxidation reaction tends to precipitate as sulfates, oxides or oxyhydroxides which is accompanied by the re-adsorption and coprecipitation of As. Under very acidic conditions (pH ~ 1.5) and high arsenic/iron ratios, As may be removed from the fluid by the formation of scorodite ($FeAsO_4 \cdot 2H_2O$) (Langmuir *et al.* 1999), however breakdown to iron oxide can occur. This re-immobilization of As reduces the efficiency of sulfide oxidation for As mobilization. However, if geochemical conditions become favorable (e.g., due to mixing or in contact with other ground-, soil, and surface water sources), the formed metal oxides and hydroxyoxides (especially those of Fe) become unstable (under very acidic conditions; see 2.6.1), or they desorb As from their surfaces (generally at pH >8; see 2.6.2), resulting in increased As concentrations in the fluid. So, Fe oxides/oxyhydroxides are the principal source of As resulting from sulfide oxidation.

Sulfide oxidation also plays a role during the weathering of coal, bitumen and peat deposits. These rocks/organic minerals contain vary variable amounts of sulfides, in particular pyrite, whose oxidation can release important amounts of As into the hydrosphere. There are several areas, where coal deposits crop out at the surface or are found near to it and leaching may occur but mining accelerates the process.

Sulfide oxidation can be important in sedimentary aquifers, which contain layers or lentils of coal and peat black shales, phosphorites, (also some clays and claystones), which contain original or diagenetically formed authigenic pyrite or other As-bearing sulfides, whose oxidation can release As into the groundwater. Sediments derived from mineralized areas may incorporate sulfides through diagenesis. This oxidation process can explain the elevated As concentrations in several sedimentary aquifers, where the As source can be related to the content of these components/products when coming into contact with oxidizing groundwater or soil air. Sulfide minerals seem to be a source of arsenic in groundwater in several aquifers used for drinking water purposes, e.g., (1) groundwater in some bedrock units that underlie an area extending from Massachusetts into Maine (Peters 2008, Zuena and Keane 1985, Boudette *et al.* 1985, Marvinney *et al.* 1994, Ayotte *et al.* 1998, Peters *et al.* 1999) contains high arsenic concentrations, (2) Interior Plains of northeastern Wisconsin (Simo *et al.* 1996, Schreiber *et al.* 2000) and (3) Interior Plains of eastern Michigan (Westjohn *et al.* 1998, Kolker *et al.* 1998). In the Michigan example, authigenic pyrite in consolidated sandstone containing ≤6.5% As (Kolker *et al.* 1998) appears to be an important source of arsenic for the groundwater in the overlying glacial aquifer. However, pyrite has not been identified in the glacial deposits. The example from Wisconsin will be described separately in section 2.4.2 followed by an example from the Franconian Upper Triassic sandstone aquifer, where a hypothesis relates the occurrence of dissolved As in the sandstone aquifer to the presence of clay layers (2.4.3).

2.4.2   *Example: Arsenic mobilization by sulfide oxidation in the near-neutral sandstone aquifer of northeastern Wisconsin, USA*

High concentrations of dissolved As have been reported in the sandstone aquifers from eastern Wisconsin, USA (Schreiber *et al.* 2000; Table 2.4), where As concentrations as high as 12,000 μg L^{-1} were recorded in a confined aquifer of Ordovician sandstone (Schreiber *et al.* 2000). The correlation of dissolved As concentrations with Fe, Cd, Zn, Mn, Cu and SO_4^{2-} have led these authors to the hypothesis that the As release mobilization is due to the oxidation of authigenic sulfides (pyrite and marcasite) which are present in a layer with secondary cementation. These authors further suggest that sulfide oxidation became favored by groundwater exploitation. A groundwater table decrease brought the cemented layer under the influence of atmospheric oxygen. Iron oxyhydroxides are stated as another additional As source (Schreiber *et al.* 2000). However, these minerals may be formed secondarily after sulfide oxidation and release of Fe and As.

2.4.3   *Example: Franconian Upper Triassic sandstone aquifer, Germany*

Geogenic enrichment of As in aquifers is well known from parts of southern Germany. In the Franconian Upper Triassic sandstones, As occurs at concentrations of more than 100 μg L^{-1} in groundwater, and at low concentrations of <4 μg kg^{-1} (mostly <2 μg kg^{-1}) in the sandstone (Heinrichs and Udluft 1999). Sandstone aquifers have generally low concentrations of dissolved As. However, in some cases, such as the aquifer system of the Keuper sandstones (Triassic; Heinrichs and Udluft 1999), which is used for the public water supplies in northern Bavaria, high As concentrations varying from 10 to 150 μg L^{-1} have been reported. The authors relate the presence of As to the geochemical characteristics of specific lithofacies comprising sediments of continental origin. Analyses of core samples from boreholes suggest that As contained in the rock is probably the source of As in the groundwater. Marked As accumulations occur in mudstone layers, where the As content reaches up to 35 mg kg^{-1}. Finally, they concluded that the As contained in the rocks was deposited in a fluvial-marine transitional environment and that the high As content could be linked to mineralized regions along fracture zones, where further groundwater from the deeper Buntsandstein aquifer may rise up and mix with the water of the Upper Triassic sandstone aquifer. This may explain the locally high As concentrations detected in the groundwater.

2.5   ARSENIC INPUT DUE TO LEACHING IN GEOTHERMAL RESERVOIRS: THE ROLE OF GEOTHERMAL FLUIDS

From many sites around the world, geothermal activities, especially geothermal waters (as per definition, >37°C) are known as important As contributors to surface water bodies and cold aquifers (<37°C). These ascending geothermal waters may either discharge at the earth's surface in the form of hot springs, or the uprising geothermal water mixes underground with the cold groundwater of aquifers, which makes its identification as an As source difficult.

Recent geothermal systems are analogous to fossil hydrothermal ore deposits and the linkage between the As release into the environment (e.g., due to hot spring activity), at present or in fossil hot spring systems (which resulted in hot spring stockwork gold deposits) becomes obvious. This analogy explains why the study of recent geothermal systems as showcases and laboratories for fossil systems is of such great importance.

Geothermal activities are mainly found along tectonic plate boundaries. We can distinguish geothermal systems associated with (1) active volcanism and tectonics (e.g., the circumpacific volcanic belt with the examples of Taiwan and the geothermal fields of New Zealand, Philippines, Indonesia, and Middle America); (2) continental collision zones (e.g., Himalayas; Alps); (3) continental rift systems associated with active volcanism (e.g., East Africa); (4) continental rifts (e.g., Italy: Larderello; Germany: Rhine river valley; India: West coast and Gujarat, Rajasthan and SONATA geothermal provinces) (Chandrasekharam and Bundschuh 2008).

In the area belonging to the first and third group with active volcanism, other additional geothermal surface manifestations, such as fumaroles, solfatares, etc., must be considered as potential sources of As. However, they are generally less important when compared to geothermal waters.

The presence of As in geothermal systems, principally in geothermal waters and their impact on cold aquifers, surface waters, and other surface environments are important in many geological settings around the world and will be discussed in an overview in section 2.5.1. In section 2.5.2, some typical case studies will be presented.

2.5.1   *Arsenic input from geothermal waters and other geothermal manifestations*

High As concentrations have been reported in geothermal waters since the 19th century. Some of these historical references can be found in Clarke (1908). Gooch and Whitfield (1888) reported As concentrations of the thermal waters of Yellowstone National Park. Since then elevated As concentrations in geothermal reservoir waters have been reported by hundreds of investigators (see e.g., Stauffer *et al.* 1980, Stauffer and Thompson 1984, Criaud and Fouillac 1989, Welch *et al.* 2000, Romero *et al.* 2000, Bundschuh *et al.* 2009a) from different parts of the world, comprising all habited continents. Geothermal As may degrade drinking water resources, such as the Los Angeles California Aqueduct (average As concentration of 20.2 μg L^{-1}), where the As has been attributed to geothermal activity in the Long Valley area (Mono County) (Wilkie and Hering 1998).

Arsenic is leached from the host rocks of the geothermal reservoir where high residence time, temperature and pressure, and reducing conditions (presence of As(III)), together with the undersaturation of most reservoir fluids with respect to arsenopyrite and other As minerals favor the dissolution of As (Webster and Nordstrom 2003). This leaching occurs along with other elements, such as antimony (Sb), boron (B), fluoride (F), lithium (Li), mercury (Hg), selenium (Se), thallium (Tl), and hydrogen sulfide (H_2S), making these elements good fingerprints for identifying components of geothermal waters where mixing with cold aquifers or surface waters has occurred. In many rocks of geothermal reservoirs, As is mainly concentrated in pyrite, whereas arsenopyrite (FeAsS) appears to be uncommon (Webster and Nordstrom 2003). At temperatures of 150–250°C, As occurs predominantly as As-bearing pyrite, with As concentrations up to 3.7 wt% (Ballantyne and Moore 1988, Ewers and Keyas 1977) or associated with Fe oxides (Christensen *et al.* 1983). At temperatures >250°C, As is mostly found as arsenopyrite (FeAsS), As-bearing pyrite (FeS_2), and other arsenides, such as lollingite ($FeAs_2$) (Heinrich and Eadington 1986, Pokrovski *et al.* 2002, Scott, 1983). How-ever, the exact chemical environment of As in pyrite is still unclear (Kolker and Nordstrom 2001, Simon *et al.* 1999), and the formation of a temperature-dependent solid-solution reac-tion between pyrite and arsenopyrite is assumed (Pokrovski *et al.* 2002).

Table 2.5 gives a worldwide overview of As concentrations in fluids of geothermal wells and geothermal springs, together with a simplified geological characterization of the rock type of the geothermal reservoir and its temperature. Generally, the highest As concentrations are found in the fluids of geothermal reservoirs in volcanic rocks (typical range thousands to tens of thousands of µg kg^{-1}, but as high as 162,000 µg kg^{-1} at Los Humeros, Mexico (González *et al.* 2001, Arellano *et al.* 2003; Table 2.5). Exceptions are the geothermal reservoirs of Hawaii and Iceland, where low As concentrations are found in the fluids of volcanic reser-voir rocks (<0.1 µg kg^{-1}), which is explained by the presence of fresh basaltic reservoir host rocks low in arsenic (Webster and Nordstrom 2003). However, the low residence time may be the key reason rather than the low As concentration in the reservoir rocks. Compared to geo-thermal reservoirs hosted in volcanic rocks, much lower As concentrations (<3–2010 µg kg^{-1}; Table 2.5) are found in geothermal reservoirs located in sedimentary rocks comprising both high (e.g., Cerro Prieto, Mexico: 250–500 µg kg^{-1}) and low *enthalpy geothermal reservoirs* (SE Mexican oil fields: <3–2010 µg kg^{-1}, Table 2.5) (Birkle *et al.* 2002, 2003, 2009a,b, Birkle 2004, Birkle and Bundschuh 2009).

Geothermal waters change their chemical composition during their ascent from the geo-thermal reservoir to or near to the earth's surface due to several physical and chemical proc-esses. Fluids, which ascend towards the earth's surface, without loss of heat (marginal loss of heat due to conductive cooling), will emerge as NaCl-type water with near neutral pH, high silica content and a Cl^-/SO_4^{2-} ratio >1. Magmatic CO_2 and H_2S are the major gas phases in these waters. These NaCl waters generally show the highest As concentrations since the water corresponds to the original reservoir waters. Uprising geothermal waters, when mixed with the near surface groundwater rich in HCO_3^- become a HCO_3-type water. Geothermal waters with a high content of H_2S gas, which condense near to the earth's surface, form pools of water with high SO_4^{2-} and low Cl^- contents. Due to oxidation, such pools give rise to low pH and high SO_4^{2-} water at the surface. Acid sulfate-chloride water evolves when the oxidation of volcanic H_2S which is sometimes contained in large amounts in NaCl waters occurs. This forms bisulfate ions (HSO_4^-), a reaction which is promoted by decreasing temperature, that means in the zone near to the earth's surface, resulting in low pH water (Nordstrom *et al.* 2009). Arsenic and Cl remain in the fluid phase during subsurface boiling and phase separa-tion (Webster and Nordstrom 2003). Consequently, many geothermal waters show a positive correlation between As and Cl. However, this correlation does not prove a common source of As and Cl^-. Although As is derived predominantly from the leaching of the host rocks in the geothermal reservoir, there are different potential Cl^- sources: (1) host rock leaching, (2) seawater component (see the example of the Mexican oilfields in section 2.5.2 where a linear relation between As and Cl^- corresponding to that of seawater was observed, or the Taiwan example, where one of the theories for the genesis of the arseniferous aquifers is seawater and evaporative concentration increase), (3) gaseous HCl from magmatic components.

During the adiabatic cooling of uprising geothermal waters (through boiling due to lower pressure) without a loss of heat to the wall rock, the pH increases due to CO_2 loss and base metals (e.g., Cu, Pb, Zn) precipitate as the first ore deposits. However, As, together with

Table 2.5.   Arsenic concentrations of fluids from wells and hot springs of selected geothermal systems.

Geothermal field/area	Arsenic (μg kg^{-1})	Reference
Geothermal wells		
Chile		
El Tatio, Chile (VOL; H)	30000–40000	Cusicanqui *et al.* 1976
	45000–50000	Ellis and Mahon 1977
Costa Rica		
Miravalles (VOL; H)	11900–29100	Hammarlund and Piñones 2009
Rincón de la Vieja (VOL; H)	6000–13000	Hammarlund and Piñones 2009
Mexico		
Cerro Prieto (SED; H)	250–1500	Lippmann *et al.* 1999,
		Mercado *et al.*1989
Los Azufres (VOL; H)	5100–49600	Birkle 1998,
		González *et al.* 2000
Los Humeros (VOL; H)	500–162000	González *et al.* 2001,
		Arellano *et al.* 2003
Cactus-Sitio Grande (SED; L)	<3–47	Birkle and Portugal 2001,
		Birkle and Angulo 2005
Luna-Sen (SED; L)	<3–548	Birkle *et al.* 2002
Jujo-Tecominoacán (SED, L)	<3–1900	Birkle 2004,
		Birkle *et al.* 2009a,b
Pol-Chuc-Abkatún (SED, L)	90–2010	Birkle 2003
New Zealand		
Broadlands (VOL; H)	5700–8900	Ewers and Keays 1977
Kawerau (VOL; H)	539–4860	Browns and Simmons 2003
Orakei Korako (VOL; H)	599–802	Ellis and Mahon 1977
Waiotapu (VOL; H)	2900–3100	Ellis and Mahon 1977
Wairakei (VOL; H)	4100–4800	Ellis and Mahon 1977
	1000–5200	Ritche 1961
Philippines		
Tongonan (VOL; H)	20000–34000	Kingston 1979
	28000 (mean)	Darby 1980
USA		
Lassen National Park (VOL; H)	2000–19000	Thompson 1985
Russia		
Kamchatka (VOL; H)	2000–30000	Goleva 1974
Ebeko volcano, Kuril Islands (VOL)	190–28000	Khrarnova 1974
Geothermal springs		
Chile		
El Tatio, Chile (VOL)	47000	Cusicanqui *et al.* 1976
	45000–50000	Ellis and Mahon 1977
Ecuador		
Tambo river area (VOL)	1090–7850	Cumbal *et al.* 2009,
		Bundschuh *et al.* 2009.
Costa Rica		
Miravalles (VOL; H)	5–4650	Hammarlund and Piñones 2009
Rincón de la Vieja (VOL; H)	5–10900	Hammarlund and Piñones 2009
Nicaragua		
Tipitapa (VOL)	262	Lacayo *et al.* 1992
Mexico		
Los Azufres (VOL; H)	<3900	Birkle and Bundschuh 2009
USA		
Salton Sea (granite intrusion; H)	30–12000	White 1968
Kilahuea, Hawaii (VOL)	60–105	de Carlo and Thomas 1985
Soda Dam/Valles Caldera, NM	1700	Reid *et al.* 2003
(VOL; H)	21–2400	Criaud and Fouillac 1989
Yellowstone, WY (VOL; H)	157–15000	Stauffer and Thompson 1984
		Nordstrom 2009, *pers. commun.*

(*Continued*)

Table 2.5.   (*Continued*)

Geothermal field/area	Arsenic ($\mu g\ kg^{-1}$)	Reference
Russia		
Kamchatka (VOL; L, H)	794–944	Belkova *et al.* 2004
Dominica (Lesser Antilles) (VOL)	7–90	McCarthy *et al.* 2005
Papua New Guinea	817–952	Pichler and Veizer 1999
Tutum Bay (VOL)		
New Zealand		
Broadlands (VOL; H)	996	Ellis and Mahon 1977
Waiotapu (VOL; H)	712–6470	Jones *et al.* 2001
	710–6500	Webster 1990
Orakei Korako (VOL; H)	307–382	Papke *et al.* 2003
Wairakei (VOL; H)	3740–5110	McKenzie *et al.* 2001
	230–3000	Ritchie 1961
Philippines		
Mt. Apo (VOL; H)	3100–6200	Webster 1999
Japan		
Tamagawa (VOL)	2300–2600	Noguchi and Nakagawa 1969
Russia		
Kamchatka (VOL, H)	2000–3600	Karpov and Naboko 1990
Taiwan		
Geothermal Spring Valley	1070–4210	(this book)
Italy		
Phlegraean Fields (VOL)	12–12600	Celico *et al.* 1992

Geological characterization of geothermal system after Chandrasekharam and Bundschuh 2008. VOL: volcanic rocks; L: low temperature reservoir, T < 150°C; H: high temperature reservoir, T > 150°C.

other elements, such as Au, Sb, and Hg, remain dissolved as oxyanions under the higher pH conditions and precipitate later in the zone near to the earth's surface. Recent geothermal systems are analogous examples for fossil epithermal ore deposits, and the linkage between arsenic release into the environment due to the mining of epithermal gold deposits becomes evident. The most common As-bearing minerals encountered in epithermal deposits are orpiment (monoclinic As_2S_3 and its amorphous analog), realgar (monoclinic AsS), and elemental arsenic (As^0). Their deposition occurs in response to acid reactions with metal-bearing waters, such as the acidification of hot spring waters with acid-sulfate waters, subaerial cooling of the fluid or increased H_2S concentrations.

Reducing conditions prevail along the pathways of ascending geothermal water until near the earth's surface zone, where atmospheric O_2 becomes increasingly available, either during mixing of the geothermal water with cold water of oxidizing aquifers or when it discharges at the earth's surface. At the surface, the exposure to oxygen results in a fast oxidation of As(III) to As(V) and the precipitation of different mineral phases (Alsina *et al.* 2007, Bundschuh *et al.* 2008) making it to an As sink, which removes As from the fluid to a variable extent. Overviews on precipitated minerals and mechanisms are given in Alsina (2007), and Webster and Nordstrom (2003). However, in waters with a high sulfide content, this oxidation does not occur (Stauffer *et al.* 1980, Langner *et al.* 2001).

Available studies indicate As(V) is the dominant oxidation state for As in many hot spring deposits, whereas As(III) is present in mineral deposits where orpiment and realgar form. Arsenic sorption by Fe oxyhydroxides controls solid-phase As speciation in a wide range of geothermal systems where Fe is present, such as hydrothermal waters that are either about neutral NaCl waters or acidic sulfate- and chloride-rich waters (Alsina *et al.* 2007). Arsenic sorption on Fe oxyhydroxides and the related association between As and Fe oxyhydroxides in geothermal systems is essential to understand the mobility of geothermal As. It is well-known that microbes in geothermal systems may catalyze the fast oxidation of As(III) to

As(V), but it is little known how microbial catalysis and/or biomineralization control the formation of Fe oxyhydroxides.

The predominant role of Fe oxyhydroxides as primary As sequestration agents in hot spring deposits, and possibly also in the shallow cool oxidizing aquifer, indicates that they constitute the principal As source for re-mobilization under changing geochemical conditions, as we will discuss in section 2.7. We also need to consider that if geothermal water discharges into shallow aquifers, or surface waters, significant amounts of As will remain in solution, either as As(V) or as As(III). The As species distribution depends on residence time of the groundwater in the aquifer and on the kinetics of the As(III) oxidation, which strongly depends whether and to which extent microbes, which can catalyze the As oxidation are present. If discharging in surface waters, in most cases, the dissolved As is found as As(V), and this is also true in our case study from Ecuador (see section 2.5.2). However, if the geothermal water discharges directly into aquifers, oxidation may be a slow and purely inorganic chemical reaction.

The Fe-oydhydroxides and other mineral phases can be further eroded by aeolic processes and surface run-off after rainfall events and reach different environments. Thereby As may be remobilized into aqueous environments.

However, we need to consider that we can expect a significant release of As only from those geothermal waters which contain a significant component of geothermal reservoir water, which correspond generally to NaCl type waters, and not from those geothermal waters that correspond exclusively to shallow groundwater, which was heated by conduction or steam as we can see in the case examples from Costa Rica (see 2.6.2).

For evaluating the importance of geothermal fluids as As sources, we consider that geothermal surface manifestations, such as geothermal springs, where local conditions as morphology or near surface geology favors these natural outlets, are rare events compared to subsurface discharges of geothermal waters into cold aquifers. This, together with the wide distribution of geothermal activities, lets us assume that geothermal waters may be much more important As sources for groundwater resources than previously thought.

2.5.2   *Examples of arsenic input from geothermal waters*

One of the best investigated areas is the geothermal springs in the Yellowstone National Park (USA), with volcanic host rocks as the geothermal reservoir(s), which have As concentrations of mostly >1000 µg L^{-1} (Staufer and Thompson 1984, Ball *et al.* 1998) and up to 15,000 µg L^{-1} (Nordstrom 2009, *pers. commun.*) Their discharge into the Madison river results in As concentrations of 360 µg L^{-1}, and 19 µg L^{-1} in the Missouri river, 470 km downstream of the springs (Nimick 1994, Nimick *et al.* 1998). The discharging water contains predominantly As(III), which is oxidized quickly to As(V) (Nordstrom *et al.* 2005). Eccles (1976) and Wilkie and Hering (1998) consider geothermal waters also in several other areas of the USA as the primary source of arsenic, including some surface waters in the eastern part of the Sierra Nevada.

In northern Chile, high As concentrations have been found in geothermal waters in the El Tatio geothermal system in the Antofagasta region (100–1000 µg L^{-1}) (Caceres 1999, Queirolo 2000). However, much higher As concentrations of up to 50,000 µg kg^{-1} have been reported by Cusicanqui *et al.* 1976 from Tatito springs, whereas concentrations in well water are 30,000–40,000 µg kg^{-1}. Ellis and Mahon (1977) mentioned similar arsenic concentrations of 45,000–50,000 µg kg^{-1} at El Tatio. These numbers have recently been confirmed by some unpublished data from Bennet and others (Nordstrom 2009, *pers. commun.*). These geothermal discharges contribute to As enrichment in Loa river (Cusicanqui *et al.* 1976), the principal source of drinking and irrigation water for a population of the Loa river basin (~500,000). However, Tatio springs are not the main source of As in the Loa river. The principal sources are located in the volcanic mountain chain of the Andes, from where As is released by weathering processes (Bundschuh *et al.* 2006, 2008a,b, 2009, Bundschuh and García 2008).

In Ecuador, geothermal As (113–844 µg L^{-1}) has been determined in the northern Andean region from different localities (El Carchi, Imbabura, Pichincha, Cotopaxi, and Tungurahua provinces; Cumbal *et al.* unpubl. data). El Angel river (El Carchi province) receives thermal waters and shows As concentrations in the range of 64 to 113 µg As L^{-1} (Cumbal *et al.* unpubl. data).

The discharge of geothermal springs into surface waters was studied by Cumbal *et al.* (2009) in the Andean region (Quijos county, Napo province, NE Ecuador). Here seven geothermal springs discharge at an average altitude of 3360 m a.s.l. into the Tambo river. The water of the studied springs contained 1090–7852 µg L^{-1} of As. Arsenic species were only determined in four of them. In two of the springs, As(III) dominates (74.4 and 61.2% of total As, amounting 3152 and 6120 µg L^{-1}, respectively). This result is in agreement with the low redox potential of the discharged water (–112.2 and –103.8 mV, respectively). Abundant Fe(III) precipitates around the spring outlets indicate a fast oxidation of the water and explain why As found in the surface water is As(V). In contrast, As(V) is predominant in the other two springs (67.8% and 66.5% of their total As concentrations, amounting to 3555 and 7852 µg L^{-1}, respectively). This predominance of As(V) is in agreement with the higher measured redox potential compared to the previous two springs (+9.2 and +7.3 mV, respectively). The Tambo river flows 13 km further downstream, together with 6 other smaller tributaries with low As concentrations (6–13 µg L^{-1}), into Papallacta lake, causing high As concentrations in the lake, which undergo significant seasonal fluctuations (86–369 µg L^{-1}) (Cumbal *et al.* 2009).

In Costa Rica, at Rincón de la Vieja geothermal area, geothermal springs are characterized by low As concentrations with the exception of one site, which is the only site where a natural surface discharge of the Na-Cl geothermal reservoir fluids exists (two closely located springs; Salitral Norte 1 and 2) (Birkle and Bundschuh 2007). As the only surface discharges of the geothermal reservoir (predominantly andesites), their As concentration is high (10,600 and 10,900 µg L^{-1}, respectively; Hammarlund and Piñones 2009) reflecting the high As concentrations found in the geothermal reservoir (6000–13,000 µg L^{-1}; mean from 4 wells 9900 µg L^{-1}). All other geothermal springs correspond to predominantly conductively heated shallow meteoric waters (Birkle and Bundschuh 2007) and As concentrations are consequently lower (5.2–132 µg L^{-1}; mean from 16 springs: 25 µg L^{-1}; Hammarlund and Piñones 2009); those with the higher As concentrations indicate that the spring water has a subordinate component of reservoir water.

From the springs of Miravalles geothermal reservoir area (Costa Rica), two adjacent springs show an As concentration, which by far exceeds the As concentrations of the other springs of the areas. These springs (Salitral Bagaces 1 and 2; As: 4564 µg L^{-1}) are also the highest mineralized springs of the area (TDS 6386 to ~6656 mg kg^{-1}) and contain high concentrations of Cl$^-$ and Na$^+$ (2600 and 2100 mg L^{-1}, respectively) and its water can be characterized as Na-Cl-HCO$_3$-type water (Birkle and Bundschuh 2007). This, together with a surface temperature of about 60°C, indicates that this water is a product of the mixing of cold local meteoric waters, which dilute the water that is uprising from the andesitic geothermal reservoir, which is of Na-Cl type and contains As concentrations of 11,900–29,100 µg L^{-1} (13 wells; average 24,400 µg L^{-1}; Hammarlund and Piñones 2009). All other geothermal springs have much lower As concentrations (5–280 µg L^{-1}; mean from 14 springs: 281 µg L^{-1}; Hammarlund and Piñones 2009). They are predominantly steam or conductively heated, and those with higher As concentrations indicate a greater contribution of the geothermal reservoir component. Three prevailing steam-heated fumaroles/mud pots described in Miravalles and Rincón de la Vieja area (temperature: 87–91°C; pH: 2–3.8; rich in S^{2-} and SO$_4^{2-}$), show low As concentrations in the fluids (5.9–194.9 µg L^{-1}, 2.7–141.4 µg L^{-1}, respectively), whereas the concentrations of dissolved Fe, Mn and Al are high (Hammarlund and Piñones 2009; Birkle and

Bundschuh 2007). In contrast, the extremely acidic geothermal fluids of Poás volcano (pH <1) (Rowe *et al.* 1992ab) contain high As concentrations (e.g., Naranjo fumarole: 14,710 µg L^{-1}; Bundschuh *et al.* unpublished data).

In Mexico, Birkle and Bundschuh (2009) performed a comparative study of As in the thermal fluids of high enthalpy geothermal (>150°C) and low enthalpy petroleum reservoirs (<150°C). However, the lack of chemical similarity between both systems indicated a distinct origin for As. Oilfield waters from sedimentary basins in SE-Mexico show maximum As concentrations of ~2000 µg L^{-1} (depth: 2900–6100 m b.s.l.) (Cactus-Sitio Grande: <3–47 µg L^{-1}; Luna-Sen <3–548 µg L^{-1}; Jujo-Tecominoacán <3–1890 µg L^{-1}; Pol-Chuc-Abkatún: 90–2010 µg L^{-1}). The linear Cl$^-$/As correlation for oilfield waters indicates that As input occurs during the mixing between meteoric water and evaporated seawater, and that there is only minor As derived from interaction with the carbonate host rock, even at relative high temperatures of about 130°C. In the Los Azufres area, As concentrations of up to 3900 µg L^{-1} found in surface manifestations (hot springs, fumaroles), are probably due to the vertical ascent of convective fluids to the surface (Birkle and Merkel 2000). Arsenic concentrations in the volcanic host rocks of the reservoir range from 45–49,600 µg L^{-1}. The reservoir water of Los Humeros geothermal field shows a wide range of As concentrations (500–162,000 µg L^{-1}). The latter is the highest As concentration detected to date in geothermal reservoir fluids in Latin America. In both of these geothermal fields, the lack of correlation between As and salinity reflects the importance of secondary water-rock interaction processes between the fluid and the volcanic host rocks. In contrast, the dominance of sandstones in the sedimentary basin of the Cerro Prieto geothermal field (Baja California state, NW Mexico) explains the relatively low As concentrations in the reservoir fluid (250–1500 µg L^{-1}) even though temperatures reach up to 370°C. The sedimentary origin of the reservoir rocks also explains the similar As concentrations in Cerro Prieto and in the oil reservoir fluids.

We have already mentioned that it is often difficult to identity the geothermal waters as the principal As source if they mix underground with cold aquifers, especially if the background chemistry of the cold groundwater and the supposed mixing geothermal water are similar or if the detailed hydrochemical composition is unknown. There are many reported examples of the occurrence of As in warm aquifers (e.g., Hurtado-Jiménez and Gardea-Torresdey 2009: Los Altos de Jalisco, Mexico; Nicolli *et al.* 2009: Burruyacú basin, Tucumán province, Argentina; García *et al.* 2009: northwestern Chaco-Pampean plain, Argentina) where groundwater temperatures do not correspond to the regional geothermal gradient, which is either caused by conductive heating due to local heat anomalies or uprising deep geothermal water. However, in most cases it remains unclear whether the increased temperatures may increase As mobility in the aquifer itself, or if the inflowing geothermal water is the source of the As (or whether both processes coexist).

For example, García *et al.* (2009) performed a hydrochemical study of As in deep geothermal groundwater from the northwestern Chaco-Pampean plain, close to the Andean mountains range from where the aquifer is recharged (geothermal area of Las Termas de Río Hondo, SE Tucumán province; Argentina) to identify the processes that control its distribution in the aquifer. Two aquifers were identified in the geothermal area: (1) a shallow unconfined aquifer with prevailing loess-type sediments (depth: 0.05–30 m), and (2) a confined multilayered aquifer composed of alternating layers of gravel, sand, silt and clay. The presence of a fault system allows the upflow of hot deep groundwater into this deep aquifer system. Samples from 10 wells (depths: 203–474 m; average: 334 m) show moderate oxidizing and alkaline conditions (Eh: 144–311 mV, average: 189 mV; pH: 7.9–9.1, average 8.0) and As concentrations ranging from 27–76 µg L^{-1} (mean 52 µg L^{-1}). Groundwater temperatures range from 34.2–44.5°C (average 38.8°C), confirming the inflow of deep geothermal water, but the contribution of As to the groundwater can not be confirmed since other sources, such as the aquifer material, may explain the main and trace element composition of the groundwater.

2.6   THE ROLE OF Fe, Mn, AND Al OXIDES AND OXYHYDROXIDES AS SOURCES AND SINKS FOR DISSOLVED ARSENIC

We have seen in previous sections that metal oxides and oxyhydroxides, especially those of Fe, Mn and Al, can be formed as secondary As-bearing minerals from many geogenic processes such as sulfide oxidation, geothermal activities, and general dissolution/leaching of rocks and minerals followed by precipitation of these minerals as a function of local hydrochemical conditions (especially of redox and pH), e.g., on sediment particles of sedimentary aquifers. Arsenic can thereby be coprecipitated (within the crystal lattice) or adsorbed on the surface of the metal oxides and oxyhydroxides. Oxides and oxyhydroxides of Fe, Mn, and Al form the principal secondary minerals which can act—depending on the present geochemical conditions—as sources and sinks for dissolved As. The retention/release of As by oxides and oxyhydroxides is predominantly controlled by sorption and precipitation/dissolution. These processes can explain the As concentrations of many areseniferous aquifers, especially those formed by unconsolidated and consolidated sediments.

To understand the As mobilization processes from metal oxides and oxyhydroxides, we need to consider under different geochemical conditions: (1) the stability of these minerals, (2) their capacity for As adsorption, and (3) the role of As species present in the fluid.

2.6.1   *Arsenic release by dissolution of metal oxyhydroxides*

There are two principal mechanisms for the dissolution of metal oxides and oxyhydroxides: (a) oxides and oxyhydroxides of Fe and Mn dissolve in strongly acid environments, such as acid mine drainage, and acidic fumaroles or acidic hot spring deposit environments, resulting in a release of As, (b) dissolution of Fe and Mn oxides and oxyhydroxides under strongly reducing conditions at near-neutral pH (reductive dissolution) and release of As. The reductive dissolution, especially that of $Fe(OH)_3$ and the related As release into the aqueous phase, involves iron oxide and organic carbon, which comes from organic matter contained in the sediments (or anthropogenic organic compounds). This process can explain the release of As, which is either present in a coprecipitated form or adsorbed on the surface into the groundwater for several important aquifers, such as the Bengal delta aquifer. Here, reductive dissolution of amorphous Fe(III) oxyhydroxides (present as coatings on sediment particles in combination with organic matter) is the principal process responsible for the elevated concentrations of dissolved As. However, there are still uncertainties about the detailed steps and reductants, including organic matter and anaerobic bacteria involved in the reaction. The same process also seems to be the predominant As mobilization mechanism in the sedimentary aquifers of SE Asia, such as the Menkong delta, Indus plain, Irrawady delta and Red River delta whose sediments and As have its origin in the Himalayas and its southeastern continuation, and which are deposited under similar sedimentological conditions. Reductive dissolution seems also to be the responsible primary As release mechanism in several aquifers.

In the USA, arsenic release into the groundwater from the dissolution of Fe-oxides has been described from several alluvial and glacial aquifers of the upper Midwest of the USA (Interior Plains), whose groundwater contains high concentrations of As (Voelker 1986, Holm and Curtiss 1988, Panno *et al.* 1994, Holm 1995, Roberts *et al.* 1985, Kanivetsky 2000, Ziegler *et al.* 1993, Korte 1991). In some parts, additional As desorption due to elevated pH is reported (Roberts *et al.* 1985, Kanivetsky 2000). Due to the widespread association between glacial deposits and high As concentration in groundwater, Welch *et al.* (2000) also suggested that other parts of the upper Midwest may have high As concentrations. Other areas in the USA, where As release due to the dissolution of Fe oxides has been reported, comprise the Intermontane Plateaus and parts of the Pacific Mountain System where As evaporative concentration together with dissolution of Fe oxides explain the high As concentrations found in shallow aquifers of the Carson Desert, Nevada (Welch and Lico 1998), the southern San Joaquin Valley (Fujii and Swain 1995, Swartz 1995, Swartz *et al.* 1996), the Appalachian

Highlands, NE Ohio (Matisoff *et al.* 1982) and in some areas of the Pacific Mountain System (Hinkle 1997, Hinkle and Polette 1999).

Example 1: The Bengal delta plain comprises the state of West Bengal (India), large parts of Bangladesh and the northern and western regions of the Indo-Gangetic Alluvium plain in India and Nepal (Bhattacharya *et al.* 1999, 2003). The source of the sediments and the As is attributed to erosion in the Himalayas and sedimentation in the Bengal delta plain by the rivers Ganges, Brahmaputra, and Megna (Nickson *et al.* 2000, Bhattacharya *et al.* 2002a).

The concentration of As in the groundwater of the Bengal delta shows a patchy distribution which is caused by the variability of hydrogeochemical aquifer conditions (redox potential, pH, temperature, water type, etc.). The analyses of ~125,000 groundwater samples (Mandal *et al.* 1996, Chakravorti *et al.* 2002, 2004, Chowdhury *et al.* 2000) report an As concentration range of <1–1300 µg L^{-1}; organic As compounds were not detected.

Dissolved As found in the reducing aquifers is predominantly present as As(III). The predominant As(III) species is the neutral arsenious acid (H_3AsO_3). Charged As(III) species only become important in alkaline waters. At $pH_w \approx 9$, $H_2AsO_3^-$, and H_3AsO_3 at higher pH, $H_2AsO_3^-$ is the prevailing As(III) species. Due to its neutral form, As adsorption is not important in reducing aquifers and in the case of water that is moving from oxidizing to reducing conditions (e.g., along a groundwater flow path), As(V) may be reduced and desorbed. The release of coprecipitated As in the Fe oxyhydroxides by reductive dissolution is postulated to be the principal process in the aquifers that trigger As mobility through the dissolution of Fe(III) oxyhydroxides and transfer substantial amounts of As into the aqueous phases (Bhattacharya *et al.* 1997, 2002c, Smedley and Kinniburgh 2002, Bauer and Blodau 2006).

Bhattacharya (2002) postulated the following scheme of principal reactions as mechanisms of arsenic release in the groundwater of the Bengal delta plain:

$$CH_2O + O_2 \rightarrow CO_2 + H_2O \qquad \text{(oxidation of organic matter by } O_2\text{)}$$

$$CO_2 + H_2O \rightarrow H_2CO_3 \rightarrow H^+ + HCO_3^- \qquad \text{(dissociation and hydrolysis)}$$

$$5CH_2O + 4NO_3^- \rightarrow 2N_2 + 4HCO_3^- + CO_2 + 3H_2O \qquad \text{(denitrification)}$$

$$2CH_2O + SO_4^{2-} \rightarrow 2HCO_3^- + H_2S \qquad \text{(sulfate reduction)}$$

$$4Fe^{III}OOH + CH_2O + 7H_2CO_3 \rightarrow 4Fe^{II} + 8HCO_3^- + 6H_2O \qquad \text{(reductive dissolution of Fe-oxides)}$$

In Bangladesh, it has been found that the groundwater composition and redox conditions are strongly correlated to the sediment color (von Brömssen *et al.* 2007, Bundschuh *et al.* 2009). Groundwater extracted from the black sediments is mostly reduced, followed by white, off-white and red, which are less reduced. Consequently, neither Fe nor As were found at elevated levels in the groundwater extracted from the less reduced white, off-white and red sediments. If this field color is correlated with local geological conditions, it could be used to target safe aquifers.

In southern Taiwan, As and Fe concentrations in the Chianan plain groundwater at two different times (2005 and 2006) were variable and could be linked to the reductive dissolution of FeOOH or the formation of biogenic carbonate and sulfide minerals (Nath *et al.* 2008). The PHREEQC modeling results demonstrated that the siderite becomes comparatively more saturated, while goethite is comparatively less saturated in 2006, revealing that the Fe phase transformation controls the dissolved Fe and As concentrations in groundwater (Nath *et al.* 2008).

The concurrent increases in concentrations of Fe and Mn in 2006 may be caused by bacterial Fe(III) and Mn(IV) reduction, which releases dissolved Fe(II), Mn(II), and other trace metals (e.g., Sr and Ba) that have been adsorbed or coprecipitated by the oxides (Nath *et al.* 2008). Arsenic concentration was slightly lowered or showed little change during 2006, suggesting that As released by bacterial Fe reduction is immobilized by precipitation with biogenic carbonate or sulfide minerals which may act as local sinks (Nath *et al.* 2008). Bacterial sulfate reduction and the precipitation of iron sulfides could strip Fe(II), SO_4, H^+, and perhaps other trace elements (by coprecipitation) from solution in different ratios (Kirk *et al.* 2004, Lowers *et al.* 2007).

In the Hetao Basin, Inner Mongolia, China, microbial processes in the sediments rich in organic matter create a favorable reducing environment and thus facilitate the mobilization of As in the aquifers of the Hetao Basin (Guo *et al.* 2008). Microbial processes also play an important role in the mobilization of As in Bangladesh groundwater (Islam *et al.* 2004, van Geen *et al.* 2004, Anawar *et al.* 2006). The anaerobic bacteria can reduce As(V) to As(III) and thus are able to mobilize As in shallow reducing aquifers (Islam *et al.* 2004) including those in Cambodia (Pederick *et al.* 2007, Lear *et al.* 2007). The incubation study by Guo *et al.* (2008) demonstrated that high concentrations of dissolved Fe are found in the amended suspensions (up to 98.3 mg L^{-1}) and thus confirmed that Fe-reducing bacteria (IRB) are present in the original Hetao aquifer material. Iron-reducing bacteria could promote the reductive dissolution of Fe(III) oxides/oxyhydroxides and stimulate the reduction of As(V) to As(III) (Guo *et al.* 2008). Many studies showed that both NO_3 reducers (NRB) and SO_4 reducers could also use As(V) as a terminal electron acceptor. NRB can respire and also use As(V) as a terminal electron acceptor (Macy *et al.* 1996). In the Hetao basin, the organic substances and the anaerobic environment are the preliminary factors accelerating the microbially mediated release of As (Guo *et al.* 2008). Akai *et al.* (2008) suggested that the combined microbial processes of sulfate reduction to generate anaerobic conditions and Fe reduction to co-reduce As, resulted in the release of As in the Bangladesh sediments.

In Cambodia, arsenic levels are particularly high in the Kandal province (average: 250 µg L^{-1}, n = 175) where reductive dissolution of arsenic-bearing minerals took place under anoxic conditions (Berg *et al.* 2007).

In Vietnam, several studies (BGS and DPHE 2001, Korte and Fernando 1991, McArthur *et al.* 2001, Nickson *et al.* 2000, Eiche *et al.* 2008, Buschmann *et al.* 2008, Berg *et al.* 2001, 2007, 2008) have suggested that elevated arsenic levels in groundwater are caused by the reductive dissolution of arsenic-rich iron oxyhydroxides.

In the sediment and water at the Mokrsko-West gold deposit, Central Bohemia, Czech Republic, the highest concentrations of dissolved As were found in groundwater (up to 1141 µg L^{-1}) (Drahota *et al.* 2009). The main processes releasing dissolved As in a redox transition zone, where neither sulfide minerals nor Fe oxyhydroxide are stable, are believed to be the reductive dissolution of Fe oxyhydroxides and arsenate minerals, resulting in a substantial decrease in their amounts below the groundwater level.

2.6.2   *Arsenic release/sequestration due to sorption by Fe, Mn and Al oxides and oxyhydroxides*

Depending upon the geochemical conditions which prevail, e.g., in an aquifer, As can be adsorbed or desorbed from metal oxides and oxyhydroxides, especially from those of Fe, Mn, and Al, which are found mostly as secondary minerals in many aquifers. Desorption is the principal control of dissolved As concentrations in many aquifers composed of consolidated and unconsolidated sediments. To understand the principal mechanism of adsorption and desorption of As from different Fe, Mn, and Al oxides and hydroxides, and its control especially by pH, redox potential and the presence of ions that are competing with the As for adsorption sites, we will give a short overview followed by some examples.

Adsorption of dissolved As on the metal oxide and oxyhydroxide surfaces may occur due to physical interactions (electrostatic forces) and/or the formation of complexes on the surface of the solid particle. We can distinguish between aqueous complexes (outer-sphere complexes) and the inner-sphere complexes, which have covalent bindings and hence a stronger binding affinity. Spectrometric studies (EXAFS) have shown that As(V) and As(III) predominantly form inner-sphere complexes on Fe, Mn and Al though the structural details are still being discussed. Evidence shows that (1) As(V) predominantly forms bidentate binuclear surface complexes on Fe and Al oxide surfaces and (2) in the case of high surface coverage, additional bidentate mononuclear complexes may be of importance (Waychunas *et al.* 1993, 1995, 1996, Fendorf *et al.* 1997, Inskeep *et al.* 2004, Randall *et al.* 2001, Sherman and Randall 2003, Arai *et al.* 2001). For As(III), evidence for the formation of bidentate binuclear complexes was also found (Manning *et al.* 1998, Arai *et al.* 2001). The formation of weaker outer-sphere complexes, which become important at low ionic strength, was described for Al minerals (e.g., Arai *et al.* 2001, Goldberg and Johnston 2001).

If we consider reducing waters, the prevailing As(III) can be strongly sorbed by ligand exchange depending on the type of oxide and the presence of other anions (Dixit and Hering 2003). In inner-sphere complexes, bidentate, binuclear species of As and oxide/oxyhydroxide (S) were identified and a 3-layer model, which permits the elucidation of the following principal reactions was established (Sverjensky and Fukushi 2006):

$$2 =SOH + As(OH)_3 \leftrightarrows (=SO)_2 \cdot As(OH) + 2H_2O$$

$$=SOH + As(OH)_3 \leftrightarrows =SOH_2^+ - AsO(OH)_2^-$$

2.6.2.1 *Influence of redox potential and pH on adsorption capacity*

The adsorption of ions on the solid phase of a surface depends on the surface charge of the solid. Generally, metal oxides and oxyhydroxides have a variable surface charge, which is a function of its mineral-specific pH point of zero charge (pH_{pzc}), where the mineral surface has no net charge on its surface and the pH of the water (pH_w). Hence, the adsorption affinity for As oxyanions depends on both of these parameters. If $pH_w < pH_{pzc}$ the mineral surface is positively charged and can adsorb As if this is present as oxyanions (adsorption capacity increases with decreasing pH_w). Considering the pH range of most ground- and surface waters (5–9), As(V) is predominantly present as oxyanions ($H_2AsO_4^-$, $HAsO_4^{2-}$), whereas the predominant As(III) species is the neutral arsenious acid (H_3AsO_3). Charged forms of As(III) only become important in alkaline waters; at $pH_w \approx 9$, $H_2AsO_3^-$ and H_3AsO_3 have the same concentration. At higher pH values, $H_2AsO_3^-$ is the prevailing As(III) species (Smedley and Kinninburgh 2002, Cullen and Reimer 1989).

The highest adsorption affinity by As(V) on metal oxides and oxyhydroxides, varies according to their pH_{zpc} which is generally between 7 and 9 (Dzombak and Morel 1990, Sigg and Stumm 1981, Davies and Kent 1990, Pierce and Moore 1982, Fuller *et al.* 1993). Iron oxides and oxyhydroxides are considered as principal adsorbents of As(V) in many sand and silt aquifers (Smedley and Kinninburgh 2002). The maximum retention capacities for As(V) were found for goethite and lepidocrocite at pH = 6, and for hematite at pH = 7–8 (Bowell 1994). The retention capacity for As(III) is lower and less pH-dependent compared to those of As(V), with maxima at pH values of about 6, 7 and 8 for goethite, hematite and lepidocrocite, respectively. The kinetics of these adsorption processes is fast (range of minutes), which allows a fast adaptation, e.g., if water flows within an aquifer with variable geochemical conditions (Fuller *et al.* 1993).

Amorphous oxyhydroxides of Al have a pH_{pzc} of ~8.6, which is similar to those of the Fe oxyhydroxides and hence they adsorb As(V) over the same pH range (Edwards 1994). In contrast, oxyhydroxides of Mn have a low pH_{pzc} (~2) so their net surface charge is negative in the pH range of most natural waters, and sorption of As oxyanions is not facilitated. However, Mn oxyhydroxides may adsorb dissolved bivalent cations, which results in a decrease in the negative surface charges (the pH_{pzc} increases) and adsorption of As(V) may become possible (Takamatsu *et al.* 1985).

Geochemical processes, which control the pH of groundwater, may indirectly control As mobility. So, cation exchange reactions influence the physicochemical properties of the water, e.g., by a pH increase due to the exchange of Ca^{2+} from a fluid by Na^+ from solid phase, correspondingly decreasing the As adsorption capacity of the solid phase. Such ion exchange processes and other processes which cause a pH increase, explain why higher concentrations of dissolved As are found in Na-HCO$_3$ type groundwater. As an example we can again use the arseniferous aquifers of the Chaco-Pampean plain (see 2.6.2.3). Here, the zones of groundwater with high As concentrations could be related to zones with the presence of Na-HCO$_3$ waters and high pH which are interpreted to occur as a result of cation exchange (Ca^{2+} by Na$^+$). This ion exchange results in a pH increase causing the release of As previously adsorbed on Mn and Al, and to a lesser extent on Fe oxyhydroxides. So in this area, the zones with low As concentrations in groundwater correspond to near-neutral waters of Ca-HCO$_3$ type (Smedley *et al.* 2005, 2008, Bundschuh *et al.* 2004, Bhattacharya *et al.* 2006b, 2008a,b, 2009) (see example 2.6.2.3).

Arsenic desorption from Fe-oxides at high pH values (>8) also explains the association of groundwater with high As concentrations and high pH value from several areas in the USA. Examples are found in the Interior Plains in South Dakota (Carter *et al.* 1998), central Oklahoma (Schlottmann and Breit 1992, Norvell 1995), the Intermontane Plateaus in Oregon (Owen-Joyce and Bell 1983, Owen-Joyce 1984, Robertson 1989), the Pacific Mountain System in Arizona (Goldblatt *et al.* 1963, Nadakavukaren *et al.* 1984) and NW Washington (Goldstein 1988, Ficklin *et al.* 1989, Davies *et al.* 1991).

2.6.2.2   *Influence of competing ions on arsenic adsorption capacity*

The adsorption capacity of Fe, Mn, and Al oxides and oxyhydroxides further depends on the presence of other ions in the fluid. Hence, As may be released from the solid phase in the presence of competing ions, such as phosphate, bicarbonate or silica (Smedley and Kinninburgh 2002, Reynolds *et al.* 1999). Phosphate is a well-known strong competitor with both As(V) and As(III) for available adsorption sites (Hingston *et al.* 1971, Manning and Goldberg 1996, Jain and Loeppert 2000, Smith *et al.* 2002), and silicic acid competes with As(V) and As(III) for adsorption sites on Fe oxides (Swedlund and Webster 1999, Roberts *et al.* 2004). Carbonate competes with As(V) for adsorption on ferrihydrite (Appelo *et al.* 2002) and on hematite (Arai *et al.* 2004). Additionally, it is also supposed that SO$_4^{2-}$ competes with both As(V) and As(III) (Gustafsson 2001). The presence of Ca^{2+} promotes As(V) adsorption (Smith *et al.* 2002), as result of the increased positive charge resulting from calcium adsorption (Rietra *et al.* 2001), which causes an increase in the pH$_{pzc}$ and favors As(V) adsorption.

The release of As from the solid phase in the presence of competing ions is a ligand exchange process and the mobilization of As from the solid phase to the liquid phase is favored under saline conditions, where a large number of anions are competing in the ion exchange processes. Consequently, As(V) is predominantly released due to the competitive exchange reactions of ligands. In particular, phosphate anions, which have similar chemical properties, are an important control of the solubility and mobility of As. Phosphate anions can rapidly substitute for As(V) oxyanions on the surface of Fe oxyhydroxides (Jackson and Miller 2000), which is especially of importance if we consider aquifers contaminated by fertilizers (Acharyya *et al.* 1999) or natural phosphates. Thus, the addition of phosphate can release As oxyanions into solution (Gao *et al.* 2004, Melamed *et al.* 1995, Peryea and Kammereck 1997).

Phosphate and As(V) anions compete for the active groups of the oxide surfaces (Jain and Loeppert 2000). Both are forming inner-sphere complexes. Phosphate adsorption is stronger than that of As, especially in sediments rich in Fe and Al oxides (Wauchope and Mc Dowell 1984). The adsorption of As(V) and phosphate anions on goethite, and gibbsite is similar over a wide pH range (Manning and Goldberg 1996). However, some active groups exist which are selective either for phosphate or for As(V) anion exchange. Arsenic mobilization by ion exchange (As replacement by phosphate) depends on the phosphate concentration, the pH of the fluid, and the fluid-solid contact time (e.g., groundwater residence time) (Barrow 1992, Darland and Inskeep 1997, Loeppert and Inskeep 1996). An association between dissolved As and phosphate been found in addition to the association with pH, in the groundwater of several areas in the USA, e.g., in the Carson Desert, Nevada (Welch and Lico 1998) and the Pacific Mountain System (Hinkle 1997, Hinkle and Polette 1999, Goldblatt *et al.* 1963, Nadakavukaren *et al.* 1984).

2.6.2.3 *Example: Chaco-Pampean plain, Argentina*

The approximate one million square kilometers covering the Chaco-Pampean plain of Argentina is one of the largest identified areas with high As concentrations in the groundwater (e.g., Nicolli *et al.* 1989, 2001, 2009, Bhattacharya *et al.* 2006b, Uriarte *et al.* 2002, Farías *et al.* 2003, Bundschuh *et al.* 2004, 2008a,b, 2009). The As concentrations of up to several mg L^{-1} are among the highest observed in the world in non-geothermal cold aquifers (<37°C). In many parts of the predominantly semiarid Chaco-Pampean plain, groundwater is the only available source of drinking and irrigation water. Several studies were performed to identify occurrence, sources and mobility controls of As in the groundwater of the Chaco-Pampean plain (e.g., Smedley *et al.* 2005, 2008, 2009, Bundschuh *et al.* 2004, 2008a,b, 2009, Bhattacharya *et al.* 2006b, Nicolli *et al.* 2009).

Arsenic in groundwater from the Quaternary loess-type aquifer of the Chaco-Pampean plain, which has prevailing oxidizing conditions, has concentrations in the range <3–5300 µg L^{-1}, predominantly As(V). Other anions and oxyanions (B, F, Mo, V, U, Mo) also have high concentrations.

The sedimentary aquifers affected by groundwater with high As concentrations consist of Tertiary and Quaternary aeolian loess-type deposits in the Pampean plain and of aeolian and fluvial sediments in the Chaco plain. They form a cover of the Chaco-Pampean plain having a thickness of <1 to several tens of meters, and in some areas reach several hundreds meters in thickness. The loess-type sediments contain 5–25% volcanic ash in a dispersed form, and individual volcanic ash beds are intercalated.

The principal source of the loess-type sediments is the weathering of volcanic rocks of the Andes, such as andesites and basalts, but also from weathering of the peri-Pampean mountains (granites, gneiss, schists). The loess sediments typically have silica and alkali contents comparable to dacite and the volcanic ash is of rhyolitic composition. Calcretes are also well-developed under the semi-arid climate conditions and are found as components of the matrix or as nodules, veins or extensive sheets.

It is assumed that the volcanic ash, which contains over 90% rhyolitic glass, and generally has As concentrations of 5–8 mg kg^{-1} (on occasion up to 20 mg kg^{-1}) is the original source of the groundwater arsenic (Nicolli *et al.* 1989, Smedley *et al.* 1998, Bundschuh *et al.* 2004). This assumption is despite the fact that loess and the volcanic glass have similar concentrations of As, and is explained by the much higher solubility rate of amorphous glass compared to the crystalline sediment minerals. In geological time spans, this As was dissolved under favorable conditions and precipitated or adsorbed on Fe, Al, and Mn oxyhydroxides from where desorption can occur as a fast process, depending upon changing geochemical conditions.

The As concentration in groundwater of the Chaco-Pampean plain shows a very patchy distribution (e.g., Bundschuh *et al.* 2004, Bhattacharya *et al.* 2006b), as we have also found in other areas, such as in the Bengal delta (Bhattacharya *et al.* 2002b) and the Sébacco valley (Altamirano Espinoza and Bundschuh 2009). This is a result of the spatial variability of hydrogeochemical aquifer conditions, and therefore hydrogeochemical processes are the principal control generating zones with low and high concentrations of As in the groundwater.

Common features of As-enriched hotspots comprise: loess-type sediments with volcanic ash/glass and the presence of Na-HCO$_3$ type groundwater with a high pH, high electrical conductivity and often together with high concentrations of the trace elements B, F, Mo, V, U, and Mo which indicate a common origin in the volcanic ash. Low concentrations of dissolved As are found in zones with a Ca-HCO$_3$ type of groundwater with a neutral pH. The pH increase can be explained by ion exchange (due to longer residence time in these sections of the aquifer) where Na$^+$ in the solid phase is exchanged by Ca^{2+} from the groundwater, and further by the dissolution of carbonates, resulting in Na-HCO$_3$ type groundwater and a pH increase (pH of water becomes more positively depleted versus pH_{pzc} of the adsorbent), which favors the desorption of As from oxyhydroxides of Fe, Al, and Mn. The control of the individual oxyhydroxides varies in the different regions, sometimes with Fe and sometimes with Al being the principal oxyhydroxide.

However, other processes must be considered as possible additional As mobility controls, such as the presence of other ions which compete for the As adsorption sites, e.g., anions and oxyanions of V, Mo, PO_4^{3-} and HCO_3^-. This process is favored, since the groundwater As hot spots often correspond to zones with highly mineralized groundwater (concentrations of As often correlate with those of V, Mo, and HCO_3^-). Additionally, the evaporation of water contributes, especially in small internal drainage systems, to As concentration increases in the water, but is not the principal control for the observed high As concentrations. The accumulation of As (dissolved and sorbed) through flow towards depressions, which are very common all over the area, and a lack of flushing are likely controls. Consequently, the principal controls of the As concentration in groundwater are sorption processes involving Fe, Al, and Mn oxyhydroxides.

2.6.2.4   *Example: Molasse trough sand aquifer, Southern Germany*
High As concentrations are found in the Tertiary Badenian sands of the Southern German Molasse trough (Bayer and Henken-Mellies 1998). Arsenic is present at concentrations of up to 1900 $\mu g\ g^{-1}$ adsorbed to iron oxyhydroxide coatings on the sand. However, As is effectively immobile in the oxic environment of this aquifer with concentrations of only 0.3–1.8 $\mu g\ L^{-1}$ (1st quartile to 3rd quartile), but with a few exceptions in anoxic areas (Wagner *et al.* 2003).

2.7   ADSORPTION PROCESSES AND CAPACITY OF CLAY MINERALS

Depending on the state of protonation of the oxygen and the deprotonation of the oxyhydroxyl groups on the mineral surface, clay minerals show no constant mineral-specific charge but have variable surface charges. A wide range of pH_{pzc} values (4–8) have been reported, which favors in many cases As adsorption in the acid range (Sadiq 1995, Manning and Goldberg 1997, Frost and Griffin 1976), where high As(V) adsorption capacity is reported for chlorite and haloisite, compared to kaolinite, montmorillonite and illite (Lin and Puls 2000). In the alkaline range, the highest As(V) adsorption was reported for illite, followed by kaolinite and montmorillonite (Manning and Goldberg 1997). Arsenite adsorption occurs predominantly on montmorillonite and kaolinite (Manning and Goldberg 1997). However, the adsorption capacity of clay minerals is much smaller when compared to metal oxyhydroxides (Smedley and Kinniburgh 2002) and consequently these processes generally play a subordinate role in As mobilization/demobilization in aquifers.

2.8   PRECIPITATION/DISSOLUTION AND SORPTION PROCESSES OF CALCITE

Arsenic adsorption may occur on the surface of calcite, which is found in many sediments though the adsorption capacity is much smaller compared to metal oxyhydroxides (Smedley and Kinniburgh 2002).

The pH_{pzc} values reported for calcite vary from 7–10.8, which is explained by the variable material properties, such as crystallinity, grade of hydration, purity, precipitation of new mineral phases on the crystals' surfaces and variable experimental conditions (e.g., variable $p_{CO2(g)}$) (see e.g., Sadiq 1995, Manning and Goldberg 1997, Frost and Griffin 1976, Stumm 1992, Van Cappellen *et al.* 1993). This pH_{pzc} range indicates that calcite may adsorb As in the alkaline range (Zachara *et al.* 1993), and a maximum adsorption capacity was found at pH 10 under laboratory conditions (Goldberg and Glaubig 1988). A study by Romero *et al.* (2004) reports a retention of As(V) by the material of a limestone aquifer (Zimapán aquifer, Mexico) over the range of pH 7–9 due to a coprecipitation and adsorption on Fe oxyhydroxides, calcite and clays.

There exist doubts whether the dissolution of carbonate minerals (e.g., calcite, dolomite, siderite) can significantly contribute to the As release into water (Smedley and Kinniburgh 2002). The As contents of most calcites are 1–8 mg kg^{-1}, those of dolomite and siderite are <3 mg kg^{-1}. Arsenic may occupy the positions of carbon in the crystal latter of calcite by substitution of CO_3^{2-} by AsO_3^{3-} and this mechanism may be a limiting factor for the mobilization of As in conditions where the immobilization by adsorption on Fe and Mn oxyhydroxides is not efficient (Di Benedetto *et al.* 2006). The same doubt was confirmed by Birkle and Bundschuh (2009), who found that the distribution of As in a limestone aquifer in SE Mexico even at high temperatures of 130°C does not contribute to the concentration of dissolved As.

2.9 INTERACTIONS BETWEEN ARSENIC AND HUMIC SUBSTANCES

Anion forming organic acids, such as humic substances, compete with As for adsorption sites on metal oxide surfaces, which is generally explained by the strong affinity of the carboxylic and phenolic functional groups of the organic acids for oxide surfaces (Xu *et al.* 1988, 1991, Bowell 1994, Gustafsson and Jacks 1995, Redman *et al.* 2002, Simeoni *et al.* 2003). Humic substances with high-molecular-weights and multiple functional groups are preferentially adsorbed on oxide surfaces (Vermeer and Koopal 1998) and desorption only occurs at high pHs due to the increased negative charge of the oxide surface (Avena and Koopal 1998).

Gustafsson and Bhattacharya (2007) used a simplified model to simulate As adsorbed humic functional groups R–COO$^-$ to the concentration of singly coordinated surface sites with [R–COO$^-$]/[= FeOH] ratios varying from 0.05 to 0.4. They found that humic substances have a small competitive effect when only a small part of the oxide surface is covered, but at large surface coverage the effect is important.

CHAPTER 3

History of blackfoot disease*

Blackfoot disease (BFD) is a unique, advanced peripheral vascular disorder, which was endemic in a limited area on the southwest coast of Taiwan. Clinically the disease starts with numbness, and/or coldness of extremities. In the end stages, the diseased extremities usually progress to ulceration, black discoloration, gangrenous change, or spontaneous amputation of the distal parts of the affected extremities. The disease was first found in four neighboring townships in southwestern Taiwan in the early 20th century. Starting in the mid-1950s, a series of studies has been carried out to elucidate the natural history, epidemiological characteristics, pathological changes, and etiology of the disease. This chapter will review the discovery history, clinical manifestations, diagnostic criteria, prognosis and prevalence of BFD. The cause and associated health hazards of BFD will be reviewed in chapter 4.

3.1   PROLOGUE: A MYSTERIOUS DISEASE

Blackfoot disease, endemic to southwestern Taiwan (Fig. 3.1), was once called "raw dry snake", "black dry snake" and "spontaneous gangrene" by local inhabitants to denote the mysterious discoloration of the affected extremities, which become black and well demarcated from the adjacent unaffected tissues. The feet of a male adult patient affected by BFD are shown in Figure 3.2a. The characteristic gangrene resulting in obvious demarcation occurred in both feet. The disease usually affects the feet, occasionally the legs, and rarely the hands. The dry gangrene in toes and fingers of another female adult patient are shown in Figure 3.2b, c. Blackfoot disease is usually accompanied by lancinating, gnawing or burning pain, which leaves patients unable to sleep or rest in ease. Another dreadful characteristic is the progression of the disease even after the affected limb has been surgically or spontaneously removed, sometimes spreading as far as the upper leg. Repeated removal of affected parts has caused severe physical disability and psychological trauma to the patients. Some patients cannot tolerate the excruciating process of the disease, finally committing suicide.

Cases of BDF occurred sporadically during the Japanese occupation of Taiwan from 1895 to 1945. However, it did not attract much attention until the 1950's when the disease occurrence increased and the miserable physical and mental suffering of victims were widely reported in newspapers. The disease ("spontaneous gangrene") was first reported in the *Journal of Formosan Medical Association* in 1954 (Kao and Kao 1954). The clinical, pathological and epidemiological characteristics as well as etiological factors of BFD had been separately investigated since the mid-1950s by two major investigation teams from National Taiwan University and US Naval Medical Research Unit 2.

3.2   CLINICAL CHARACTERISTICS OF BLACKFOOT DISEASE

Blackfoot disease is a peripheral vascular disease with an insidious onset and initial symptoms of numbness and/or coldness in one or more extremities (Tseng *et al.* 1961). The disease

---

* This chapter was prepared in collaboration with Chih-Hao Wang Ph.D., Department of Cardiology, Cardinal Tien Hospital, College of Medicine, Fu-Jen Catholic University and Department of Internal Medicine, School of Medicine, Taipei Medical University, Taipei, Taiwan.

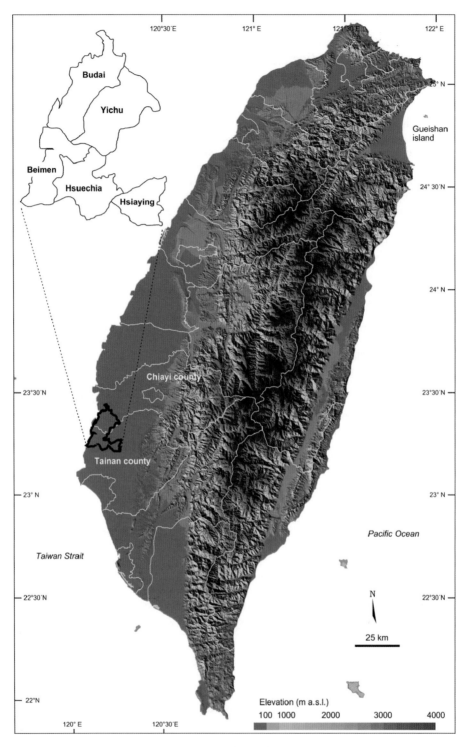

Figure 3.1.   Geophysical setting of Taiwan and location of the blackfoot disease area in SW Taiwan.
Most blackfoot disease cases were clustered in the townships Hsuechia, Beimen, Hsiaying
(Tainan county) and Yichu and Budai townships (Chiayi county).

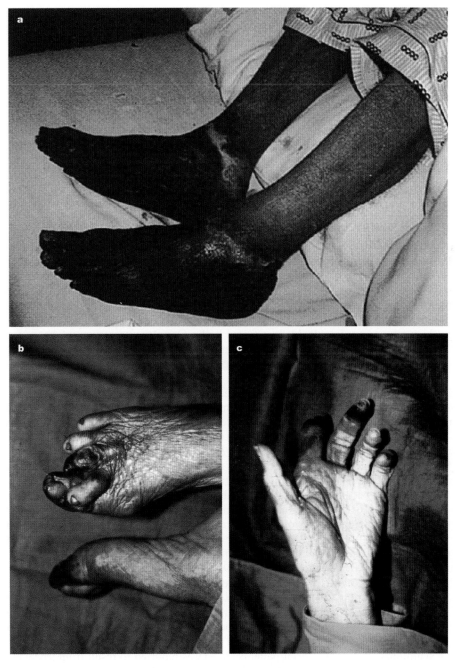

Figure 3.2.   (a) Feet of a male adult patient affected with blackfoot disease; (b) and (c) Dry gangrene in toes and fingers of a female adult patient affected with blackfoot disease.

usually progresses to ulceration, black discoloration, dry gangrene, or all these changes simultaneously. A small ulcer following a trivial trauma tends to extend and is associated with severe pain. In the end stage, the dry gangrene leads to spontaneous and/or surgical amputations of distal parts of affected extremities.

The diagnosis of BFD depends on: (1) objective signs of ischemia (absence or diminution of arterial pulsations, pallor on elevation or rubor on dependency of ischemic extremities, and various degrees of ulceration and gangrene), and (2) subjective symptoms of ischemia (intermittent claudication, pain at rest, and ischemic peripheral neuropathy). The diagnostic procedures, well described by Tseng (1974), included examinations of skin temperature and discoloration, varicose veins, peripheral arterial pulse, and blood flow. Blackfoot disease patients often have cold and pale limbs after rest for 10 minutes at a warm room temperature. The skin temperature of affected extremities is lower than unaffected extremities. The pale skin color of affected extremities depends on postural changes. The pale color of affected hands and feet becomes more apparent when they are lifted above the level of the heart for approximately one minute. The skin color might regain its original color when the lifted extremities are lowered for about 15 seconds. The longer the time interval needed for the lowered extremity to recover its original skin color, the more severe is the occlusion of peripheral circulation. The skin color of affected feet might become pale after walking. When affected feet hang downward over a period of time, the bottom of soles might appear purple-red as shown in Figure 3.3a.

In addition to the examination of skin color and temperature, varicose veins and peripheral arterial pulse are also important signs of BFD. In normal persons when extremities are lowered after being lifted for about one minute, varicose veins can be observed within 15 seconds, as shown in Figure 3.3b. It usually takes a much longer time for the lowered extremities to show varicose veins in BFD patients. As shown in Figure 3.3c, d, palpation of peripheral artery pulse may be needed in the evaluation of BFD. Peripheral artery pulse might be weak or even impalpable in affected limbs. Doppler ultrasonograph examination is the method of choice to diagnose or evaluate the disease (Tseng 1978). Although artery angiography may help to identify the precise occlusion site of affected extremities, it is rarely used in this setting.

The clinical manifestation of blackfoot disease is classified into the following stages: (1) *Symptomatic stage*: Patient has complaints of numbness, coldness, intermittent claudication, skin discoloration, and pressing pain of affected extremities. The color of affected extremities is white, red or purple rather than black as shown in Figure 3.4a, b. Patients at this stage may not pay much attention to the disease. (2) *Pain stage*: Severe pain of affected extremities appears. Most patients seek medical care at this stage. (3) *Necrosis stage*: Gangrene and necrosis occur at this stage. The sensory and motor functions of affected extremities are gradually lost, and the pain becomes more severe. As shown in Figure 3.4c, the affected extremities gradually turn black, the defining characteristic of the disease. (4) *Spontaneous amputation stage*: The black gangrenous lesions of affected extremities fall off without special sensation as shown in Figure 3.4d. (5) *Relapse stage*: New lesions develop at the proximal part of amputated extremities or in unaffected extremities.

In a study on 1070 BFD patients (Tseng 1989), the initial symptoms of BFD were reported to include numbness of extremities (75%), coldness of extremities (57%), intermittent claudication (42%), ruborous feet (15%), burning sensation in soles (15%), pale feet (15%), weakness of extremities (10%) and itching sensation in soles (7%). The symptoms in all patients were obviously due to the ischemia of extremities. Although the clinical onset of BFD is usually insidious, some patients (14%) had sudden onset without initial symptoms. The onset of initial symptoms of some blackfoot patients (33%) was reported to result from trivial injury or exposure to cold. The progression from intermittent claudication to pain at rest may take several months to several years with a wide variation. The rest pain was often described as burning, gnawing, shooting and lancinating. It was localized at toes or the base of toes in the area of metatarsal bone. Frequently, a progressively enlarging ulcer would develop at the site of a minor trauma.

In affected extremities, dry gangrene first developed, progressing to moist gangrene when the site became infected, ultimately becoming necrotic. In most cases, infection and

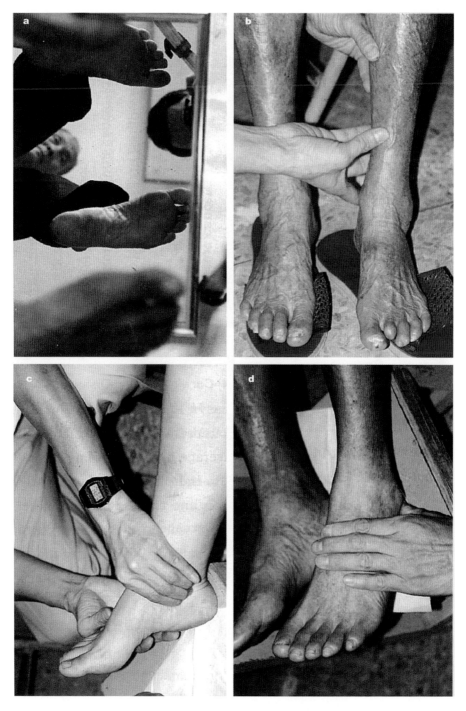

Figure 3.3. (a) Purple-red discoloration in the sole of a male adult blackfoot disease patient with his feet hanging downward; (b) Venous expansion examination of blackfoot disease; (c) and (d) Peripheral arterial pulse examination of blackfoot disease.

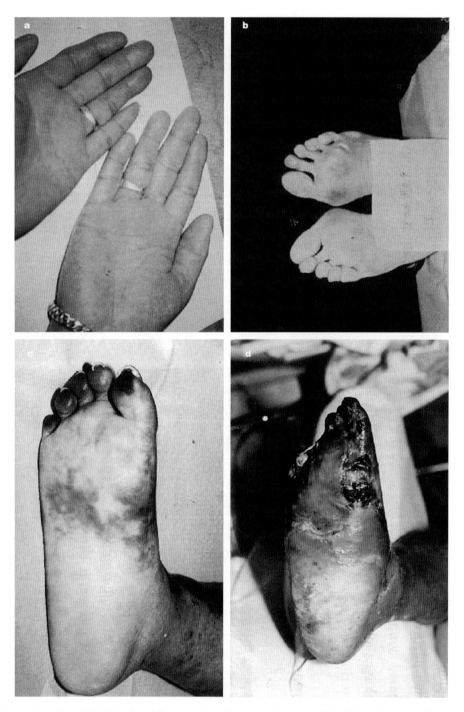

Figure 3.4.   (a) and (b) Discolored fingers, sole and toes at early stage of blackfoot disease; (c) Gangrene of toes at middle stage of blackfoot disease; (d) Spontaneous amputation of affected extremities at end stage of blackfoot disease.

inflammation were limited to the tissues immediately surrounding and proximal to areas of skin gangrene. In the final stage, the distal parts of affected extremities eventually underwent spontaneous or surgical amputation in a majority (68%) of patients (Tseng 1989). Some patients had a limited involvement of the tip of a digit, but most patients had entire extremities involved. Among 1300 BFD patients, 65% of them were affected in a single extremity, 30% in two extremities, 3% in three extremities, and 2% in all four extremities. Most of them (91%) had lower extremities affected only, 2% with only upper extremities involved, and 7% with the involvement of both lower and upper extremities. Prior to 1958, in patients who experienced spontaneous amputation the duration of gangrenous stage prior to the amputation below the transmetatarsal bone ranged from one month (20%) to more than one year (4%) with a median duration around three months. Surgical amputation became more frequent after 1959.

As the gangrene progressively extended from distal to proximal parts of affected extremities, the level of amputation could correspondingly change over time. Of the 1233 amputations among 830 patients (Tseng 1989), the final level of amputation included toe (36%), below-the-knee (35%), above-the-knee (13%), fingers (7%), transmetatarsal (4%), and others (5%). The incidence rates of major amputation per 100 patient-years, by age at onset, are shown in Figure 3.5a. The incidence rates increased significantly with increasing age at onset from 0.69 per 100 patient-years for age at onset <10 years to 21.48 per 100 patient-years for age at onset of 70 or more years (Tseng 1989).

The five-year, ten-year, twenty-year and thirty-year survival rates after onset of BFD were 76, 60, 38 and 29%, respectively (Tseng 1989). The median survival time was 13.5 years after the disease onset. The all-cause mortality rates per 1000 patient-years by age at onset of BFD are shown in Figure 3.5b. The mortality rates increased with increasing age at onset from 11 per 1000 patient-years for age at onset <10 years to 160 per 1000 patient-years for age at onset of 70 or more years. Among the causes of death, BFD patients had a higher mortality from cardiovascular disease, cerebrovascular disease, peripheral vascular disease (BFD), and cancers of the lung, bladder, liver, skin and kidney than the general population in the endemic area (Tseng 1989).

In a 15-year follow-up study on cause-specific mortality of 789 BFD patients, BFD patients were found to have a statistically increased mortality from peripheral vascular disease,

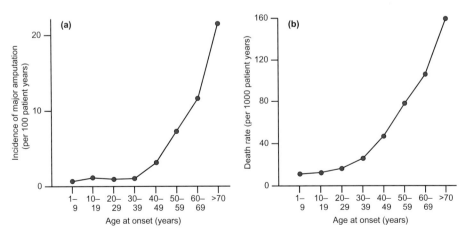

Figure 3.5.   (a) Incidence rates of major amputation and (b) mortality rates by age at onset of 1300 blackfoot disease patients.

cardiovascular disease, and cancers of the lung, liver, bladder and skin compared with the general population in the endemic area (Chen *et al.* 1988b). The standardized mortality ratio was 4.5, 3.5, 2.8, 2.6, 2.5, and 1.6 for skin cancer, peripheral vascular disease, lung cancer, bladder cancer, liver cancer, and cardiovascular disease, respectively.

3.3   PATHOLOGICAL FINDINGS OF BLACKFOOT DISEASE

The pathological features of BFD have been extensively examined in 63 amputated parts of extremities from 51 BFD patients (Yeh and How 1963). The disease was compatible with two distinct histological types including thromboangitis obliterans (30%) and arteriosclerosis obliterans (70%), although all showed the same appearance of dry gangrene of the affected extremities with or without sharp demarcation, depending on the extent of secondary infection.

The most striking features of the thromboangiitis obliterans type of BFD included: segmental occlusion of both arteries and veins, occlusive lesions of various stages in different vessels or in the same vessel, fibrinoid degeneration of intimal connective tissue in larger vessels, fibrinoid necrosis of the whole vessel wall in arterioles or venules, proliferation and activation of vascular endothelium, marked arteriosclerotic changes, and frequent medial calcification. In the arteriosclerosis obliterans type of BFD, the occluding lesions were mostly older than those in arteriosclerotic gangrene in the general population. Significant and intensive arteriosclerotic changes with marked atheroma formation, calcification, bone formation, hyalinization, and increase in elastic fibers in intimal coat were observed. Medial calcification and "pipe-stem" artery commonly occurred, while marked adventitial fibrosis was also seen.

In the same study, generalized atherosclerosis was observed in all three autopsy cases of BFD. The atherosclerosis was especially marked in the coronary arteries, even in a 24-year-old woman who had thromboangiitis obliterans. The conclusions were that "the fundamental vascular changes in common in both groups of BFD is an unduly developed severe arteriosclerosis, leading to pure arteriosclerotic gangrene of extremities in 69.23% and the thromboangiitis obliterans is another disease superimposed on pre-existing or co-existent arteriosclerosis." (Yeh and How 1963).

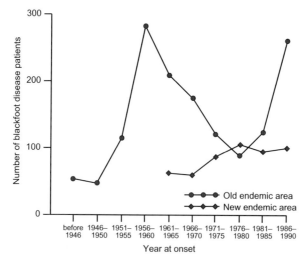

Figure 3.6.   Year at onset of 1300 blackfoot disease patients reported from old (Hsuechia, Beimen, Hsiaying of Tainan county and Yichu and Budai townships of Chiayi county) and new (other townships in southwestern Taiwan) endemic areas.

3.4   EPIDEMIOLOGICAL CHARACTERISTICS OF BLACKFOOT DISEASE

Sporadic cases of blackfoot disease were first identified in southwestern Taiwan in the early 20th century. Most BFD patients reported before 1961 were clustered in the endemic area including Hsuechia, Beimen, Hsiaying townships of Tainan county and Yichu and Budai townships of Chiayi county. After an intensive survey extended to other townships in southwestern Taiwan, there were more BFD patients identified after 1961. A total of 2252 BFD patients were identified from 1912 to 1990. Figure 3.6 shows the distribution of onset year of 1992 BFD reported from the old endemic area and new endemic area (Taiwan Provincial Department of Health, 1993). In the old endemic area, the number of BFD patients increased

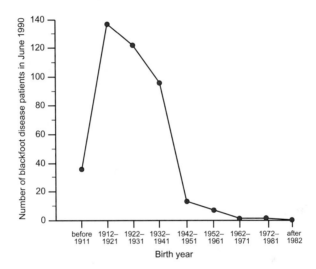

Figure 3.7.   Year at birth of 413 blackfoot disease patients living in Tainan and Chiayi counties in June 1990.

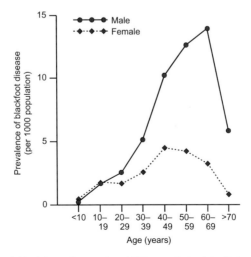

Figure 3.8.   Prevalence of blackfoot disease (per 1000 population) in Beimen, Hsuechia, Budai and Yichu townships in 1960.

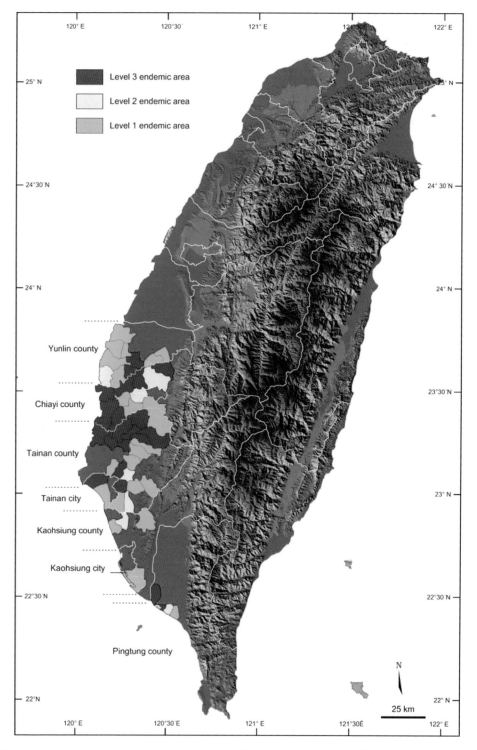

Figure 3.9.   Map of townships by three levels of blackfoot disease endemicity in Taiwan (county
boundaries are in white).

gradually before 1946 up to 1950. It increased rapidly after 1950, peaked in 1956–1960, and gradually decreased after 1961. There was a surge of newly reported BFD patients after 1980 resulting from the implementation of Mobile Medical Care Program by the Preventive Center of Blackfoot Disease. The program gave priority to internal medicine rather than surgery for patient care, which attracted many early cases to the program. The number of BFD in the new endemic area increased from before 1961 to 1976–1980. Incident disease rate peaked in the new endemic area 24 years after the peak in the old endemic area.

Based on a series of 1300 BFD patients mostly in the old endemic area, the distribution of onset age was similar for men and women. More than 70% patients had an age at onset at least 40 years. The onset age of BFD peaked at 50–59 for both males and females, however 8% of patients had an age at onset <20 years old. Despite the wide range of age at onset of BFD in the old endemic area, most patients who lived in the area in June 1990 were born between 1912 to 1941 as shown in Figure 3.7. In other words, the onset year was more widely distributed than the birth year. This indicated that exposure to some environmental factors in the endemic area might play an important role in the induction of BFD.

In an epidemiological survey of BFD in the old endemic area from February 1958 to August 1960, a total of 327 BFD patients was diagnosed. The prevalence of BFD per 1000 people was highest in Beimen (5.57), followed by Hsuechia (3.87), Budai (2.02), and Yichu (0.64), and lowest in Hsiaying township (0.14). The prevalence of BFD varied significantly in 109 villages of the five townships. The number of villages with a BFD prevalence of 0, 0.1–1.9, 2.0–4.9, 5.0–9.9, 10.0–14.9, 15.0–19.9 per 1000 population was 53, 19, 13, 12, 6 and 6, respectively. There were 83% (15/18) villages in Beimen, 75% (21/28) village in Hsuechia, 48% (12/25) villages in Budai, 30% (7/23) villages in Yichu, and 7% (1/15) villages in Hsiaying had BFD patients. Twenty four villages with a BFD prevalence of 5 or more per 1000 population were located in Hsuechia, Beimen and Budai. Figure 3.8 shows the age-sex-specific prevalence of BFD in Beimen, Hsuechia, Budai and Yichu townships. The prevalence increased significantly with increasing age in both males and females. Males had a higher prevalence than females in age groups older than 20 years old. In addition to age and sex, occupation was also associated with the prevalence of BFD. In the four endemic townships, the prevalence per 1000 population was 19.7, 9.9 and 3.4 for salt-field workers, fishermen, and farmers, respectively (Wu *et al.* 1961). However, it was suggested that the source of drinking water rather than occupation was responsible for the development of the disease. All villages where salt-field workers lived had used artesian wells until 1956.

Based on an island-wide study on BFD, the endemicity of BFD in various townships was classified into three levels, according to the high arsenic concentration in groundwater (>350 µg L^{-1}), the occurrence of patients affected with BFD, and the symptoms of chronic arsenic (As) poisoning (hyperpigmentation and hyperkeratosis) in children (Taiwan Provincial Department of Health 1993). Level 1 township was defined as a township with the highest As concentration in groundwater >350 µg L^{-1} but without any BFD patient or children with chronic As poisoning. Level 2 township was defined as a township with the highest As concentration in groundwater >350 µg L^{-1} and either some BFD patients or children with chronic As poisoning. Level 3 township was defined as a township with the highest As concentration in groundwater >350 µg L^{-1}, BFD patients, and children with chronic As poisoning. Figure 3.9 shows the map of townships at three levels of BFD endemicity. Most level 3 townships clustered in Chiayi and Tainan county, while level 2 and 1 townships were scattered in Yunlin, Chiayi, Tainan, Kaohsiung and Pingtung counties.

CHAPTER 4

Cause of blackfoot disease: Arsenic in artesian well water*

Blackfoot disease (BFD) is an endemic peripheral vascular disease occurring in a limited coastal area in southwestern Taiwan. The clinical manifestation, diagnostic method, prognosis and epidemiological characteristics of BFD are described in chapter 3. This chapter describes the studies on the sources and characteristics of drinking water in the BFD endemic area, the association between BFD and consumption of artesian well water, and the dose-response relationship between arsenic (As) in drinking water and risk of BFD, skin and internal cancers, microvascular and macrovascular diseases, carotid atherosclerosis, diabetes, hypertension, and erectile dysfunction.

4.1   TYPES OF WELLS IN BLACKFOOT DISEASE-ENDEMIC AREA

As shown in Figure 4.1, the endemic area of BFD in the early 1960s included Beimen and Hsuechia in Tainan county and Budai and Yichu townships in Chiayi county. These townships are located on the southwestern coast of Taiwan. Both Beimen and Budai township are situated along the seaside, and were noted for fishing and sun-dried salt production, with most residents engaged in fishing, salt production and farming. Hsuechia and Yichu are typical agricultural townships. The lifestyle and living condition of residents in the endemic area

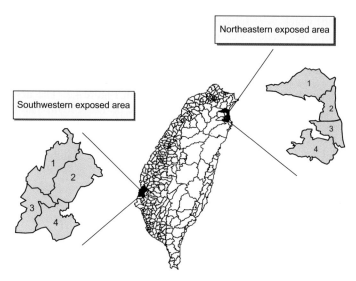

Figure 4.1.   Map of southwestern and northeastern arsenic-exposed area in Taiwan. There are four townships in the southwestern exposed area: Budai (1), Yichu (2), Beimen (3) and Hsuechia (4), and four townships in the northeastern exposed area (Lanyang plain): Jiaosi (1), Jhuangwei (2), Wujie (3) and Dongshan (4).

* This chapter was prepared in collaboration with Chih-Hao Wang Ph.D., Department of Cardiology, Cardinal Tien Hospital, College of Medicine, Fu-Jen Catholic University and Department of Internal Medicine, School of Medicine, Taipei Medical University, Taipei, Taiwan.

Figure 4.2.   Wells in the blackfoot disease endemic area: (a) Shallow well; (b) Artesian well.

were very similar to those of residents of southern Taiwan. However, the soil of the endemic area has a high salt content, which was not suitable for crops, especially rice. The majority of the residents consumed a mixture of rice and dried sweet potato chips as the staple food. The lower the socioeconomic status of a household, the higher the proportion of dried sweet potato chips consumed in their diet. In the early 1990s, a new area of chronic As poisoning was identified in northeastern Taiwan.

In the BFD endemic area, residents in most villages were using well water as the main drinking water source. Two types of wells were used in this area: shallow wells and artesian wells. They were remarkably different in their outer appearance and depth. As shown in Figure 4.2a, a shallow well was an ordinary well (4–12 m deep) with walls usually made of bricks or stones which were sometimes coated with cement. The well water was drawn by bucket with rope. As shallow well water in some villages of the endemic area was salty, residents had started to use water from artesian wells in the 1920s. As shown in Figure 4.2b, an artesian well was an open cement tank (2 m long, 1.5 m wide, 1–2 m high) partly located underground, to store the water spontaneous coming up by natural ground pressure. The cross-sectional diagram of an artesian well is shown in Figure 4.3. The depth of artesian wells varied from 100 to 280 m with 80% in the range of 120–180 m. The water came up automatically to the water tank through bamboo pipes. The lower part of the bamboo pipes was full of small holes for a length of 6 m to let water flow in, and the outside of the pipes was covered with linen to prevent sand and gravel from entering the pipes through the holes. Usually every four to five years it was necessary to dig another artesian well hole and put in new bamboo pipes.

4.2   CHARACTERISTICS OF WELL WATER IN BLACKFOOT DISEASE ENDEMIC AREA

The first study of the physicochemical characteristics of drinking water samples from 130 wells in the BFD endemic area was carried out in 1959 (Chen *et al.* 1962). It included 18 physicochemical analyses: turbidity, bicarbonate, carbon dioxide, pH value, chloride, fluoride, ammonia nitrogen, nitrite nitrogen, nitrate nitrogen, total hardness, total solids, calcium, magnesium, iron, manganese, oxygen absorption, silica and arsenic. The As was tested by Gutzeit's procedure. There were 41 samples from artesian wells in the endemic area, 13 samples from artesian wells in a non-endemic area, 15 samples from shallow wells in the endemic area, and 61 samples from shallow wells in a non-endemic area. One of the remarkable differences between artesian and shallow wells was the growth of algae in artesian well water,

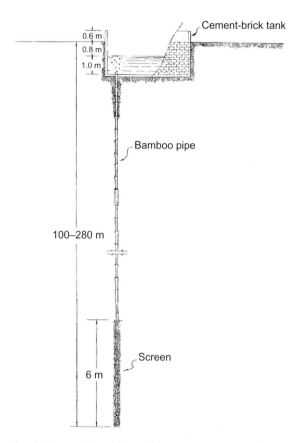

Figure 4.3.   Cross-sectional diagram of artesian well.

which was favored by direct sunshine and fresh water. The green algae was floating on the surface of the water, or fixed to the surrounding wall and bottom of the tank. The amount of green algae varied from time to time and from one well to another, depending on the tank cleaning frequency, daily water usage, age of the well, and water temperature.

The medians of 18 physicochemical characteristics in four groups of wells are shown in Table 4.1. No remarkable differences were found for pH value, bicarbonate, carbon dioxide, nitrite nitrogen, nitrate nitrogen and fluoride. Water samples from shallow wells had higher concentration of calcium, magnesium, manganese, total hardness, total solids and chloride than artesian well water. Median concentration of silica was higher in artesian well water than in shallow well water, but there was no significant difference in water samples from artesian wells in endemic and non-endemic areas. The median levels of turbidity, ammonia nitrogen, iron, dissolved oxygen (uptake of oxygen may have occurred during exposure of the well water to the atmosphere increasing the high values of oxygen content) and As were higher in artesian well water than in shallow well water, as well as higher in artesian well water samples from the BFD endemic area compared with those from a non-endemic area. Based on the possible toxic effects of the chemicals, it was concluded that high As content in artesian well water was most likely to be a main causal factor of BFD. Arsenic levels in water samples from artesian and shallow wells in BFD endemic and non-endemic areas in southwestern Taiwan are shown in Table 4.2. The As concentrations of all water samples from artesian wells in the endemic area were equal to or greater than 350 μg L^{-1}, while all water samples from shallow wells in endemic and non-endemic areas were less than 300 μg L^{-1}.

Table 4.1.   Physicochemical characteristics of water samples from 130 wells in the blackfoot disease endemic area in southwestern Taiwan (Chen *et al.* 1962).

Physicochemical characteristics	Artesian wells		Shallow wells	
	Endemic area	Non-endemic area	Endemic area	Non-endemic area
pH value	7.4	7.6	7.4	7.2
Bicarbonate (mg L^{-1})	575	467	483	533
Carbon dioxide (mg L^{-1})	32	15	30	50
Nitrite nitrogen (mg L^{-1})	trace	0.005	0.002	trace
Nitrate nitrogen (mg L^{-1})	0.02	0.02	0.02	0.02
Fluoride (mg L^{-1})	0.25	0.13	0.20	0.24
Calcium (mg L^{-1})	33	23	72	85
Magnesium (mg L^{-1})	17	13	48	65
Manganese (mg L^{-1})	0.05	0.10	0.20	0.46
Total hardness (mg L^{-1})	168	150	425	500
Total solids (mg L^{-1})	900	607	1000	1438
Chloride (mg L^{-1})	81	32	150	290
Silica (mg L^{-1})	17.3	18.5	7.5	12.5
Turbidity (mg L^{-1})	10	4	5	2
Ammonia nitrogen (mg L^{-1})	1.53	1.00	0.04	0.09
Iron (mg L^{-1})	1.01	0.25	0.65	0.58
Oxygen (mg L^{-1})	9.5	4.75	2.25	1.86
Arsenic (µg L^{-1})	780	380	40	30

Table 4.2.   Arsenic in water samples from 130 wells in the southwestern endemic area of Taiwan and of 3894 wells in the northeastern endemic area of Taiwan with chronic arsenic poisoning.

Arsenic in water (µg L^{-1})	Southwestern Taiwan				Northeastern Taiwan
	Artesian well		Shallow well		
	Endemic area	Non-endemic area	Endemic area	Non-endemic area	
<50	–	–	10	60	2489
50–140	–	–	3	–	801
150–240	–	3	1	–	160
250–340	–	2	–	1	117
350–440	4	1	–	–	91
450–540	5	4	–	–	52
550–640	3	–	–	–	24
650–740	5	–	–	–	29
750–840	3	–	–	–	30
850–940	6	–	–	–	19
950–1040	5	–	–	–	8
>1050	3	–	–	–	74

4.3   ARSENIC LEVELS IN WELL WATER IN LANYANG BASIN

In northeastern Taiwan, the As exposure areas are located in Jiaosi, Jhuangwei, Wujie and Dongshan townships in the Lanyang basin of Yilan county (Fig. 4.1). These are typical agricultural townships. Residents in this northeastern endemic area had a much better socioeconomic status and nutrition intake than those in the southwestern endemic area.

Dried sweet potato chips were not a common staple food in the northeastern endemic area. Because of the abundance of groundwater in the Lanyang basin, residents there had been using water from household shallow wells (<40 meters in depth) since the late 1940s. Each household in the northeastern endemic area had a shallow well in its backyard, while only a few wells were shared by many households in the villages of the southwestern endemic area. The community water system was implemented in the northeastern endemic area in the early 1990s. In a comprehensive survey of As concentrations in water of 3894 wells in the northeastern endemic area from 1991–1994 (Chiou *et al.* 1997), As was analyzed by hydride generation combined with flame atomic absorption spectrometry. The As levels of water samples from wells in the northeastern endemic area ranged from undetectable to >1000 µg L^{-1} as also shown in Table 4.2, with a wide variation in median As concentrations ranging from undetectable to 140 µg L^{-1} in various villages. The variation in As levels in well water was more striking in the northeastern endemic area of Taiwan compared to that in the southwestern endemic area.

4.4 ASSOCIATION BETWEEN BLACKFOOT DISEASE AND ARTESIAN WELL WATER

An epidemiological study of BFD was carried out in five townships in the southwestern endemic area of Taiwan from February 1958 to August 1960 (Wu *et al.* 1961, Chen and Wu 1962). A total of 327 patients affected with BFD were identified in 109 villages of the five townships where 906 wells were used for drinking water. These 109 villages may be classified into four groups according to their drinking water sources: 39 villages used artesian wells only (Group I with 195 artesian wells), 30 villages used both shallow and artesian wells (Group II with 136 artesian wells and 223 shallow wells), 38 villages used shallow wells only (Group III with 352 shallow wells), and two villages used surface water (Group IV). Among them, 39 (89.5%) Group I villages and 19 (63.3%) Group II villages had BFD patients, but no BFD patients were found in Group III and IV villages. BFD patients were found among users of 89 (45.6%) artesian wells in Group I villages and 29 (21.3%) artesian wells in Group II, while no BFD patient was found among users of shallow wells in Group II and III. The prevalence of BFD was highest (4.88 per 1000 population) in Group I villages, followed by Group II villages (2.03 per 1000 population), with no BFD patients found in Group III and IV villages as shown in Table 4.3. The increased prevalence of BFD with the use of artesian well water was observed in all five townships. There was a correlation between the BFD prevalence and percentage of villages using artesian wells in the five studied townships. The BFD prevalence per 1000 population in Beimen, Hsuechia, Budai, Yichu and Hsiaying was 5.57, 3.87, 2.02, 0.64 and 0.14, respectively; while the corresponding percentage of villages using artesian wells

Table 4.3. Prevalence of blackfoot disease per 1000 population in 109 villages in southwestern Taiwan by types of wells for drinking water (Wu *et al.* 1961).

Township	Artesian wells only (Group I)	Both artesian and shallow wells (Group II)	Shallow wells only (Group III)	Surface water (Group IV)	Total
Beimen	7.94	5.70	0	0	5.57
Hsuechia	6.19	3.70	0	NA	3.87
Budai	3.54	0.09	0	NA	2.02
Yichu	1.01	1.45	0	NA	0.64
Hsiaying	3.31	0	0	NA	0.14
Total	4.88	2.03	0	0	2.10

NA: surface water was not used for drinking water.

was 83.6%, 32.4%, 61.4%, 22.6% and 13.3%. It was thus concluded that the BFD was caused by the consumption of artesian well water (Chen and Wu 1962).

In a community-based study of 353 BFD cases and 353 matched unaffected controls (Ch'i and Blackwell 1968), the length of residence in the endemic villages was significantly associated with the risk of developing BFD in a dose-response relationship. While all 353 (100%) BFD cases had used artesian well water as the principal source of drinking water during the 15 years before onset, only 233 (66%) matched controls had consumed artesian well water. In another community-based study of 241 BFD patients and 759 matched healthy controls (Chen *et al.* 1988b), there was a significant dose-response relationship between the risk of BFD and the duration of consumption of artesian well water. The longer the period of subjects consuming artesian well water as the main drinking water source, the more likely they were to be affected with BFD.

Based on the above-mentioned studies, the consumption of high-As artesian well water was considered the most important risk predictor of BFD. As artesian wells were widely used in early twentieth century in the endemic area, the peak occurrence of BFD was observed in the late 1950s. The induction period of developing BFD after the consumption of artesian well water was estimated to be as long as 30–40 years, with individual variation.

4.5   ARSENIC IN DRINKING WATER: THE CAUSE OF BLACKFOOT DISEASE

In a community-based cross-sectional survey of 40,421 residents in 37 villages of BFD endemic area in southwestern Taiwan, the prevalence of BFD associated with As in well water was examined (Tseng 1977). A significant dose-response relationship was found between the As in drinking water and the prevalence of BFD as shown in Figure 4.4. The higher the As in drinking water was, the higher the prevalence of BFD, and the older was the age group, the higher the prevalence of BFD. The age effect may reflect the increased cumulative exposure to As in drinking water as well was the aging process.

In an analysis of age-adjusted mortality from peripheral vascular disease (mainly BFD) during 1973–1986, among residents in 42 villages of the endemic area of BFD (Wu *et al.*

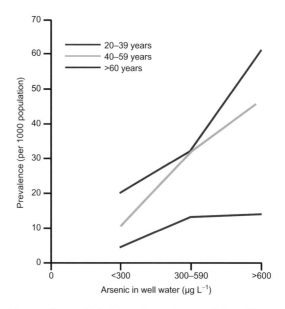

Figure 4.4.   Age-specific prevalence of blackfoot disease by arsenic in well water.

1989), a significant dose-response relationship between As in drinking water and age-adjusted mortality from peripheral vascular disease was observed: the higher the As level in drinking water, the higher the age-adjusted mortality from severe peripheral vascular disease.

In another comprehensive health survey in three hyperendemic villages of BFD, the subclinical peripheral vascular disease of 263 male and 319 female adult residents was examined by Doppler ultrasonography (Tseng *et al.* 1996). The prevalence of the disease was found to be associated with the cumulative As exposure from drinking water in a dose-response relationship after adjustment for age, gender, body mass index, cigarette smoking, and serum levels of total cholesterol and triglycerides. This study used a more objective and sensitive tool to define peripheral vascular disease rather than clinical diagnosis based on severe manifestation of BFD.

These studies have clearly documented the main cause of mild, severe and lethal peripheral vascular disease, mainly BFD, in the endemic area to be As in drinking water. The increased risk of peripheral vascular disease has been observed in other countries where high-As drinking water was consumed (Chen and Lin 1994, Wang, C.-H. *et al.* 2007). However, overt BFD, a severe type of peripheral vascular disease, has rarely been reported in other studies (Borgono and Greiber 1972, Borgono *et al.* 1977, Salcedo *et al.* 1984, Cebrian 1987). The extremely high As level in well water of the BFD-endemic area compared with the intermediately high As levels in drinking water of other countries might explain in part the varying severity of peripheral vascular disease in different As-exposed areas. Host and environmental co-factors of BFD are discussed below.

Fluorescent substances, humic acids, and silica in artesian well water have been hypothesized to be the cause of BFD. However, there has never been any epidemiological study to assess the association between exposure to these chemicals in well water and the occurrence of BFD. Neither hazard identification nor dose-response assessment has been done for these chemicals.

4.6   CO-MORBIDITY OF UNIQUE ARSENIC-INDUCED SKIN LESIONS AND BLACKFOOT DISEASE

Residents in the BFD endemic area had unique skin lesions induced by chronic As poisoning including hyperpigmentation and/or hypopigmentation, hyperkeratosis, Bowen's disease, and squamous cell carcinoma and basal cell carcinoma of the skin. As shown in Figure 4.5, there were multiple Bowen's disease and skin cancers in the trunk of a male adult resident in the endemic area. In a community-based cross-sectional survey on As-induced skin lesions of 40,421 residents in 37 villages of BFD endemic area (Tseng *et al.* 1968), a significant dose-response relationship was found between the As in drinking water and the prevalence of skin

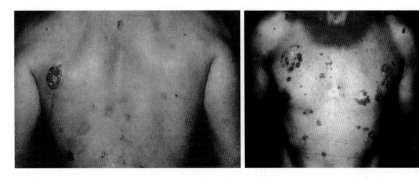

Figure 4.5.   Arsenic-induced skin lesions of a male adult resident in the blackfoot disease endemic area.

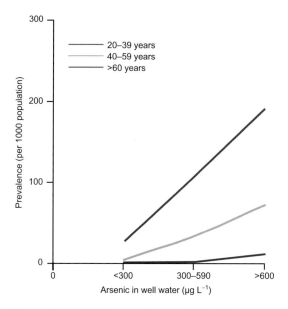

Figure 4.6.   Age-specific prevalence of skin cancer by arsenic in well water.

cancer as shown in Figure 4.6: the higher the As in drinking water, the higher the prevalence of cancer; and the older the age group, the higher the prevalence of skin cancer. The skin lesions are unique characteristics of chronic As poisoning. They have been used to define the chronic toxicity of As exposure.

The co-morbidity of BFD and As-induced skin lesions has been very well documented (Tseng 1977). In other words, BFD patients were found to have a significantly higher prevalence of As-induced skin hyperpigmentation, palmoplantar hyperkeratosis, and skin cancer than unaffected residents in the endemic area. The risk of developing BFD was 9.7, 14.0, 62.4 times higher for residents who were affected with As-induced skin hyperpigmentation, palmosolar hyperkeratosis, and skin cancer, respectively, compared to residents without any As-induced skin lesions. The co-existence of BFD and unique As-induced skin lesions further supports the important role of As in the development of BFD.

Chen *et al.* (1990) evaluated the toxic effects of As on cultured human umbilical vein endothelial cells and found that As inhibited endothelial cell proliferation and glycoprotein synthesis at levels higher than 100 ng mL^{-1}, but only inhibited the proliferation at levels lower than 50 ng mL^{-1}. They believed the findings suggested that As may damage endothelial cells, and such damage may play an important role in the pathogenesis of BFD. Because destruction of vascular endothelial cells takes place at an early stage in the limbs affected by BFD, Yu *et al.* (1998) examined the factors related to endothelial cell damage in BFD and had the following three findings: (1) Endothelial cell binding activity of IgG was significantly higher in BFD patients than that in unaffected controls, (2) IgG in BFD patients at a concentration >100 µg mL^{-1} induced concentration-dependent endothelial cell cytotoxicity, but not for IgG in unaffected controls, and (3) IgG in BFD patients at a concentration of 100 µg mL^{-1} stimulated neither the release of von Willebrand factor nor the expression of intercellular adhesion molecule-1 by EC. These findings suggested a significant role of immunological mechanisms in the development of BFD, which might account for the individual susceptibility to BFD. They believed that only persons who produce the IgG anti-endothelial cell antibody are potential victims of BFD.

One of the main pathologic features of BFD is atherosclerosis, and Lee *et al.* (2007) studied eight BFD patients and four age-matched healthy controls to assess the peripheral adrenergic responses. They found BFD patients had persistently decreased skin perfusion, while controls had a transient decrease in skin perfusion after iontophoresis of phenylephrine. Because increased peripheral alpha-adrenergic response and decreased beta-2-adrenergic response are related to increased vascular tone and result in atherosclerosis, they believed that their findings of accentuated alpha-adrenergic response in microcirculation and decreased lymphocyte beta-2-adrenoceptor response in BFD patients play an important role in the pathogenesis of atherosclerosis in BFD.

4.7   HOST AND ENVIRONMENTAL CO-FACTORS FOR BLACKFOOT DISEASE

Since only a fraction of people consuming artesian well water were affected with BFD, host factors including nutrition deficiency were suggested to modify the effect of artesian well water consumption on BFD. Two case control studies were carried out to explore co-factors of BFD in addition to As in artesian well water. Blackfoot disease patients were reported to have a lower educational level and socioeconomic status than matched controls (Ch'i and Blackwell 1968). Undernourishment was also associated with the development of BFD independent of the consumption of artesian well water and As-induced skin lesions.

4.8   ARSENIC IN DRINKING WATER AND CIRCULATORY DISEASES OTHER THAN BLACKFOOT DISEASE

Blackfoot disease patients were reported to have a significantly increased mortality from ischemic heart disease (Chen *et al.* 1988b). In a study conducted in 42 villages of the endemic area of BFD (Wu *et al.* 1989), the age-adjusted mortality from cardiovascular diseases during 1973–1986 was found to be significantly higher among residents in the endemic area than the general population in Taiwan. Furthermore, there was a significant dose-response relationship between age-adjusted mortality from cardiovascular disease and As in drinking water (Wu *et al.* 1989). For residents living in villages with the median As concentration in well water of <100, 100–340, 340–590 and >600 $\mu g\ L^{-1}$, their life-time (0–79 years old) risk of dying from ischemic heart disease was 3.4, 3.5, 4.7 and 6.6%, respectively. Residents with cumulative As exposures of 100–990, 10000–19900 and ≥20,000 $\mu g\ L^{-1}$-years, respectively, were found to have 2.5, 4.0 and 6.5 times the risk of dying from ischemic heart disease compared with those without As exposure after adjustment for age, sex, cigarette smoking, body mass index, serum cholesterol and triglyceride level, and disease status of hypertension and diabetes (Chen *et al.* 1996).

In another study of 66,667 residents living in the BFD-endemic area and 639,667 residents in a non-endemic area (Wang *et al.* 2003), the age-gender-adjusted prevalence of microvascular disease in endemic and non-endemic areas was 20.0 and 6.0%, respectively, for diabetes, and 8.6% and 1.0%, respectively, for non-diabetes. The corresponding prevalence of macrovascular disease was 25.3 and 13.7% for diabetes, and 12.3 and 5.5% for non-diabetes.

A dose-response relationship was also observed between the prevalence of ischemic heart disease based on Minnesota codes of probable and possible coronary heart disease on resting electrocardiogram and long-term As exposure from drinking well water in the same survey (Tseng *et al.* 2003). The higher the cumulative exposure to As in drinking water, the higher was the risk of being affected with ischemic heart disease.

In a study of carotid atherosclerosis assessed by duplex ultrasonography among 199 male and 264 female adult residents in the BFD-hyperendemic villages (Wang *et al.* 2002), significant associations of carotid atherosclerosis were observed for three indices of long-term exposure to As from drinking artesian well water, including the duration of consuming artesian well water, the average As concentration in consumed artesian water, and the cumulative

As exposure showing a dose-response relationship. The prevalence of carotid atherosclerosis increased with increasing exposure to As in drinking water after adjustment for age, gender, hypertension, diabetes mellitus, cigarette smoking, alcohol consumption, waist-to-hip ratio, and serum levels of total cholesterol and low-density lipoprotein cholesterol. Prolonged QT interval in electrocardiogram was significantly associated with ischemic heart disease and carotid atherosclerosis. Three indices of long-term exposure to As in drinking water were also significantly associated with the risk of QT prolongation showing dose-response relationships (Wang *et al.* 2009).

The association between As in drinking water and circulatory diseases other than BFD was also observed in the As-exposed area in northeastern Taiwan (Chiou *et al.* 1997). A dose-response relationship between prevalence of cerebrovascular disease, especially cerebral infarction, and As level in well water was reported. After adjustment for age, gender, hypertension, diabetes mellitus, cigarette smoking, and alcohol consumption, the risk of cerebral infarction was 3.4, 4.5 and 6.9 times, respectively, for those who consumed well water with As level of 0.1–50.0, 50.1–299.9 and ≥300 µg L^{-1} compared with the level <0.1 µg L^{-1} as the reference group.

In a cross-sectional study aimed to assess the association between ingested As and erectile dysfunction, the prevalence of erectile dysfunction was greater in the As-exposed (83.3%) group than the unexposed (66.7%). After adjustment for age, cigarette smoking, diabetes mellitus, hypertension, and cardiovascular disease, the group with exposure to well water containing >50 µg L^{-1} As had a three-fold risk of erectile dysfunction compared with the As level in drinking water ≤50 µg L^{-1} (Hsieh *et al.* 2008). The relative risk was even higher (7.5-fold) for severe erectile dysfunction as the outcome.

These studies have documented that clinical and subclinical atherosclerotic diseases including ischemic heart disease and cerebral infarction are significantly associated with As in drinking water showing a significant biological gradient. The associations between As in drinking water and risk of cardiovascular diseases have also been reported in many countries (Wang C.-H. *et al.* 2007). As atherosclerosis is the underlying pathological change of BFD, all these findings further support that As in drinking water is the major cause of BFD.

4.9   ARSENIC IN DRINKING WATER AND PREVALENCE OF DIABETES AND HYPERTENSION

As both diabetes and hypertension are important risk factors for atherosclerotic diseases, it is essential to examine the associations between ingested As and risk of these two diseases. Residents in three BFD-hyperendemic villages were found to have a significantly increased (around two-fold) prevalence of diabetes mellitus (Lai *et al.* 1994) and hypertension (Chen *et al.* 1995) compared to residents in non-endemic areas in Taiwan. Furthermore, there was a significant dose-response relationship between the cumulative As exposure and prevalence of diabetes mellitus and hypertension. A follow-up study also found a significantly higher incidence of diabetes in residents of hyperendemic villages than those in non-endemic areas, and a significant association between cumulative As exposure and diabetes incidence (Tseng *et al.* 2000).

In another study of diabetes prevalence of residents living in the BFD-endemic and non-endemic areas, the age-sex-adjusted prevalence of diabetes derived from the Taiwan National Health Insurance Database from 1999 to 2000 was higher in the BFD endemic area (7.5%) than the non-endemic area (3.5%) (Wang *et al.* 2003).

The associations with As in drinking water for diabetes and hypertension have also been reported in other countries (Chen *et al.* 2007). Arsenic in drinking water may induce BFD through its effect on the development of diabetes and hypertension, but there seems to be an independent effect of As on the development of BFD as well.

4.10   REDUCTION IN MORTALITY OF ARSENIC-INDUCED DISEASES AFTER IMPLEMENTATION OF PUBLIC WATER SUPPLY SYSTEM IN THE ENDEMIC AREA OF BLACKFOOT DISEASE

The implementation of a public water supply system, which used uncontaminated surface water from distant reservoirs, in the BFD endemic area of southwestern Taiwan was started in the early 1960s and completed in the 1970s. The As-induced health hazards identified from above-mentioned studies could thus be validated through the comparison of the secular changes in morbidity and mortality of As-induced diseases in the endemic and non-endemic areas. In a series of studies, cause-specific standardized mortality ratios of residents in the BFD endemic area from 1971 to 2003 were calculated using the general population in Taiwan as the standard population. Significant declines in mortality from ischemic heart disease (Chang *et al.* 2004), renal disease (Chiu and Yang 2005), and peripheral vascular disease (Yang 2006) were observed for both males and females in the endemic area of BFD. Based on the reversibility criterion, it can be concluded that association between exposure to As in drinking water and various vascular diseases including BFD is causal.

CHAPTER 5

Non-vascular health effects of arsenic in drinking water in Taiwan

5.1 INTRODUCTION

The study of the health effects of arsenic (As) in Taiwan was triggered by blackfoot disease (BFD), an obstructive peripheral vascular disease named after its most striking clinical feature—the black discoloration of extremities, mostly feet, resulting from the development of gangrene (Yeh and How 1963, Tseng 1988). Various vascular effects of As were discovered after epidemiological studies were conducted in the endemic area of BFD, including other peripheral vascular diseases (Wu *et al.* 1989), cardiovascular diseases (Tseng, 1989, Wu *et al.* 2003), ischemic heart disease (IHD) (Chen *et al.* 1996, Hsueh *et al.* 1998, Chang *et al.* 2004), and cerebrovascular accidents (Chiou *et al.* 1997). In addition, diabetes mellitus (DM), which is related to vascular changes, was also found to be prevalent in the BFD area (Lai *et al.* 1994, Tseng *et al.* 2000, Wang *et al.* 2003). Furthermore, As was found to be associated with many other diseases that are not directly related to vascular changes.

5.2 SKIN CANCER

The study of BFD triggered the study of the association between As in drinking water and skin cancer in Taiwan. The skin cancers observed in the BFD area were those typically seen among patients intoxicated with As, especially Bowen's disease (Yeh, 1963, Yeh *et al.* 1968, Guo *et al.* 2001). Yeh *et al.* (1968) studied 238 arsenical skin cancers in 153 patients and found 46 of them were epidermoid carcinomas, 36 were basal carcinomas, 139 were intraepidermal carcinomas (including 121 classical Bowen's lesions and its variants), and 17 were combined forms. In the survey of 40,421 inhabitants of the BFD area (Fig. 4.1) by Yeh (1973), the prevalence of skin cancer was found to be 10.6 per 1000; of the 303 skin cancer cases in 184 patients, 57 were epidermoid carcinomas, 45 were basal cell carcinomas, 176 were intraepidermal carcinomas (including 153 classic Bowen's lesions and its variants), and 25 were combined forms. The high prevalence among BFD patients of other skin changes typically seen in As intoxication, such as hyperpigmentation and hyperkeratosis (Yeh 1963a), indicates that As is a major cause of the skin cancers observed in the BFD area. In addition, Tseng *et al.* (1968) conducted an ecologic study in the BFD area and observed a positive dose-response relationship between the As level in well water and prevalence of skin cancer, thus, providing further supporting evidence.

In the study by Tseng *et al.* (1968), 37 villages with a total of 40,421 inhabitants were examined, and skin cancers were observed in 428 individuals. Data from a survey conducted by Kuo (1968) were adopted to assess the exposure, and the survey covered 114 wells, including 110 artesian wells and 4 shallow wells. However, only 33 of the 37 villages had data on As levels in groundwater. According to the mean As level observed in each village, Tseng *et al.* (1968) categorized the villages into four exposure groups: "low" (below 0.30 mg L^{-1}), "mid" (0.30–0.60 mg L^{-1}), "high" (>0.60 mg L^{-1}), and "undetermined". Age- and sex-specific prevalence was calculated for three age groups: 20–39 years, 40–59 years, and ≥60 years. For comparison, the study also included 4978 residents of Matsu island (in the Strait of Taiwan, 30 km from mainland China), where As was not detectable in the drinking water from shallow wells, and 2522 residents in five villages outside the BFD area, where As levels in the drinking water from shallow wells ranged from 0.001 to 0.017 mg L^{-1}. As a result, Tseng *et al.* observed a positive dose-response relationship between the As level in well water and the prevalence of skin cancer in 33 BFD endemic villages. Brown *et al.* (1989) re-analyzed the data, estimating the numbers of persons at risk over three dose intervals and four exposure durations and then applying the method of maximum likelihood to a multistage-Weibull time/dose-response model. They assumed a constant exposure level since birth for each of the exposure categories and found that the cumulative hazard increases as a power of three in age, and is linear or quadratic (with a linear coefficient) in dose. They verified the model using data from a smaller epidemiologic study in Mexico and estimated that an American male would have a lifetime risk of developing skin cancer of 1.3×10^{-3} or 3.0×10^{-3} if exposed to 1 µg kg^{-1} day^{-1} for a 76-year lifespan (median lifespan in the U.S.). Because the examiners were not blind to the exposure status of the participants and no skin cancer patients were found among the 7500 participants of the comparison populations, it has been speculated that this study may have a certain level of information (observer) bias (Risk Assessment Forum 1988).

Nearly two decades later, Chen *et al.* (1985) conducted another ecological analysis and calculated the sex-specific age-standardized mortality ratios (SMR) from 1973 to 1986. They found the SMR for skin cancer was significantly higher in the BFD area: 534 for men and 652 for women. In a following dose-response analysis (Chen *et al.* 1988a), they calculated the sex-specific age-standardized mortality rates from 1973 to 1986 for 42 villages in the BFD area on the basis of 899,811 person-years and 1031 cancer deaths. Again, they used the median well water As level observed by Kuo (1968) in each village to categorize the villages into the same three exposure groups— <0.30 mg L^{-1}, 0.30–0.59 mg L^{-1}, and ≥0.60 mg L^{-1}. In comparison with the general population in Taiwan, higher SMR's were observed in the BFD area for skin cancers in all three exposure groups, and a positive dose-response relationship was observed. Because cancer cases below 20 years of age were unlikely to be attributable to As exposure from drinking water, a re-analysis of the data on residents 20-year old or older were conducted later (Wu *et al.* 1989). The re-analysis on the basis of 889,806 person-years and 1152 cancer deaths revealed a significant positive dose-response relationship between the As level in drinking water and SMR for skin cancer in both men and women. Chen *et al.* (1992) re-analyzed the data again using the Armitage-Doll multistage model to estimate the cancer risks attributable to As in drinking water. On the basis of new data, they reassigned one village in the "high" exposure group into the "low" exposure group. In addition, unlike the previous studies in which a cancer death was defined as a person whose underlying cause of death was cancer (Chen *et al.* 1985, 1988a, Wu *et al.* 1989), the re-analysis identified any case with primary cancers listed on the death certificate as a cancer death. Still, they observed significant dose-response relationships between the As level in drinking water and skin cancer. The data were re-analyzed further by Morales *et al.* (2000) using several variations of the generalized linear model and the multistage-Weibull model. As a result, they found that the risk estimates were sensitive to the model choice, to whether or not a comparison population was used to define the unexposed disease mortality rates, and to whether the comparison population was from all of Taiwan or just from the southwestern region. Nonetheless, they believed the data suggested that the previous standard of 0.05 mg L^{-1} was associated with a substantial increased risk of cancer and thus was not sufficiently protective of public health. Accordingly, the US EPA lowered the Maximum Contaminant Level (MCL) for As in drinking water to 0.010 mg L^{-1} (US EPA 2001).

Chen *et al.* (1988b) studied 241 BFD patients and 759 age-sex-residence-matched healthy community controls and found the BFD patients had a significantly higher mortality from skin cancer. Between September 1988 and March 1989, Hsueh *et al.* (1995, 1996) studied 1571 residents of three villages in the BFD area who were 30 or more years of age. They interviewed the participants using a structured questionnaire to collect data on risk factors for skin cancer. Among the participants, 1081 (68.8%), including 468 men and 613 women, also received a physical examination. The prevalence of skin cancer among the examined subjects was 6.1%. Using duration of residence in the endemic area, duration of consumption of high-As artesian well water, average As exposure (divided into three groups: 0 mg L^{-1}, 0.0–0.70 mg L^{-1}, and >0.71 mg L^{-1}), and cumulative As exposure (divided into three groups: ≤4, 5–24, and ≥25 mg L^{-1}-years) as indicators of As exposure, the investigators observed significant dose-response relationships between As exposure and skin cancer. Again, the As exposure was determined by the median level observed by Kuo (1968) in the village of residence. Hseuh *et al.* (1997) studied a subset of the study population (654 participants) to evaluate the role of serum β-carotene and observed a negative association between the serum β-carotene level and prevalence of skin cancer. Still, after adjusting for gender, age, and serum β-carotene level, they observed a significant positive dose-response relation between skin cancer prevalence and cumulative As exposure.

About three decades after the initiation of studies in the BFD area, researchers began to study the health effects of As in drinking water in another endemic area in the northeastern region of Taiwan. Using the median As levels from the survey of 6986 wells by the Taiwan Provincial Institute of Environmental Sanitation (Lo and Lin 1982), Lin *et al.* (1986) assigned 43 villages in seven townships of the Yilan county (Lanyang plain, Fig. 4.1) into the following three exposure groups: I (≤0.05 mg L^{-1}), II (0.051–0.125 mg L^{-1}), and III (0.126–0.2 mg L^{-1}). They analyzed the mortality from 1972 to 1983 and found that in comparison with the general population in Taiwan, men in category II had a decreased mortality for skin cancer. Because such a decrease was not observed in women and for other cancers, it is likely that some other factors affected the associations between As in drinking water and skin cancer in this area.

Chen and Wang (1990) conducted the first nation-wide study on As and cancers in Taiwan, which was an ecological analysis of the 314 townships surveyed by the Taiwan Provincial Institute of Environmental Sanitation (Lo *et al.* 1977). They calculated the SMR's between 1972 and 1983 for cancers of 21 sites and found a positive association between the As level in drinking water and mortality of skin cancer. The study population was around 15 million, and the mean As level in water in each township was used as the exposure indicator. Using multivariate regression, they found an increase in age-adjusted mortality per 100,000 person-years for every 0.1 mg L^{-1} increase in As level of well water was 6.8 and 2.0, 0.7 and 0.4, 5.3 and 5.3, 0.9 and 1.0, 3.9 and 4.2, as well as 1.1 and 1.7, respectively, in males and females for cancers of the liver, nasal cavity, lung, skin, bladder and kidney. When they performed separate analyses of the data on 170 townships in the southwestern region of Taiwan, a similar association was observed.

Another nation-wide ecological analysis was conducted on the 243 townships where individual well measurement reports from the survey by the Taiwan Provincial Institute of Environmental Sanitation (Lo *et al.* 1977) were available (Guo *et al.* 1998). In the analysis, As levels were grouped into the following 10 categories based on the standard solution concentrations used in the measurements: 0 mg L^{-1} (undetectable), trace (between the 0.001 mg L^{-1} detection limit and 0.01 mg L^{-1}), 0.01 mg L^{-1}, 0.02 mg L^{-1}, 0.03–0.04 mg L^{-1}, 0.05–0.08 mg L^{-1}, 0.09–0.16 mg L^{-1}, 0.17–0.32 mg L^{-1}, 0.33–0.64 mg L^{-1}, and >0.64 mg L^{-1}. Incidence rates of bladder and kidney cancers between 1980 and 1987 were assessed using data from the National Cancer Registration Program, and standardized incidence rates (SIR's) were calculated for each township. The study covered a population of about 11.4 million people, and 1547 cases of skin cancer were identified. The data were analyzed by regression models using a series of variables to describe the exposure status, and each variable denoted the proportion of wells in a specific As exposure category in each township. This approach, which can produce unbiased risk estimates when the dose-response relationship is not linear (Guo *et al.* 1998) did show such a dose-response relationship: a positive association with the incidence of skin cancer was observed for the highest As exposure group (>0.64 mg L^{-1}) in both men and women, but not for the other exposure groups. A further analysis of data between January 1, 1980 and December 31, 1989, which included 2369 skin cancer patients, showed that among the three major cell types of skin cancer, squamous cell carcinoma and basal cell carcinoma appeared to be associated with ingestion of As, but such an association was not observed for malignant melanoma (Guo *et al.* 2001). Merkel cell carcinoma is a rare type of skin cancer. Lien *et al.* (1999) reviewed all 11 cases of Merkel cell carcinoma diagnosed at two medical centers in Taiwan and found 6 of them were residents of the BFD area. These findings suggested that the carcinogenic effect of As is cell-type specific.

Because some researchers raised concerns about effects of possible incomplete reporting of the cancer registry (Abernathy *et al.* 1996), Guo conducted a study on skin cancer to compare results obtained from different sources, including cancer registry, death certificates, and physical examinations (Guo 1997). The results showed that although the registry's ascertainment rate was about 30%, the relative risk estimates generated on the basis of the registry data were compatible with those generated on the basis of the other two data sources, indicating that the information bias, even if existed, was not large enough to affect the study results.

Tsai *et al.* (1998) assessed the cancer mortality trends in the BFD area following the establishment of the tapwater system in 1956, which delivered As-safe water using two methods to estimate the age-adjusted mortality rate ratio (SM ratio). The first method used the first time interval (1971–1973) as the standard, and the SM ratio was estimated from this interval through the last interval (1992–1994) using Poisson regression. The combination of Chiayi and Tainan counties, excluding the BFD area, was used as a local reference, and the general population of Taiwan was used as a national reference. The second method compared the SM ratio for the BFD area to that for the local and national references for the same time intervals for each disease category. The results showed significantly declining trends for SM ratios all malignant tumors, including skin cancer, with 1971–1973 as the standard for the BFD area, especially in women. Compared to either local or national reference, a decrease of SM ratios of malignant cancers, including skin cancer, was found in the age group of >40 years for both genders. The findings suggest that the As exposure from artesian well water is associated with the mortality from cancers in the BFD area. They also calculated SM ratios of all death causes in the BFD area from 1971 to 1994 and compared them to the two reference groups (Tsai *et al.* 1999). In comparison with the local reference, residents of the BFD area had higher risks of dying from various cancers including skin cancer.

To clarify the role of As methylation capacity in the development of As-related skin cancer, Yu *et al.* (2000) recruited 26 patients with matched controls who had been exposed to similar high concentrations of As in drinking water. They found that the patients had higher percentage of inorganic As ($13.1 \pm 3.7\%$ *vs.* $11.43 \pm 2.1\%$) and monomethylarsonic acid ([MMA(V)] $16.4 \pm 3.2\%$ *vs.* $14.6 \pm 2.6\%$), a lower percentage of dimethylarsinic acid ([DMA(V)] $70.5 \pm 5.8\%$ *vs.* $73.9 \pm 3.3\%$), and a higher ratio of MMA(V) to DMA(V) [MMA(V)/DMA(V) 0.24 ± 0.06 *vs.* 0.20 ± 0.04] in urine. They also found that individuals with a higher percentage of [MMA(V)] (>15.5%) had an odds ratio (OR) of 5.50 (95% CI, 1.22–24.81) for developing skin cancer, which indicated that As biotransformation including methylation capacity may have a role in the development of As-induced skin disorders. The hypothesis was supported by a study of 76 cases of skin cancer and 224 controls from the BFD area, which found associations between the secondary As methylation index (SMI), defined as [DMA(V)/MMA(V)] and skin cancer (Chen *et al.* 2003a). Specifically, a low SMI (≤ 5) with cumulative As exposure of >15 mg L^{-1}-years was associated with an OR of 7.48 (95% CI, 1.65–33.99) for developing skin cancer in comparison with cumulative As exposure of ≤ 2 mg L^{-1}-years. Further support came from a study on 64 residents of the Lanyang plain on the northeast coast of Taiwan (Fig. 4.1), which found that the As concentration in whole blood had a negative association with the level of plasma antioxidant capacity ($r = -0.30$, $p = 0.014$), while the primary As methylation capability had a positive association with the level of plasma antioxidant capacity ($p = 0.029$) (Wu *et al.* 2001). The researchers also found that the As concentration in whole blood had a positive association with the level of reactive oxidants in plasma ($r = 0.41$, $p = 0.001$) and believed that persistent oxidative stress in peripheral blood may be a mechanism underlying the carcinogenesis and atherosclerosis induced by long-term As exposure.

5.3 INTERNAL CANCERS

As early as the 1950s, the occurrence of urinary cancers, including cancers of the bladder, kidney, ureter, and urethra, was noted among As intoxicated patients (Sommers and McManus 1952). Studies in Taiwan also observed associations between the As level in drinking water and the occurrence of cancers of the lung, urinary bladder, kidney, liver, and probably colon and prostate (Guo *et al.* 1994a).

Tseng (1977) followed-up on 1108 patients of BFD and found that the most common cause of death in the patients with skin cancer and BFD was carcinoma of various sites. In the ecological analysis of 42 villages in the BFD area by Chen *et al.* (1985), in addition to skin cancer, the SM ratio was significantly higher in the BFD area for cancers of urinary bladder, kidney, lung, liver, and colon. Specifically, the SM ratio for cancers of bladder, kidney, lung, liver, and colon were 1100, 772, 320, 170, and 160 for men, and 2009, 1119, 413, 229, and 168 for women. In a following dose-response assessment, higher SM rates were observed for cancers of liver, lung, prostate, urinary bladder, and kidney, as well as all cancers combined in all three exposure groups, except for prostate cancer in the low exposure group. Furthermore, a positive dose-response relationship was observed for each of the six types of cancers (Chen *et al.* 1988a). In the re-analysis of the data by Wu *et al.* (1989), sex-specific SM rates were calculated for the same three levels of exposure, and additional analyses were conducted for leukemia and cancers of nasopharynx, esophagus, stomach, colon, and cervix. In addition to skin cancer, the investigators observed significant positive dose-response relationships for cancers of urinary bladder, kidney, and lung, as well as all cancers combined in both genders. Such a relationship was also observed for liver and prostate cancers in men, but not for liver or cervical cancer in women, nor for esophageal cancer, nasopharyngeal cancer, stomach cancer, colon cancer, or leukemia in either gender. In the further analysis of the data by Chen *et al.* (1992), significant dose-response relationships were observed for cancers of the bladder, lung, and liver, in addition to skin cancer. In order to compare the risks of various internal cancers induced by ingested inorganic As and to assess the differences in risk between males and females, Chen *et al.* (1992) calculated cancer potency indices by applying the Armitage-Doll multistage model to analyze mortality rates observed on 898,806 person-years of residents in the BFD area. On the basis of 202 liver cancer, 304 lung cancer, 202 bladder cancer and 64 kidney cancer deaths, they estimated that the potency index of developing cancer of the liver, lung, bladder, and kidney due to an intake of 10 µg kg^{-1} day^{-1} of As was 4.3×10^{-3}, 1.2×10^{-2}, 1.2×10^{-2}, and 4.2×10^{-3}, respectively, for men and 3.6×10^{-3}, 1.3×10^{-2}, 1.7×10^{-2}, and 4.8×10^{-3}, respectively, for women. In a recent study, Liao *et al.* (2009) applied an integrated approach by linking the Weibull dose-response function and a physiologically based pharmacokinetic model to analyze data on an 8-year follow-up of 10,138 residents in arseniasis-endemic areas in southwestern and northeastern Taiwan and found positive relationships between As exposures and cumulative incidence ratios of bladder, lung, and urinary-related cancers. Accordingly, they recommended 3.4 µg L^{-1} as the reference As guideline based on male bladder cancer with an excess risk of 10^{-4} for a 75-year lifetime exposure. The likelihood of reference As guideline and excess lifetime cancer risk estimates range from 1.9–10.2 µg L^{-1} and 2.84×10^{-5} to 1.96×10^{-4}, respectively, based on the drinking water uptake rates of 1.08–6.52 L day^{-1}.

Soon after Chen *et al.* (1992) published the re-analysis of data on 42 villages, Brown and Chen (1993) reported findings from a larger ecological analysis of 60 villages, which included the 42 villages covered by the previous studies. On the basis of the median As level in drinking water observed in the survey by Kuo (1968) and the survey by Lo *et al.* (1982), they categorized the villages into 11 exposure groups (<0.01, 0.01–0.019, 0.02–0.029, 0.03–0.039, 0.04–0.049, 0.05–0.09, 0.1–0.29, 0.3–0.49, 0.5–0.69, 0.7–0.89, and ≥0.9 mg L^{-1}), and SM rates of cancers of bladder, liver, lung, and skin from 1973 to 1986 were calculated for each category. The dose-response relationships did not appear to be linear; only median As levels at 0.05–0.09 mg L^{-1} and above were associated with mortality rates of skin cancer.

Taking the uncertainties in exposure measurements into account, they re-analyzed the data and confirmed that the dose-response patterns for lung, liver, and bladder cancers were not linear (Brown and Chen 1995). The resultant dose-response patterns showed no evidence of excess risk at As levels below 0.1 mg L^{-1}.

To control for potential confounders other than gender and age and to assess the risk using personal data instead of ecological data, Chen *et al.* (1986) conducted a case-control study which included 69 bladder cancer, 76 lung cancer and 59 liver cancer deaths and 368 alive community controls group-matched on age and sex. They collected information on risk factors through proxy interviews of the cases and personal interviews of the controls using a standardized structured questionnaire. They found a dose-response relationship was between the exposure to artesian well water and cancers of bladder, lung and liver. In comparison with those who never used artesian well water, the age-sex-adjusted odds ratios (OR) of developing bladder, lung and liver cancers for those who had used artesian well water for 40 or more years were 3.90, 3.39, and 2.67, respectively. While this retrospective study had to use proxy information for the cases, Chiou *et al.* (1995) conducted a seven-year follow-up on 263 BFD patients and 2293 healthy residents in the BFD area and collected data from them through standardized questionnaire interviews. They obtained data on consumption of well water, sociodemographic characteristics, life-style and dietary habits, and personal and family history of cancers. According to the median As level observed by Kuo (1968) in the village of residence and the history of well water consumption reported by the participant, they assigned participants to four exposure groups: "0 mg L^{-1}-years," "0.1–19.9 mg L^{-1}-years," "20.0 + mg L^{-1}-years," and "unknown." Cases of cancers were identified through annual health examinations, personal interviews during home visits, household registration data checks, and national death certification and cancer registry profile linkages. As a result, they observed a dose-response relationship between the As exposure from drinking well water and the incidence of lung cancer, bladder cancer, and all cancers combined after adjustment for age, sex, and cigarette smoking. In the study of 241 BFD patients and 759 matched controls, Chen *et al.* (1988b) found a significantly higher mortality from cancers of bladder, lung, and liver among BFD patients, in addition to skin cancer.

Horng *et al.* (1995) followed up on 257 patients of BFD and on 753 healthy residents in four townships of the BFD area from 1985 to 1992 and collected data on potential confounders through questionnaire interviews. They identified 21 lung cancer deaths, including 9 among BFD patients, and estimated As exposure using the median level observed in the village of residence by Kuo (1968). After adjusting for age, gender, and cigarette smoking, they observed a significant dose-response relationship between the cumulative As exposure—which was classified by four groups: 0, <20, ≥20 mg L^{-1}-years, and unknown—and the mortality of lung cancer. Liaw *et al.* (1995) analyzed data from the same study population for cancers of the lower urinary tract but used different exposure grouping: 0.1–19.9, 20–29.9, 30 + mg L^{-1}-years, and unknown. They identified 20 deaths of urinary bladder cancer and 2 of renal pelvis cancer. After adjusting for age, gender, and cigarette smoking, they also found a positive dose-response relationship between the cumulative As exposure and the mortality related to cancers of lower urinary tract.

Study by Lin *et al.* (1986) in the northeastern region of Taiwan found that mortality increased for rectal cancer in category I, for liver cancer in category III, and for all cancers combined in categories I and III, whereas a decreased mortality for skin cancer was observed in men in category II. In women, they observed increased mortality for liver cancer in category III, but decreased mortality was observed for rectal cancer in category I and for lung cancer in category III. When the data from their survey on 391 wells were used to estimate the exposure, a decreased mortality was observed for all cancers combined in both genders. Because the water samples taken at the time of their study might not accurately represent exposure levels in the past, results from the second analysis are not reliable. From 1991 to 1994, Chiou *et al.* (2001) conducted a follow-up on 8102 residents 40 years old or older in 18 villages in the same region. They measured As in a sample of water taken from the household of each participant and categorized the participants into four As exposure groups: 0–10.0, 10.1–50.0, 50.1–100.0, and >100.0 µg L^{-1}. During the follow-up, they identified 18 cases of urinary cancers, including 9 in the bladder, 8 in the kidney, and 1 in both bladder and kidney. Among these cases, 10 were transitional cell carcinomas (TCCs). After adjustment for age, sex, and cigarette smoking, they found a significant positive dose-response relationship between As exposure and risk of urinary cancers, especially TCC.

To assess the effect of cigarette smoking on the association between As ingestion and lung cancer, Chen *et al.* (2004) followed up on 2503 residents in southwestern and 8088 in northeastern arseniasis-endemic areas for an average period of 8 years. They collected information on As exposure, cigarette smoking, and other risk factors through standardized questionnaire interviews and identified 139 new cases of lung cancer through the national cancer registry from January 1985 to December 2000 (83,783 person-years). After adjustment for cigarette smoking and other risk factors, they observed a monotonic trend of lung cancer risk by As level in drinking water of less than 10 to 700 µg L^{-1} or more ($p < 0.001$). The etiologic fraction of lung cancer attributable to the joint exposure of ingested As and cigarette smoking ranged from 32% to 55%, and the synergy indices ranged from 1.62 to 2.52, indicating a synergistic effect. While confirming the significant dose-response trend for the ingested As and lung cancer risk, on the basis of the synergistic effect, they argued that the risk assessment of lung cancer induced by ingested As should take cigarette smoking into consideration.

In the first nation-wide study conducted by Chen and Wang (1990) in 314 townships, in addition to skin cancer, the SM ratios were calculated for leukemia and cancers of nasopharynx, esophagus, stomach, small intestine, colon, rectum, liver, pancreas, nasal cavity, larynx, lung, bone and cartilage, breast, uterine cervix, ovary, prostate, bladder, kidney, and brain. Significant associations were observed for cancers of liver, nasal cavity, lung, urinary bladder, kidney, and prostate. When analyses were limited to the 170 southwestern townships, significant associations between arsenic exposure and these cancers were generally observed.

Guo *et al.* (1994b, 1997) conducted a nation-wide ecological analysis on urinary cancers in the 243 townships where individual well measurement reports from the survey by the Taiwan Provincial Institute of Environmental Sanitation (Lo *et al.* 1977) were available. The preliminary report (Guo *et al.* 1994b) categorized As levels into 10 groups as in the study on skin cancer (Guo *et al.* 1998), and assessed incidence rates of bladder and kidney cancers between 1980 and 1987 using data from the National Cancer Registration Program. The study identified 1972 bladder cancer cases and 726 kidney cancer cases. The data were analyzed by regression models using a series of variables to describe the exposure status, and each variable denoted the proportion of wells in a specific As exposure category in each township. For bladder cancers, going from the lowest to the highest As exposure groups, significant positive associations with some categories were observed, alternating with insignificant or significant negative associations, and the highest exposure group (>0.64 mg L^{-1}) was associated with the highest risks. For kidney cancers, an association was observed between high As levels and transitional cell carcinoma (239 cases in total), but not renal cell carcinoma (210 cases in total). For bladder cancers, an association was observed between high As levels and transitional cell carcinoma, but not squamous cell carcinoma. These findings supported the hypotheses generated from studies on skin cancer that the carcinogenic effects of As are cell-type specific and that the dose-response relationship is not linear.

A further analysis was conducted on all urinary cancers, including 170 cases of ureter cancer and 57 cases of urethral cancer. However, because the method used for As analysis was more reliable above 0.05 mg L^{-1}, the five lowest exposure groups were combined into a single exposure group (As < 0.05 mg L^{-1}) (Guo *et al.* 1997). This analysis found associations between high As levels in drinking water and transitional cell carcinomas of the bladder, kidney and ureter and all urethral cancers combined in both genders. Such an association was observed in adenocarcinomas of the bladder in males, but not in squamous cell carcinomas of the bladder or renal cell carcinomas or nephroblastomas of the kidney. A positive association between urbanization index and transitional cell carcinomas of the ureter was observed in males. Although the analysis accounted for the number of cigarettes sold per capita to adjust for the effects of cigarette smoking, the cigarette sale data was not a good predictor for urinary cancers. The results supported the main conclusion of the preliminary analysis: the carcinogenic effects of As are cell-type specific and the dose-response relationship was not linear. Chiou *et al.* (2001) followed up on 8102 residents of an arseniasis-endemic area in northeastern Taiwan and confirmed the cell-type specific carcinogenicity of As on urinary cancers.

The cell-type specificity and non-linear dose-response pattern of the carcinogenic effect of As were verified also for lung cancer. Guo *et al.* (2004) reviewed certificates of deaths issued between January 1, 1971 and December 31, 1990 for 138 villages in the BFD area and identified 673 male and 405 female mortality cases due to lung cancer. He found that As levels above 0.64 mg L^{-1} were associated with a significant increase in the mortality due to lung cancer in both genders, but no significant effect was observed at lower levels. He confirmed the non-linear dose-response relationship using post-hoc analyses. Guo *et al.* (2004) evaluated the cell-type specificity for lung cancers using data on 37,290 lung cancer patients from 243 townships in Taiwan, including 26,850 men and 10,440 women, who were diagnosed between January 1, 1980 and December 31, 1999. They compared 5 townships in the BFD area to other 238 townships and found that patients from the BFD area had higher proportions of squamous cell and small cell carcinomas, but a lower proportion of adenocarcinomas. Furthermore, through a review of literature, they found that the association between adenocarcinoma and As exposures through inhalation appeared to be stronger than that for squamous cell carcinoma and speculated that As may give rise to different mechanisms in the development of lung cancers through different exposure routes. Lamm *et al.* (2006) re-analyzed the data published by Wu *et al.* (1989) and found that As accounted for only 21% of the variance in the village SM ratios for bladder and lung cancer and that data for bladder and lung cancer mortality fit an inverse linear regression model ($p < 0.001$) with an estimated threshold at 151 µg L^{-1} (95% CI, 42–229 µg L^{-1}). Although a study of liver cancer did not identify any specific cell type that is related to As ingestion, a literature review showed that hepatic angiosarcoma, a rare primary mesenchymal malignancy of the liver, has close association to As intoxication (Guo 2003). In addition, a case of hepatic angiosarcoma with an As level of 0.12 mg L^{-1} in her drinking till 21 years of age was identified in Taiwan (Ho *et al.* 2004).

Tsai *et al.* (1998) assessed the cancer mortality trends in the BFD area following the water source being shifted to tapwater in 1956 (providing As-safe water) and observed significantly declining trends for SM ratios of all malignant tumors in the BFD area, especially in women. Compared to either the local or the national reference, a decrease of SM ratios of malignant cancers was found in the age group of >40 years for both genders. These findings suggest that the As exposure from artesian well water is associated with the mortality of cancers in the BFD area. They also calculated SM ratios of all death causes in the BFD area from 1971 to 1994 and compared them to both local and national references (Tsai *et al.* 1999). In comparison with the local reference, residents of the BFD area had higher risks of dying from cancers of urinary bladder, kidney, skin, lung, nasal-cavity, bone, liver, larynx, colon, and stomach, as well as lymphoma. Chiu *et al.* (2004a) calculated the SM ratios for lung cancer in the BFD area from 1971 to 2000 and found that mortality from lung cancer declined gradually after the improvement of drinking water supply system to eliminate As exposure from artesian well water. Based on the reversibility criterion, they argued that the association between As exposure and lung cancer mortality is likely to be causal. Yang *et al.* applied the same method to analyze the data on kidney (Yang *et al.* 2004), bladder (Yang *et al.* 2005), colon (using data from 1971 to 2006) (Yang *et al.* 2008a), and prostate (using data from 1971 to 2006) (Yang *et al.* 2008b) cancers and observed the same phenomenon. However, when they calculated the SM ratios for liver cancer, they found that mortality from liver cancer declined starting 9 years after the cessation of consumption of high-As artesian well water in women, but not in men (Chiu *et al.* 2004b).

The association between the As level in drinking water and mortality of bladder cancer was evaluated by an ecologic study in 10 townships in southwestern Taiwan, including the 4 major townships in the BFD area, from 1971 to 1990 (Guo 1999). This study categorized the As levels in drinking water into the six groups (<0.04, 0.05–0.08, 0.09–0.16, 0.17–0.32, 0.33–0.64, and >0.64 mg L^{-1}) and found that exposure >0.64 mg L^{-1} was associated with an increase in bladder cancer mortality. In contrast, no association was found for lower exposure groups. Guo and Tseng (2000) compared the results obtained using cancer registry data with those obtained using death certificates and found that the two approaches identified similar associations between high As levels in drinking water and occurrence of bladder cancer, which further justified the validity of studies based on the cancer registry.

Chow *et al.* (1997) compared the characteristics of 49 patients of As-related bladder cancers and 64 other bladder cancer patients and found a higher histological grading for the As-related group ($p = 0.04$), but no other differences in pathobiological features or prognosis. Tan *et al.* (2008) followed up on 474 patients with pathologically diagnosed transitional cell carcinoma of the genitourinary tract and found no significant differences between the groups in age, sex, tumour stage and grade between patients from the BFD area and patient from other areas, although patients from the BFD area had a significantly higher proportion of women. They found that among patients with urinary bladder transitional cell carcinoma, patients from the BFD area had a lower 5-year survival rate (58.7% *vs.* 72.4%) and that in early-stages (pTa and pT1), patients from the BFD area had a higher mortality rate with tumor progression and recurrence after transurethral resection of the bladder tumor. Similarly, Chen *et al.* (2009) studied 977 bladder cancer patients from 1993 through 2006 and found that patients with As-related bladder cancer may have decreased overall cancer specific survival because they have more unfavorable tumor phenotypes. Lu *et al.* (2004) compared 65 patients with hepatocellular carcinoma (HCC) from the BFD area to 130 age- and sex-matched HCC control patients from non-BFD-endemic areas and found the two groups had similar clinicopathological features, including hepatitis viral infection status, hepatitis activity, liver function, histological findings, computed tomography scan characteristics, and patient survival.

There are other factors that might contribute to the high risk of developing cancer among residents of the BFD area. Lu *et al.* (1986) conducted an ecological analysis in the BFD area and a neighboring township and observed a positive association between the fluorescent intensity of well water (a surrogate measurement of humic acids) and incidence of bladder cancer. Lu and Chen (1991) studied 129 BFD patients and 374 age-sex-residence-matched community controls and found similar hepatitis B surface antigen (HBsAg) carrier rates between BFD patients (20.9%) and the controls (20.1%). Residents in the endemic area had an HBsAg carrier rate (20.2%) similar to that of the general population in Taiwan. Guo (2003) reviewed several other studies and confirmed these finding, which indicated hepatitis B infection is not a major cause of the high risk of HCC in the BFD area. Similar to skin cancer, it has been speculated that As methylation capacity may have a role in the development of As-induced internal cancers. Chen *et al.* (2003b) recruited 49 new patients with bladder cancer and 224 fracture and cataract patients as controls from January 1996 to December 1999 and found that a cumulative As exposure of >12 mg L^{-1}-years with a low SMI ($\leq$4.8) was associated with an OR of 4.23 (95% CI, 1.12–16.01) for developing bladder cancer compared to 2 mg L^{-1}-years. These findings were verified in a prospective cohort study in which 1078 residents of southwestern Taiwan were followed for an average of 12 years, and 37 new patients of urothelial carcinomas were identified through the National Cancer Registry of Taiwan during a follow-up period of 11,655 person-years between January 1985 and December 2001 (Huang *et al.* 2008). Significantly higher percentages of MMA(V) and lower percentages of DMA(V) were observed in the urine of patients with urothelial carcinoma. After adjusting for age, gender, educational level, and smoking status, the researchers found that the percentage of urinary DMA(V) was inversely associated with the risk of urothelial carcinoma, with relative risks of the tertile strata of 1.0, 0.3, and 0.3, respectively ($p < 0.05$ for the trend test). The relative risk of residents with a cumulative As exposure of $\geq$20 mg L^{-1}-years and a higher percentage of MMA(V) or a cumulative As exposure of $\geq$20 mg L^{-1}-years and a lower percentage of DMA(V) was 3.7 (95% CI, 1.2–11.6) or 4.2 (95% CI, 1.3–13.4), respectively, as compared to residents with a cumulative As exposure of <20 mg L^{-1}-years and a lower percentage of MMA(V) or a cumulative As exposure of <20 mg L^{-1}-years and a higher percentage of DMA(V). Thus, they confirmed a significant association between inefficient As methylation and the development of urothelial carcinoma in the residents in the high CAE exposure strata.

5.4   EYE DISEASES

See *et al.* (2007) studied 349 residents of the BFD area to evaluate the effect of As exposure from drinking water on various types of cataract and found that cumulative exposure to As and the duration of consuming artesian well water were associated with an increased risk of all types of lens opacity, but after adjustment for age, sex, diabetes status, and occupational sunlight exposure, only the associations for posterior subcapsular opacity were statistically significant ($p = 0.014$ and $p = 0.023$, respectively).

In addition, Lin, W. *et al.* (2008) evaluated the association between As exposure through drinking water and the occurrence of pterygium in southwestern Taiwan and recruited participants >40 years of age from three villages in the BFD area and four neighboring non-endemic villages. They included 223 participants from the endemic villages and 160 from the non-endemic comparison villages and found the prevalence of pterygium was higher in the endemic villages across all age groups in both sexes and increased with cumulative As exposure. After adjusting for age, sex, working under sunlight, and working in sandy environments, they found that cumulative As exposures of 0.1–15.0 and $\geq$15.1 mg L^{-1}-years were associated with increased risks of developing pterygium; the adjusted odds ratios were 2.04 (95% CI, 1.04–3.99) and 2.88 (95% CI, 1.42–5.83), respectively.

5.5   OTHER HEALTH OUTCOMES

In addition to diabetes mellitus, goiter is another endocrine disorder reported to be prevalent in the BFD area. Chang *et al.* (1991) examined all elementary school students in the Budai and Beimen townships and divided them into two groups—endemic and non-endemic for BFD. Of the 4567 participants, 120 (2.63%) were found to have goiter of grade I or above, and the prevalence of goiters in school children from the endemic area was higher than that from the non-endemic area (3.44 *vs.* 2.08%, $p < 0.01$). However, the authors pointed out that humic substances had also been reported as a possible source of environmental goitrogen.

In addition to urinary cancers, long-term As exposure has also been associated with mortality attributed to renal diseases. Chiu and Yang (2005) conducted an ecological analysis to evaluate the association between As exposure from drinking water and mortality due to renal diseases and calculated the SMRs for renal diseases for the BFD area from 1971 to 2000. They found that the mortality from renal disease declined gradually after improvement of the drinking-water supply system to eliminate As from artesian well water, thus, supporting a positive association between As exposure and mortality attributed to renal diseases.

Tsai *et al.* (1999) calculated SM ratios for all death causes in the BFD area from 1971 to 1994 and compared them to both local and national references. In comparison with the local reference, in addition to various cancers, residents of the BFD area had higher risks of dying from vascular disease, IHD, DM, and bronchitis.

Many carcinogens are also teratogens, and so it is reasonable to speculate that As exposure during pregnancy may lead to adverse outcomes. Yang *et al.* (2003) conducted a study in the Lanyang Basin in the northeastern region of Taiwan to evaluate associations between As exposure and preterm delivery and birthweight. After adjustment for potential confounders, As exposure from drinking well water was associated with a reduction in birth weight of 29.05 g (95% CI, 13.55–44.55 g). The odds ratio for preterm delivery was 1.10 (95% CI, 0.91–1.33), which was not statistically significant. These findings provide evidence for a potential role for As exposure in increasing the risk of adverse pregnancy outcomes.

Tseng (2003) examined 85 residents of the BFD area and 75 external normal controls without exposure and measured the current perception threshold (CPT) at the trigeminal, median, and superficial peroneal nerves. He found that 36 of the 85 (42.4%) BFD area residents had at least one abnormal measurement, and after adjusting for age, sex, body height and body weight, residency in the BFD area was significantly associated with higher CPT values. In addition, he found that the longer nerves (superficial peroneal and median nerves) were involved more commonly than the shorter (trigeminal) nerve, and the lower frequencies (5 and 250 Hz) were more commonly involved than the higher (2000 Hz) frequency.

5.6 SUMMARY AND CONCLUSIONS

Studies in Taiwan found associations between As in drinking water and cancers of the skin, urinary bladder, kidney, lung, and liver. Although similar associations were also found for other types of cancers, it should be noted that the studies in Taiwan are not totally independent because many studies included participants covered by other studies. Therefore, it is not surprising that the results show some consistency, and this should be taken into account in making conclusions. Studies conducted from outside of Taiwan should be reviewed to validate the findings in Taiwan. It should also be noted that drinking water is not the only route of exposure to As. A study of rice and yams obtained from the southwestern region of Taiwan showed that a substantial portion of the As in these food items were in inorganic forms, which contradicts the general belief that almost all the As in food is in organic forms (Schoof *et al.* 1998). Furthermore, studies in the northeastern region showed some conflicting results. Whereas the findings reported by Chiou *et al.* (2001) on urinary cancers were compatible with those observed in studies in other areas, Lin *et al.* (1986) reported very different findings except for those on liver cancer. Even though an increase in mortality was observed for all cancers combined in men, the dose-response relationship was not linear, and the finding was not observed in women. For rectal cancer, an increase in mortality was observed in men, but a decrease was observed in women. In addition, a decrease in mortality was observed for skin cancer in category II in men and lung cancer in women, which is not compatible with the current knowledge of the carcinogenicity of As. A possible explanation is "ecological fallacy" (Last 2001), because Lin *et al.* (1986) used ecological exposure data while Chiou *et al.* (2001) obtained water samples from individual households.

In summary, a significant positive association was consistently observed between the As level in drinking water and cancers of skin, lung, kidney, and urinary bladder by different research groups using different study approaches on different study populations. Associations of As ingestion with liver cancer were also observed quite consistently by the same research group in different studies with different study designs and study populations. However, except for the study in the Yi-Lan county by Lin *et al.* (1986), which was judged to be unreliable as described previously, no studies by other researchers in Taiwan are available to confirm the finding. Associations between As in drinking water and prostate cancer were observed in three studies with different designs and study populations by the same research team (Chen *et al.* 1988a, Wu *et al.* 1989, Chen and Wang 1990), but no studies by other researchers in Taiwan are available to confirm the finding. The association between As ingestion and cancer of the nasal cavity was observed in only one study (Chen and Wang 1990) and, therefore, it should be regarded as inconclusive. Conflicting results on the association between As exposure and rectal cancer were observed by a study in the Yilan county between the two genders (Lin *et al.* 1986), and no association was observed by any studies conducted by another research team (Chen and Wang 1990). These results should also be regarded as inconclusive. No statistically significant associations were observed for other cancers, including colon cancer, esophageal cancer, nasopharyngeal cancer, stomach cancer, uterine cervix cancer, cancer of small intestine, laryngeal cancer, cancers of bone and cartilage, breast cancer, ovarian cancer, brain cancer, and leukemia.

Many epidemiologic studies of the carcinogenic effects of As in drinking water conducted outside of Taiwan were inconclusive, mostly due to the small sizes of study populations (Bates *et al.* 1992, US EPA 1984). Nonetheless, positive associations were observed between As ingestion and skin cancer, liver angiosarcoma, and urinary cancers regardless of the sources of exposure (Bates *et al.* 1992, Guo *et al.* 1997, 2001). Positive associations between As exposure through inhalation, mostly occupational exposures, and lung cancer were also frequently observed (US EPA 1984). Some studies showed that ingested As was also related to the occurrence of lung cancer (Hopenhayn-Rich *et al.* 1998, Smith *et al.* 1998, Ferreccio *et al.* 2000).

There is increasing evidence suggesting the dose-response relation between the As level in drinking water and occurrence of cancer is not linear (Guo 1997, 1999, Guo and Valberg 1997, Guo *et al.* 1997, 1998, 2001, Guo and Tseng 2000). Although some other studies found increased risk associated with low-level exposures, the mean or median level of each unit population (a village or a township) was used as the exposure indicator, and therefore the results were not necessarily incompatible (Guo *et al.* 1998). In addition, evidence of the cell-type specificity of the carcinogenic effects of As is also increasing (Guo *et al.* 1997, 1998, Chiou *et al.* 2001). While studies in Taiwan provide the best available data on ingested As and cancer, all key studies, except for the study in the Lanyang Basin (Chiou *et al.* 2001), used ecological exposure data. Because of the limitations inherent in ecologic studies, further studies, such as case-control or cohort studies with exposure data at individual level, are necessary to confirm hypotheses generated from the previous studies in Taiwan. In particular, further studies on the dose-response relationship and cell-type specificity may help the understanding of the mechanism of the carcinogenic effects and the determination of regulatory standards.

CHAPTER 6

Arsenic sources, occurrences and mobility in surface water, groundwater and sediments

6.1 INTRODUCTION

Blackfoot disease (BFD) cases caused by drinking groundwater were first reported in 1954 in an endemic area of arsenic intoxication, along the southwestern coast of Taiwan (Kao and Kao, 1954, Kao et al. 1954, Tseng 1977), but were most prevalent between 1950~1956. After 1956, some of the residents started to utilize public surface water supplies instead of groundwater for drinking purposes. From 1970, most residents drank tapwater instead of groundwater, which resulted in a major reduction in the number of BFD cases. Since 1990, almost no new BFD cases were found. Whereas BFD is generally believed to be caused by As-containing groundwater (Jean 1999), the specific health effects depend on the presence of other substances and on the hydrogeological conditions (Jacobson 1998, Lu 1975, 1989, Chen et al. 1988, Chen and Wu 1962, Chen et al. 1962, Blackwell 1961, Yang and Tsai 1961). The details on health and toxicology can be referred to in chapters 3, 4, and 5.

   In the early 1950s, the Division of Health of the Taiwan Provincial Government set up the Taiwan Blackfoot Disease Prevention and Control Center (TBFDPCC) in the Beimen township of Tainan county to look after the BFD patients and allocated funds for projects to study the mechanism that caused BFD and the possible etiological agents of BFD from medical and environmental points of view. The first BFD report by the TBFDPCC was published in Chinese in 1974 and the last BFD report was published in 1994. Some of these reports were published in domestic journals (e.g., Journal of Formosan Medical Association) and foreign journals. Most of these reports dealt with medical aspects (e.g., pathology, toxicology, and epidemiology), with only one report by Wu (1978) detailing the environmental aspects of groundwater and geological characteristics in the endemic BFD areas. Since then, the BFD issue has drawn the interest of environmental scientists and hydrogeologists in Taiwan. This led to a great increase in the number of SCI (Science Citation Index) journal papers concerning the environmental aspects.

   Although no new BFD cases have been found since 1990, the etiological agents that caused BFD are still unclear despite many studies that have been reported. Arsenic-affected groundwater in Taiwan has still not been treated or removed from the water supply. The inhabitants in these arsenic-affected regions may be at risk from BFD and arsenicosis if arsenic-containing groundwater is utilized for drinking purposes. These regions are notably in the Chianan plain in SW Taiwan, Lanyang plain in NE Taiwan, Guandu plain in Taipei city, and the Chinkuashi gold mine area in NE Taiwan. It is important to detect the possible recurrence of BFD and investigate the geological/hydrochemical settings of groundwater arsenic in these regions, which will be elucidated in the following sections of this chapter.

6.2 HYDROGEOLOGY AND SEDIMENTOLOGY OF ARSENIC IN AQUIFERS

6.2.1 *Chianan plain*

The Chianan plain is located in the southwestern part of Taiwan and is surrounded by the Taiwan Strait to the west and the Central Mountain Range to the east. The Chianan plain is bounded by the Peikang river to the north and the Erjen river to the south, and covers an area of about 2400 km² (Nath *et al.* 2008) (Fig. 6.1). It extends 40 km from east to west and 60 km from north to south. Two major rivers, the Pachang river and the Tsengwen river (Figs. 6.1 and 6.2), flow from the NE to the west through the northern part and the southern part of the plain, respectively. The BFD areas are situated in the coastal alluvial Chianan plain (Fig. 6.1) with an altitude less than 50 m above sea level (a.s.l.) and are covered by alluvial deposits derived from the eastern foothill area through fluvial transportation. Fast

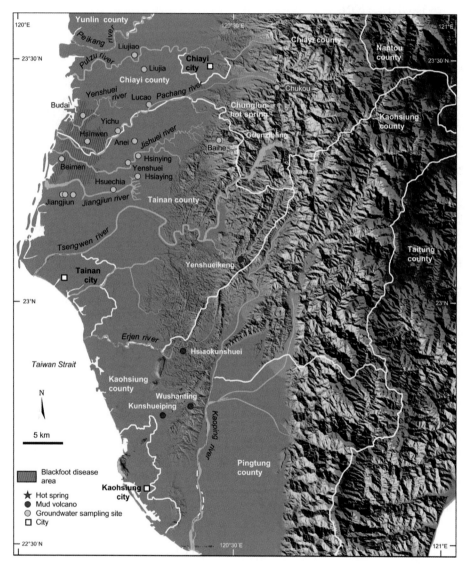

Figure 6.1.   The BFD areas in Chianan plain, SW Taiwan. Sampling sites are Beimen (comprises the well Beimen-2A), Jiangjiun (comprises the wells Jiangjiun-1B, Jiangjiun-1C, Jiangjiun-1D), Liujiao (comprises the wells Liujiao-1, Liujiao-2, Liujiao-2B), Lucao (comprises the wells Lucao-1A, Lucao-1B), Yenshuei (comprises the wells Yenshuei-1, Yenshuei-2, Yenshuei-3, Yenshuei-4, Yenshuei-5A), Yichu (comprises the well Yichu-1A), and Hsuechia (comprises the wells Hsuechia-1, Hsuechia-2).

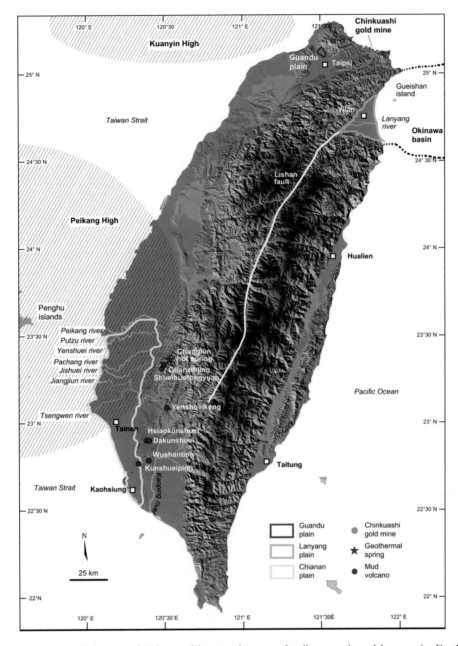

Figure 6.2.   Principal areas of Taiwan with groundwater and soil contaminated by arsenic. Further
shown are the locations of the Chinkuashi gold mine and the geothermal springs and mud
volcanoes as arsenic sources.

erosion of surface water upstream and a sharp change of river gradient have caused thick,
clay-rich muddy sediments to be deposited in the upper plain and have led to poor hydraulic
properties in the aquifers. Regional geologic studies revealed that the sedimentation rate since
the Pliocene has been very high. Alluvial deposits of clay, silt and fine-grained sand have
extensively covered the coastal plain. The deep Chianan plain groundwater is characterized

by high salinity due to NaCl, highly reducing conditions, and arsenic concentrations in the range of 4.9 to 704 µg L^{-1} (median: 402 µg L^{-1}, n = 5) (see section 6.4). Wells in these areas penetrate to a depth of 100–280 m through a multiple aquiclude-aquifer system (maximum thickness; 280 m) belonging to the "Liushuang" and "Erzhong river" formation of Pliocene age, which is composed of shallow to deep sea sediments (gray color mudstone and siltstone intercalated with three fine to thick sandstone layers of 4–15 m, 54–79 m, and 130–143 m) (Stach 1957, Ho 1975) with abundant organic matter and a very high humic acid concentration (Wu 1978). Moreover, the hydrogeological profiles (Fig. 6.3) of the Chianan plain exhibit no obvious layer structure. The regional groundwater flows, in general, from the mountain area in the east towards the Chianan plain in the west (Fig. 6.4). The groundwater recharge in the Chianan plain is from the mountainous coarser-grained materials of Chiayi Hill (Fig. 6.5).

6.2.2   *Lanyang plain (Yilan plain)*

The Lanyang plain is located in Yilan county in northeastern Taiwan (Fig. 6.2), and represents the alluvial fan of the Lanyang river (Fig. 6.6). The area has a triangular shape, with the Pacific Ocean to the east, the Snow Mountain Range located to the northwest, and the Central Mountain Range located to the southwest. The main river, the Lanyang river, flows through the middle of the area from west to east (Fig. 6.6). Lanyang plain covers approximately 400 km², with each side being about 30 km. The groundwater flows from west to east in the north of the plain, and flows to the northeast in the south. The northwestern and southwestern parts of the plain near the mountains form the recharging area of the groundwater (Fig. 6.7). The thickness of the alluvial cover ranges from ~100–400 m. Lanyang plain consists of Holocene alluvial deposits which include sands, gravels, clays, detrital slates, quartz sandstone, and crystallized gneiss (Lin and Kao 1997, Water Resources Bureau 2003). The alluvial deposits are coarse-grained in the upper stream areas and decrease in grain size to fine sand and silt in the lower stream areas. The surface layer is covered by sediments from the Quaternary Period, including silty sand, silty clay, slate, metamorphic sandstone, schist, and shale, and is partitioned into upstream, midstream, and downstream areas. The alluvial deposits consist of a multiple aquifer system with the unconfined aquifer on the

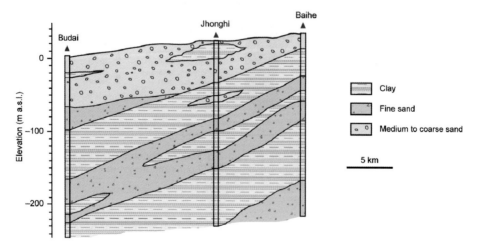

Figure 6.3.   Hydrogeological cross-section in Chianan plain (the distance between Budai and Baihe is about 28 km).

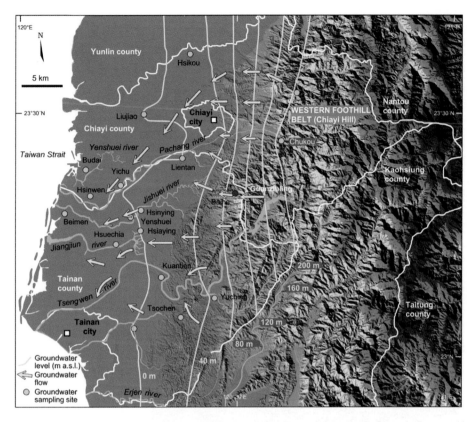

Figure 6.4.   Regional groundwater flow in Chianan plain. Groundwater levels from shallow aquifers (referred to Fig. 6.3) were measured in 2008.

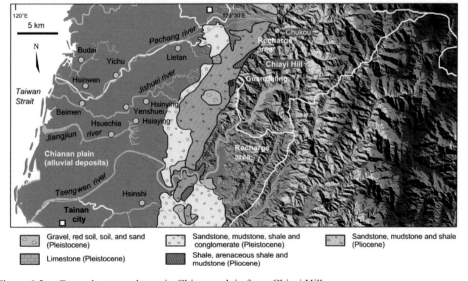

Figure 6.5.   Groundwater recharge in Chianan plain from Chiayi Hill.

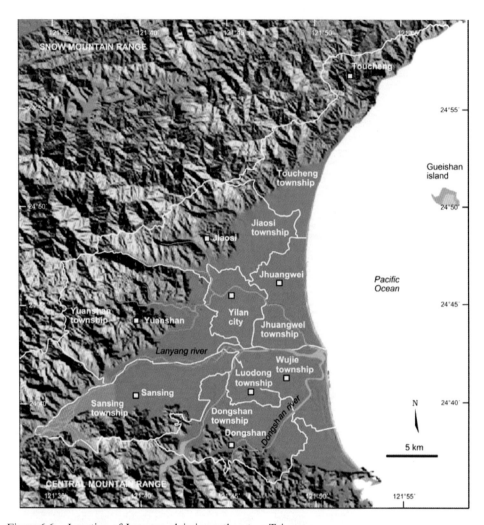

Figure 6.6.   Location of Lanyang plain in northeastern Taiwan.

top and several deeper confined aquifers. Arsenic and humic substances in these aquifers may originate from the detrital slate and argillite in the northern part of the Central Mountain Range. The Lanyang plain groundwater is characterized by low salinity, a calcium to sodium bicarbonate water type, moderately reducing conditions and an arsenic concentration in the range of 2.5 to 543 µg L^{-1} (median: 20.6 µg L^{-1}, n = 6) (see section 6.4).

The bedrock of the north bank of the Lanyang river is characterized by a sedimentary terrain of Eocene to Oligocene age, while the bedrock of the south bank is characterized by a sedimentary terrain of Miocene age. The bedrock, overlain by the alluvial deposits, is the Suao slate and argillite of Miocene age, occasionally with a thin layer of metamorphosed sandstone (Lin and Kao 1997). These two terrains may be separated by the Okinawa basin (Fig. 6.2) as tectonic units. The Lisan fault (Fig. 6.2) in the south of the Lanyang plain connects to the Lanyang river. Arsenic can be carried into the Lanyang river and Lanyang plain through the Lisan fault.

The groundwater recharge in the Lanyang plain is from the coarser-grained materials in the west of plain (Fig. 6.7) (Lee *et al.* 2008c). The rainfall is recharged in the mountain region

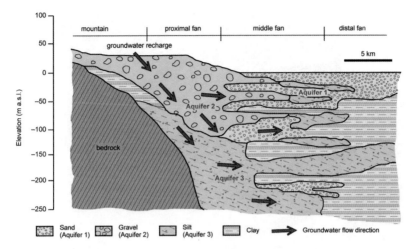

Figure 6.7.   Groundwater recharge from the mountain area in the Lanyang plain.

and then spreads eastwards into Aquifer 3, Aquifer 2, and Aquifer 1 in the proximal-fan, mid-fan, and finer-grained materials of the distal-fan. Regional groundwater flow is from west to east (Lee *et al.* 2008c) (Fig. 6.7). The As-affected regions in the Lanyang plain are near the coastal plain in the east. The median concentrations of As in these wells are <200 µg L^{-1}, whereas the maximum concentrations were 351 µg L^{-1} at Yilan city and Jiaosi township (Fig. 6.8) (Lin *et al.* 1986).

6.2.3   *Guandu plain*

The sources (Fig. 6.9) of the contamination found in the agricultural land in the Guandu plain were the Huanggang creek and Keelung river, which were contaminated mainly with As and other trace elements that were transported from the Geothermal Spring Valley (N25°8'26.11", E121°30'13") (Fig. 6.10) at Beitou district of Taipei city. There are two types of hot spring water (green sulfur spring water and white sulfur spring water) in the Geothermal Spring Valley. The water of the green sulfur spring of the Geothermal Spring Valley is a sulfate-chloride type with a pH of 1.6 and a temperature of 80–100°C which is discharged at a rate of 4000 m^3 day^{-1}, whereas the water from the white sulfur spring is a sulfate-rich water with a pH of 3.64 and a temperature of 50–60°C which is discharged at a rate of 2000 m^3 day^{-1} (Taipei Water Department 2009). The water quality analyses for the green sulfur spring reveals high concentrations of trace elements such as As (max: 4210 µg L^{-1}, mean: 1920 ± 1170 µg L^{-1} (n = 7), min: 1070 µg L^{-1}), cadmium (max: 84 µg L^{-1}, mean: 72 ± 10 µg L^{-1} (n = 7), min: 54 µg L^{-1}), chromium (max: 23 µg L^{-1}, mean: 21 ± 2 µg L^{-1} (n = 7), min: 16 µg L^{-1}), and lead (max: 3280 µg L^{-}, mean: 1680 ± 724 µg L^{-1} (n = 7), min: 1110 µg L^{-1}).

Many As-containing sulfide minerals are produced from both geothermal and hydrothermal activity (Lewis *et al.* 2007) by the discharge from the spring. Furthermore, the agricultural land soil of Guandu plain has been contaminated with arsenic-containing fluids and sediments of the precipitates from the Geothermal Spring Valley with As concentration between 25 and 272 mg kg^{-1} (mean: 135 mg kg^{-1}) in contrast to an As concentration of 6.54 mg kg^{-1} in the agricultural land soil of the other areas in Taiwan (Chang *et al.* 1999) and 10 mg kg^{-1} in the world (Das *et al.* 2002). It should be noticed that the As concentration in the groundwater of Guandu plain ranged from 26–31 µg L^{-1} (Chang *et al.* 2007), which is higher than the WHO standard (10 µg L^{-1}).

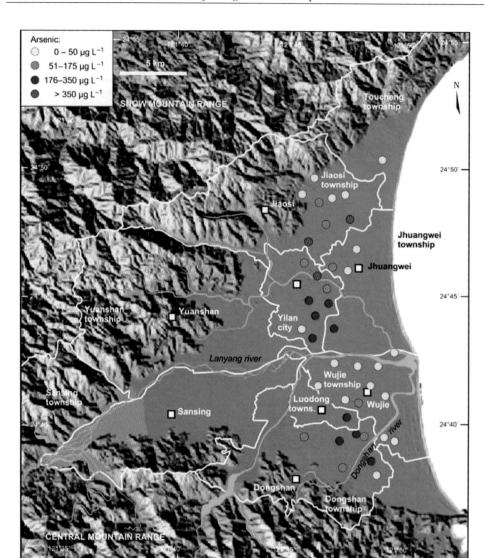

Figure 6.8.   Arsenic concentration in the groundwater of Lanyang plain (modified from Lin *et al.* 1986).

6.3   POTENTIAL ARSENIC SOURCES

6.3.1   *Geogenic sources*

There are many possible geogenic As sources and several specific mobility controls, which may occur either under oxidizing conditions (e.g., sulfide oxidation, desorption from metal oxides and oxyhydroxides at high pH) or reducing conditions (e.g., reductive dissolution of metal oxides and hydroxides and release of coprecipitated and adsorbed As), which can explain the occurrence of As in groundwater as discussed in detail in chapter 2. However, in many of the As-contaminated sites all around the world, it is difficult or impossible to identify the principal source and the mobility controls and often several sources and mobility controls may coexist (see chapter 2). The same problem also occurs in the Chianan plain

Figure 6.9. The Geothermal Spring Valley contaminates the soils and groundwater of Guandu plain with arsenic.

Figure 6.10. The Geothermal Spring Valley in Beitou district, Taipei city.

(BFD area) in SW Taiwan and Lanyang plain in NE Taiwan, where the primary As source is still unknown.

6.3.1.1 *Chianan plain*
In the Chianan plain (BFD area), black shale in the foothills of western Taiwan (called Chiayi Hill, Fig. 6.11) may be one of the As sources (Hsu *et al.* 1997, Nordstrom 2002).

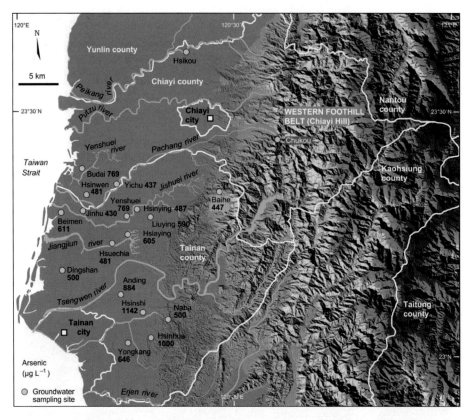

Figure 6.11.   The foothills of western Taiwan (Chiayi Hill, also called the West Mountain Belt) in the east and the arsenic concentrations ($\mu g\ L^{-1}$) in the well water of Chianan plain in the west (modified from Wu 1989).

The As concentration in black shale at Chukou (location shown in Fig. 6.4) located at the upstream part of the Pachang river (Fig. 6.12) can be up to 20 mg kg^{-1} (Yang *et al.* unpublished). *In-situ* formation of arsenopyrite in the Chianan plain, as speculated by Lewis *et al.* (2007), is possible because the As content in marine shales in the world can be up to 490 mg kg^{-1} and that of phyllite and slate from 0.5–143 mg kg^{-1} (Selinus *et al.* 2005). Aquifers with carbonaceous silts/clays found in the lower stream of the Pachang river in SW Taiwan can lead to high dissolved As concentrations in groundwater (10–1820 $\mu g\ L^{-1}$) (Nordstrom 2002). Such carbonaceous silts/clays in the Chianan plain aquifer are proposed to be a result of the eroded black shale host rock which is distributed in the upper stream of the Pachang river (Fig. 6.12). The rock debris weathered from the As-rich black shale was transported westwards along the Pachang and the Jishuei rivers (Fig. 6.11). These black shale weathering products contribute to the organic mud and clay sediments of the Chianan plain aquifer. However, high As concentrations were also found in brown sands containing iron sulfides and oxides, chlorite and illite as well as organic-rich silts, about 190 m below the ground surface in the Hsinwen drilled hole (Figs. 6.1 and 6.11) in Budai township (downstream of the Pachang river). The As concentration in core sediments was measured by ICP-MS and found to be as high as 1677 mg kg^{-1} (Huang 2009). The mean As concentration in sediments of this 250 m deep core was 44 mg kg^{-1} (max: 1677 mg kg^{-1}, min: 0.4 mg kg^{-1}, n = 57). The high As concentration of 1677 mg kg^{-1} only occurred locally in brown sand at a depth of 190 m (1 out of 57 samples). The other 56 samples at different depths in this core were mostly collected

Figure 6.12.   Chukou black shale at the upstream Pachang river.

from organic mud and clay and had As concentrations in the range of 0.4–215 mg kg^{-1}. The brown sands may be contained in the Chukou black shale, which was considered by Nordstrom (2002) to be the source of As in the Chianan plain aquifer BFD area at Budai, Yichu, Beimen, and Hsuechia (Fig. 6.1) along the downstream area of the Pachang river. However, whether or not such high As concentrations in the alluvial sediments originated from the black shale must be investigated further in the future.

There is an international debate regarding the source of As in the Chianan plain and the Lanyang plain of Taiwan as well as the Ganges (Bengal) delta plain and other areas around the world. In addition to the black shale proposed by Hsu *et al.* (1997) and Nordstrom (2002), five other potential sources of the As in the groundwater of the Chianan plain have been proposed by Lewis *et al.* (2007):

- Onshore and submarine mud volcanoes;
- Coal/peat beds;
- *In-situ* formation of pyrite and arsenopyrite;
- Hydrothermal veins associated with onshore igneous intrusions/flows;
- Offshore Miocene basalts.

Of the above five possible As sources, Lewis *et al.* (2007) considered that the onshore mud volcanoes may contribute a portion of the total As. The As contents of the Wushangting (WST) mud volcano fluids are in the range of 12.0–27.7 mg kg^{-1} (Liu, C.-C. *et al.* 2009). Although the As concentrations are high in the mud volcano fluids and muds, only a few mud volcanoes are dispersed in the east of Chianan plain (Figs. 6.1 and 6.2). These mud volcanoes are not close to rivers and thus unlikely to be able to carry and transport significant quantities of As to Chianan plain.

Coal/peat beds may also be a source of As, which are yet undiscovered beneath the BFD area (Lewis *et al.* 2007) but can be found in the intercalation of sandstone and shale of the West Mountain Belt (the foothill of western Taiwan) at Chukou, east of the BFD area (Figs. 6.1 and 6.4). However, no extensive coal/peat beds are distributed in surface and subsurface environments. Another possible As source may be the *in-situ* formation of pyrite and arsenopyrite. The As-containing pyrite/arsenopyrite may have been authigenically precipitated in sedimentary rocks and later mobilized and transported to the Chianan plain aquifer (Lewis *et al.* 2007). Arsenic in the Chianan plain might be derived from the black shale (Nordstrom 2002, Hsu *et al.* 1997) that may contain As-bearing pyrite. To date, no literature has reported the mineraological analysis of the Chukou black shale, which is worthy of detailed research. Another possible As source might be the hydrothermal veins associated with onshore igneous intrusions/flows (Lewis *et al.* 2007). Most As-containing sulfide minerals are found in the veins along faults and/or related fractures in permeable rocks. There are basalt flows in the Miocene strata present in the Chiayi and Hsinying areas (Fig. 6.1) north of the BFD area according to the seismic reflection data and cores from exploratory gas drilling by the Chinese Petroleum Corporation in Taiwan (Lewis 1985). Basalt flows and dioritic intrusions with jarosite, pyrite, iron oxides, such as hematite and limonite, and cinnabar are common in a geothermal deposit of quartz metasandstone, about 20 km east of the BFD area in the western foothills tectonic province of Taiwan (Lewis 1985). However, basalt flows are not extensively distributed and are unlikely to be widespread in the As-affected regions in Chianan plain. Another potential source of As in the BFD area is offshore Miocene basalts. Lewis *et al.* (2007) indicated that possible basaltic dikes/sills or apophyses associated with the uplift of the Peikang High (Fig. 6.2) or the Penghu islands horst could contain As that is released into the seawater in the Taiwan Strait. Eastward transport by the tidal action of seawater is possible (Lewis *et al.* 2007). However, the eastward transport of As into the aquifer system of the BFD area where As is widely distributed in the entire Chianan plain seems to be impossible due to the reverse hydraulic gradient from seawater.

In addition to the above five potential sources, arsenic in seawater can also be a possible source of As in the coastal areas of Chianan plain. Arsenic in seawater can result from natural and anthropogenic sources where it has been discharged into rivers in the Chianan plain (i.e., Peikang river, Putzu river, Pachang river, Jishuei river, Jiangjiun river, Tsengwen river, etc.). This As has been carried from upstream to downstream areas of the rivers and has been discharged into the Taiwan Strait and then migrated into the Chianan plain aquifers due to seawater intrusion. The evidence of seawater intrusion revealed that the zero meter contours of groundwater levels were shifted from the coastal area of Beimen in 1957 inwards to the east in 1974 (Fig. 6.13) due to an overdraft of groundwater. The seawater quality offshore from the Chianan plain had been polluted naturally and anthropogenically with trace elements, resulting in high As concentrations in the range of ~200 to ~1000 μg L^{-1} (highest ones at the estuaries of Peikang and Putzu rivers with 2500 μg L^{-1}) compared to 2.3 μg L^{-1} and 1.5 μg L^{-1}, respectively in the seawater offshore from other areas of Taiwan (Huang *et al.* 1975). Such high As concentrations may be due to the sediment load of the Peikang river and Putzu river (Figs. 6.1 and 6.2) as well as groundwater and wastewater discharging into the Taiwan Strait. Some of the As presence could be particulate As (i.e., As sorbed to colloids). High As concentrations at the estuaries of Peikang and Putzu rivers (Figs. 6.1 and 6.2) may be mobilized from the sediment due to the presence of high amounts of organic substances creating reducing conditions.

Arsenic and sulfate in seawater might have encroached into the Chianan plain aquifers of the BFD endemic area (Fig. 6.1) during 1957–1974. Seawater intrusion could provide the required electron acceptors of SO$_4$ for bacterial sulfate reduction and promote reducing conditions that are favorable for As mobilization from oxy(hydr)oxides (Nath *et al.* 2008), thereby reducing As(V) to As(III) in the groundwater. However, direct microbial reduction of sorbed As(V) to As(III) using arsenate as the electron acceptor may also mobilize As without necessarily reducing Fe or Mn hydroxides. Adsorbed-phase arsenate could be reduced to dissolved-phase arsenite and thus released into groundwater under reducing conditions through bacterial reduction by As(V)-reducing bacteria.

6.3.1.2   *Lanyang (or Yilan) plain*

Lewis *et al.* (2007) suggested that there are seven possible sources for the As concentration in the groundwater of the Lanyang (or Yilan) plain:

- Arsenic-containing pyrite;
- Subsurface peat/coal beds;
- Shale/slate chimneys with cinnabar, pyrite, and jarosite;
- Weathering of Chinkuashi gold-copper ores;
- Ocean floor gas vents;
- Basalt/andesitic rocks with hydrothermal veins;
- Hot springs and geothermal waters.

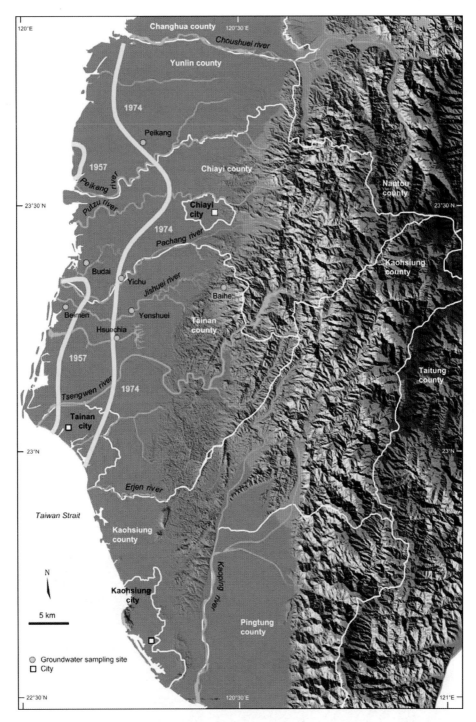

Figure 6.13. The changes in zero meter groundwater level lines in the deep arsenic-contaminated confined aquifer of Chianan plain in 1957 and 1974 (modified from Wu 1978).

Of the above seven potential As sources, Lewis *et al.* (2007) considered that the most probable ones in the Lanyang plain are As-containing pyrite. Pyrite that typically contains As may occur in the host rocks of the mountainous areas to the west of the Lanyang plain (Fig. 6.6). As the pyrite weathers, As can desorb from the pyrite detritals and be transported to Lanyang plain. Some dispersed coal deposits are found in the shales and sandstones about 30 km to the north of the Chinkuashi gold mine (Fig. 6.14) (Shen and Wang 2004) and significant quantities of coal have been extracted from surface and subsurface mines in northern Taiwan (Lewis *et al.* 2007). Arsenic concentrations in coals can be up to 35,000 mg kg^{-1} (Belkin *et al.* 2000; see also chapter 2), though arsenic concentrations in coals in Taiwan were reported to be 140 mg kg^{-1} (Liu 1995). It is likely that the shale/slate chimneys with cinnabar, pyrite, and jarosite may be the As source in the Lanyang plain. Many As minerals associated with sulfide deposits and cinnabar can occur in slate chimneys from this same area (Lewis *et al.* 2007). Several As-containing sulfide minerals that imply hydrothermal activity have been discovered in the upthrust Slate Belt of south-central Taiwan (Lewis *et al.* 2007).

Furthermore, another possible As source could result from the weathering of Chinkuashi As-containing gold-copper ores (e.g., enargite, luzonite). The Chinkuashi gold-copper mine of northern Taiwan (Fig. 6.14) is located about 30 km northwest of the Langyang plain. However, Lewis *et al.* (2007) indicated that acid mine drainage from the Chinkuashi mining district would not flow southeastwards reaching the Lanyang basin (Fig. 6.14). It is quite probable that As is released from the gold-copper ore bodies and migrates along the Lishan fault (Fig. 6.2), from where it is carried by the Lanyang river into the Lanyang plain (Lewis *et al.* 2007). Another source of As could be from the ocean floor gas vents near Gueishan island in the Okinawa basin off Taiwan's northeastern coast, about 20 km east of the Lanyang plain (Fig. 6.2). However, it is not likely that As could be so widespread throughout the whole Lanyang basin as a result of the permeation of As. Additionally, As may originate from basaltic/andesitic rocks with hydrothermal veins which are most likely tectonically linked to the rifted Okinawa back-arc basin that extends from northeastern Taiwan to southern Japan (Lewis *et al.* 2007). Outcrops of pillow basalt, chert, metasandstone, breccia and

Figure 6.14.   Location of Chinkuashi gold mine and the "Yinyang Sea" which is polluted by acid mine drainage from Chinkuashi gold mine in northern Taiwan.

slate occur in the upthrust Slate Belt of south-central Taiwan with sulfide mineralization that contains As (Smith and Lewis 2007).

Furthermore, another possible As source could be the hot springs and/or geothermal waters which commonly occur in the riverbeds of the upstream area of the Lanyang river. Arsenic can be transported to the downstream of the Lanyang river, resulting in the As contamination of the Lanyang plain aquifers.

6.3.2   *Anthropogenic sources*

Anthropogenic sourcs of arsenic in Taiwan can result mainly from mining, industrial and agricultural activities as detailed below.

6.3.2.1   *Mining activity*

During and after the extraction of mine materials, As is released and transported to As-affected regions. These mine materials include gold and coal as well as pyrite and Fe/Cu sulfides. Pyrite and other sulfides in water can be oxidized to produce an acidic solution rich in Fe and Cu. Fe(III) can be reduced to Fe(II) in acidic reducing environments and be dissolved in groundwater. However, Fe(III) may not always be produced by the oxidation, and the Fe may remain dissolved as Fe(II). Gold has been extracted commercially from the Chinkuashi gold mine (Fig. 6.14) in northeastern Taiwan. However, As and other trace elements have also been exposed, resulting in acid mine drainage (AMD) (Fig. 6.15a). When reduced groundwater seeps out to the surface and is mixed with river water, Fe(II) is oxidized to Fe(III) (Fig. 6.15b) and is transported along a creek and finally to the sea. The sea has long been severely polluted with AMD and thus is called the "Yinyang" Sea (meaning "hell and heaven" in Chinese), a reference to the resulting mix of yellow and blue colors (Fig. 6.15c). Fe(III) is insoluble in water, yielding Fe hydroxides that are suspended on the surface of the seawater and turn it yellow. To date, very few references about the contamination of aquifers and soils by this AMD have been reported. In addition, coal mines have also been exploited in northeastern Taiwan. Arsenic contamination of aquifers like Lanyang (or Yilan) plain might result from the extraction of coal and/or gold (Lewis *et al.* 2007). Since the Chinkuashi gold mine is located in a mountainous area, AMD is discharged into a creek and finally to the "Yinyang Sea" (Fig. 6.15c), but not to Lanyang plain.

6.3.2.2   *Industrial activity*

Plating and integrated circuit manufacturers in the Central Taiwan Science Park, Taichung, Central Taiwan produced wastewater with numerous contaminants, including As. However, Lin (2007) revealed that As concentrations in soils and discharged wastewater in this Science Park were respectively 11.9 ng kg^{-1} and 4.3 to ~35 μg L^{-1}. These values are much lower than the tolerance levels for soil and wastewater stipulated by the Taiwan Environmental Protection Agency (60 mg kg^{-1} and 500 μg L^{-1}, respectively; Taiwan EPA 2006). To date, the soil and groundwater of all science parks in Taiwan are not contaminated with As.

6.3.2.3   *Agricultural activity*

Pesticides and herbicides used for agricultural purposes frequently contain As and thus can be a source of groundwater contamination. Soils that are contaminated with As-contained pesticides are found at two sites; 0.99 hectares of contaminated soils in Changhua county and 0.29 hectares in Taichung county, Central Taiwan. The most serious case was at the Yuh-Tai Chemical Engineering Company in Changhua county where about 10,000 m^3 (0.99 hectares in area) of soil was contaminated according to the soil survey between 21 July 2005 and 20 July 2006 (Taiwan EPA 2006).

Figure 6.15.   (a) Acid mine drainge from Chinkuashi gold mine; (b) The oxidized iron rocks in the
Chinkuashi river; (c) The AMD-polluted "Yinyang Sea".

Although groundwater containing As in the BFD endemic area has been no longer used for drinking purpose, it is still used in fish ponds for aquaculture, whereas surface water is now used for irrigating crops. Arsenic concentrations in groundwater used in fish ponds in the BFD area are $87.8 \pm 52.5 \, \mu g \, L^{-1}$ in Beimen, $58.1 \pm 44.1 \, \mu g \, L^{-1}$ in Yichu, $53.8 \pm 21.4 \, \mu g \, L^{-1}$ in Hsuechia, and $27.5 \pm 7.90 \, \mu g \, L^{-1}$ in Budai (locations are shown in Fig. 6.1). The mean As concentration over these four areas is $54.1 \pm 36.0 \, \mu g \, L^{-1}$ (Lin *et al.* 2001). Most of these As concentrations exceeded the WHO standard ($10 \, \mu g \, L^{-1}$). The As concentration in milk fish was measured to be $1.02 \pm 1.10 \, mg \, kg^{-1}$ in Budai, the bioaccumulation factor can be up to 37, revealing that milkfish can strongly adsorb As from water and accumulate it in their body up to 37 times greater than the concentration in the immediate environment (Lin *et al.* 2001). Despite the bioaccumulation of inorganic As in milkfish, most inorganic As can be transformed into organic As which is readily taken up in the human body through the food chain. However, organic As in the human body becomes less toxic and more easily excreted. These processes are described in detail in chapter 8.

6.4   ARSENIC DISTRIBUTIONS AND MOBILITY CONTROLS

6.4.1   *Water chemistry in the Chianan and Lanyang plains*

Surface water, including river water and pond water, has been used for irrigating crops and for fish farming. If the surface water contains As, the soils and groundwater may also be contaminated. The rivers in Chianan plain such as the Pachang river and the Jishuei river (Figs. 6.1 and 6.2) increased in As concentration from the upper stream to the lower stream, from $0.83 \pm 0.54 \, \mu g \, L^{-1}$ (n = 4) to $5.5 \pm 0.54 \, \mu g \, L^{-1}$ (n = 4) in the Pachang river and from $0.92 \pm 0.33 \, \mu g \, L^{-1}$ (n = 4) to $9.4 \pm 4.1 \, \mu g \, L^{-1}$ (n = 4) in the Jishuei river, respectively (Table 6.1). The increases in As concentration in the lower streams of the Pachang river and Jishuei river are probably due to more evaporation and less rainwater in the lower streams. On the other hand, the rivers in Lanyang plain, such as the Dongshan river and the Lanyang river (Figs. 6.6 and 6.8) decreased in As concentration from the upper stream to the lower stream, from $5.4 \pm 1.9 \, \mu g \, L^{-1}$ (n = 4) to $1.3 \pm 0.69 \, \mu g \, L^{-1}$ (n = 4) in the Dongshan river and from $3.1 \pm 2.3 \, \mu g \, L^{-1}$ (n = 3) to $0.53 \pm 0.12 \, \mu g \, L^{-1}$ (n = 3) in the Lanyang river (Table 6.1). The higher As concentrations in the upper stream of the Lanyang plain and Guandu plain are due to As-rich spring waters. The As-containing surface waters are then diluted by mixing

Table 6.1.   Comparison of arsenic concentrations in the surface water of Chianan plain, Lanyang plain, and Guandu plain.

Plain	Range ($\mu g \, L^{-1}$)	Mean $\pm$ SD ($\mu g \, L^{-1}$)
Chianan plain		
Pachang river (upper stream)	0.40 – 1.60	0.83 ± 0.54 (n = 4)
Pachang river (lower stream)	3.90 – 7.60	5.50 ± 1.50 (n = 4)
Jishuei river (upper stream)	0.70 – 1.40	0.92 ± 0.33 (n = 4)
Jishuei river (lower stream)	5.80 – 15.2	9.40 ± 4.10 (n = 4)
Lanyang plain		
Dongshan river (upper stream)	3.40 – 8.00	5.40 ± 1.90 (n = 4)
Dongshan river (lower stream)	0.70 – 2.30	1.30 ± 0.69 (n = 4)
Lanyang river (upper stream)	1.20 – 5.70	3.10 ± 2.30 (n = 3)
Lanyang river (lower stream)	0.40 – 0.60	0.53 ± 0.12 (n = 3)
Guandu river		
Huang river (upper stream)	4.70 – 32.7	13.2 ± 13.3 (n = 4)
Huang river (lower stream)	0.50 – 2.90	1.30 ± 1.10 (n = 4)

with water from other rivers, especially in the wet season, and transported to the lower parts of streams.

The BFD cases were found exclusively in SW Taiwan before 1990, especially in the coastal areas of northern Chianan plain (e.g., Budai, Yichu, Beimen, Yenshuei, Hsuechia, Jiangjiun townships, Fig. 6.16) where the groundwater contains high concentrations of As (4.9–704 µg L⁻¹, median: 402 µg L⁻¹, n = 5, Reza *et al.* 2009), high concentrations of humic substances (34.8–468 µg L⁻¹, median: 93 µg L⁻¹, n = 5, Reza *et al.* 2009) and high relative fluorescence intensity (24.1–322.4 µg L⁻¹, median: 69 µg L⁻¹, n = 5, Reza *et al.* 2009) (Table 6.2). Here, 42 rural villages of six townships (Budai, Yichu, Yenshuei, Hsiaying, Hsuechia, Beimen) in the coastal region of SW Taiwan that had high median As levels in their well water were studied (Wu *et al.* 1989) (Fig. 6.16). In this area, 14 villages (33.3%) were dependent upon confined aquifer wells and 28 villages (66.7%) had alternative water source (surface or shallow aquifer). The median As concentrations in the groundwater of the six townships (Fig. 6.16) were divided into three zones: (1) ≥ 600 µg L⁻¹ (in south Budai, Beimen, etc.); (2) 300–590 µg L⁻¹ (in north Budai, north Hsuechia, etc.); (3) < 300 µg L⁻¹ (in south Hsuechia, Yichu, Yenshuei, Hsiaying, etc.) (Wu *et al.* 1989). The BFD cases were more frequent in the coastal areas, such as south Budai in Chiayi county and Beimen in Tainan county, which are at the northern and southern margins of the Pachang river (Fig. 6.17) (Ch'i and Blackwell 1968).

Interestingly, the groundwaters in Lanyang plain in NE Taiwan contained high concentrations of As with maximum values of 543 µg L⁻¹ (Table 6.2), but much weaker relative fluorescence intensities of humic substances compared to the BFD area in southwestern Taiwan (Jean *et al.* 2007, Reza *et al.* 2009). The experimental results in Reza *et al.* (2009) demonstrated that the median concentrations of As (402 µg L⁻¹, n = 5) and fluorescent humic

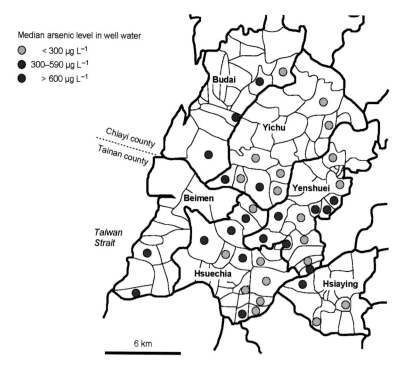

Figure 6.16.   Geographic distribution of 42 studied villages in six townships in the blackfoot disease endemic area in SW Taiwan, showing the median arsenic concentration in well water determined over the period 1964–1966.

Table 6.2.   Comparison of arsenic concentration, humic substances concentration and relative fluorescence intensity and other chemical components in deeper groundwater (>50 m) and their health effects among Chianan plain (SW Taiwan) and Lanyang plain (NE Taiwan) (modified from Reza *et al.* 2009).

Chemical composition	Chianan plain, SW Taiwan	Lanyang plain, NE Taiwan
As (μg L^{-1})	range: 4.9–704 median: 402 (n = 5)	range: 2.5–543 median: 20.6 (n = 6)
Concentration of humic substances in QSU (10 μg L^{-1})	range: 34.8–468.0 median: 93 (n = 5)	range: 16.2–211.5 median: 60.5 (n = 6)
Relative fluorescence intensity of humic substances	range: 24.1–322.4 median: 69 (n = 5)	range: 11.6–46.0, median: 41.5 (n = 6)
pH	range: 7.01–8.09 median: 7.75 (n = 5)	range: 7.27–8.58 median: 7.92 (n = 6)
EC (μS cm^{-1})	range: 1011–80300 median: 1583 (n = 5)	range: 3.23–4860 median: 166.9 (n = 6)
Na$^+$ (mg L^{-1})	range: 87–15551, median: 182 (n = 5)	range: 26.2–441.0 median: 142.1 (n = 6)
NH$_4^+$ (mg L^{-1})	range: 0–4.8 median: 0 (n = 5)	ND
K$^+$ (mg L^{-1})	range: 15–437 median: 22 (n = 5)	range: 1.01–39.7 median: 8.68 (n = 6)
Mg^{2+} (mg L^{-1})	range: 7–169 median: 23 (n = 5)	range: 3.74–17.6 median: 13.1 (n = 6)
Ca^{2+} (mg L^{-1})	range: 29–293 median: 41 (n = 5)	range: 9.39–87.8 median: 42.2 (n = 6)
Cl$^-$ (mg L^{-1})	range: 14–36540 median: 197 (n = 5)	range: 2.7–47.6 median: 3.57 (n = 6)
NO$_2^-$ (mg L^{-1})	range: 0–166 median: 0 (n = 5)	ND
SO$_4^{2-}$ (mg L^{-1})	range: 0–3375 median: 15 (n = 5)	range: 0–3.03 median: 0.37 (n = 6)
Alkalinity (mg L^{-1})	range: 163–175, median: 166 (n = 5)	range: 88–198 median: 155 (n = 6)
Fe (μg L^{-1})	range: 26–630 median: 217.1 (n = 5)	range: 15.6–235 median: 67.2 (n = 6)
Mn (μg L^{-1})	range: 9.1–345 median: 20 (n = 5)	range: 0.23–130 median: 54.3 (n = 6)
Symptoms	BFD; hyperkeratosis of hands; skin cancers on hands, arm and chest; lung cancer, liver cancer and bladder cancer	Non-BFD; skin cancer,lung cancer, bladder cancer; cardiobrain vascular diseases

ND: not detected, EC: electrical conductivity, QSU: quinine sulfate units.

substances (93 QSU, n = 5) as well as median relative intensity (69, n = 5) in the Chianan plain groundwater are higher than those (20.6 μg L^{-1}, 60.5 QSU, and 41.5, respectively, n = 6) in the Lanyang plain groundwater. The combination of high As and high concentrations of humic substances in the Chianan plain groundwater may be the cause of the blackfoot disease, which occurred only there.

The health effects in the Chianan plain are related to the consumption of water from the deep confined aquifers with reducing conditions (ORP: min: –200 mV, mean: –156.41 ± 39.17 mV

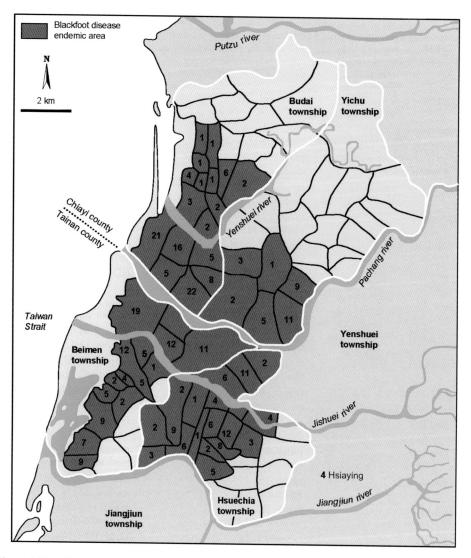

Figure 6.17.   The distribution of blackfoot disease in Chianan plain. The numerals represent the number of observed BFD cases.

(n = 12), max: –75 mV) (Table 6.3), whereas the consumption of water from the unconfined (or phreatic) aquifer wells, with much lower As concentrations did not result in As-related health effects. The water from the confined aquifer contains high concentrations of As and fluorescent humic substances, especially in the endemic BFD area of northern Chianan plain. Lu *et al.* (1985) revealed that the fluorescent humic substances in the artesian well water were positively correlated with As, pH, and TDS ($r = 0.49, 0.25$, and 0.18, respectively, $p < 0.001$, n = 1189).

6.4.2   *Arsenic in sediments*

The sequentially extracted total As content is fundamentally the same as that obtained from PHH (phosphoric acid and hydroxylamine hydrochloride) extraction from the shallow aquifer sediment samples from the Lanyang and Chianan plains and the amount of As

Table 6.3.   Mean groundwater quality data from between 4 and 13 sampling sites in arsenic-affected regions in Chianan plain during November 2005 and November 2006 based on the data from Nath *et al.* (2008). These sampling sites are Anei, Beimen-2A, Jiangjiun-1B, Jiangjiun-1C, Jiangjiun-1D, Liujiao-1, Liujiao-2, Liujiao-2B, Lucao-1A, Lucao-1B, Yenshuei-1, Yenshuei-2, Yenshuei-3, Yichu-1A, and Hsuechia-2 (for locations see Fig. 6.1).

Parameter	November 2005	November 2006
Groundwater temperature (°C)	Max: 27.2 Mean: 26.8 ± 0.77 (n = 11) Min: 24.7	Max: 29.3 Mean: 29.3 ± 0.08 (n = 13) Min: 29.1
pH	Max: 8.5 Mean: 7.61 ± 0.37 (n = 11) Min: 7.2	Max: 8.1 Mean: 7.43 ± 0.30 (n = 13) Min: 7.0
EC (μS cm^{-1})	Max: 2090 Mean: 1329.2 ± 372.5 (n = 11) Min: 841	Max: 3720 Mean: 1723.3 ± 880.5 (n = 13) Min: 824
TDS (mg L^{-1})	Max: 637 Mean: 519.9 ± 141.8 (n = 11) Min:335	Max: 1605 Mean: 1277 ± 284.0 (n = 10) Min: 796
Salinity (‰)	Max: 0.8 Mean: 0.44 ± 0.20 (n = 11) Min: 0.1	Max: 1.9 Mean: 0.71 ± 0.52 (n = 13) Min: 0.2
Alkalinity (mg L^{-1})	Max: 163 Mean: 121 ± 44.5 (n = 9) Min: 31	Max: 175 Mean: 164.8 ± 6.99 (n = 13) Min: 149
ORP (mV)	Max: -17 Mean: -45.5 ± 20.1 (n = 11) Min: -93	Max: -75 Mean: -156.4 ± 39.2 (n = 12) Min: -200
Na$^+$ (mg L^{-1})	Max: 223 Mean: 126.4 ± 76.5 (n = 9) Min: 11	Max: 494 Mean: 195.6 ± 149.5 (n = 13) Min: 41
K$^+$ (mg L^{-1})	Max: 32 Mean: 17.9 ± 8.50 (n = 7) Min: 6.4	Max: 122 Mean: 31.9 ± 31.7 (n = 13) Min: 13
Mg^{2+} (mg L^{-1})	Max: 48 Mean: 24.5 ± 17.0 (n = 9) Min: 5.3	Max: 198 Mean: 37.2 ± 49.6 (n = 13) Min: 7
Ca^{2+} (mg L^{-1})	Max: 33 Mean: 21.4 ± 6.10 (n = 9) Min: 16	Max: 122 Mean: 60.8 ± 32.1 (n = 13) Min: 21
Cl$^-$ (mg L^{-1})	Max: 230 Mean: 90.4 ± 69.1 (n = 9) Min: 14	Max: 842 Mean: 153.8 ± 228.7 (n = 13) Min: 14
SO$_4^{2-}$ (mg L^{-1})	Max: 167 Mean: 49.2 ± 63.3 (n = 8) Min:3.7	Max: 601 Mean: 120.3 ± 169.4 (n = 11) Min: 2.5
NO$_2$ (mg L^{-1})	Max: below detection limit Mean: below detection limit Min: below detection limit	Max: 166 Mean: 49.2 ± 78.0 (n = 4) Min: 7.0

(Continued)

Table 6.3.   (*Continued*)

Parameter	November 2005	November 2006
V (μg L^{-1})	Max: 0.39 Mean: 0.25 ± 0.15 (n = 11) Min: 0.07	Max: 2.5 Mean: 0.81 ± 0.71 (n = 13) Min: 0.12
Cr (μg L^{-1})	Max: 0.12 Mean: 0.09 ± 0.04 (n = 11) Min: 0.05	Max: 2.7 Mean: 0.57 ± 0.77 (n = 13) Min: 0.07
Mn (μg L^{-1})	Max: 8.5 Mean: 4.42 ± 7.10 (n = 9) Min: 0.23	Max: 710 Mean: 102.5 ± 193.9 (n = 13) Min: 9.2
Co (μg L^{-1})	Max: 0.44 Mean: 0.19 ± 0.11 (n = 11) Min: 0.05	Max: 0.76 Mean: 0.38 ± 0.23 (n = 13) Min: 0.09
Mo (μg L^{-1})	Max: 12 Mean: 4.87 ± 5.29 (n = 11) Min: 0.28	Max: 15 Mean: 3.83 ± 4.96 (n = 13) Min: 0.11
Fe (μg L^{-1})	Max: 15 Mean: 9.84 ± 3.29 (n = 11) Min: 5.1	Max: 6151 Mean: 731.4 ± 1681.8 (n = 13) Min: 11
As (μg L^{-1})	Max: 651 Mean: 236.6 ± 197.4 (n = 11) Min: 1.8	Max: 575 Mean: 223.4 ± 189.5 (n = 13) Min: 1.3
Rb (μg L^{-1})	Max: 6 Mean: 3.05 ± 1.85 (n = 11) Min: 0.26	Max: 5.1 Mean: 3.27 ± 1.45 (n = 13) Min: 0.52
Ba (μg L^{-1})	Max: 20 Mean: 16.7 ± 33.3 (n = 10) Min: 0.26	Max: 88 Mean: 30.7 ± 28.4 (n = 13) Min: 2.5
Sr (μg L^{-1})	Max: 278 Mean: 171.9 ± 79.7 (n = 11) Min: 60	Max: 1094 Mean: 433.5 ± 381.6 (n = 13) Min: 111
Cu (μg L^{-1})	Max: 5.4 Mean: 2.23 ± 1.95 (n = 10) Min: 0.39	Max: 1.6 Mean: 1.02 ± 0.51 (n = 13) Min: 0.31
Zn (μg L^{-1})	Max: 62 Mean: 18.9 ± 16.9 (n = 10) Min: 3.9	Max: 376 Mean: 106.9 ± 116.9 (n = 13) Min: 3
Ni (μg L^{-1})	Max: 1.2 Mean: 0.43 ± 0.38 (n = 11) Min: 0.1	Max: 2.9 Mean: 0.83 ± 0.86 (n = 13) Min: 0.08
Se (μg L^{-1})	Max: 0.57 Mean: 0.16 ± 0.16 (n = 11) Min: 0.01	Max: 0.42 Mean: 0.19 ± 0.16 (n = 13) Min: 0.04

extracted from organic matter is negligible when compared with that from Fe-(hydr)oxides (Chen and Liu 2007). This suggests that mobile As in the sediments of the Chianan plain was predominantly associated with clay-sized FeOOH. The amount of acid extractable As was positively correlated with the proportion of clay-sized Fe-(hydr)oxides in the sediments, whereas no correlation ($p > 0.05$) was found between the contents of As and Fe in groundwater (Chen and Liu 2007). To date, no one has done adsorption/desorption experiments to measure the concentrations of As(V) and As(III) in sediments and groundwater in the Chainan and Lanyang plains. This warrants detailed research.

The results of X-ray diffraction analysis showed that the dominant minerals in the Chianan plain and Lanyang plain sediments are quartz, mica and feldspar (Reza *et al.* 2009). Abundant quartz was found in sand, but mica (muscovite and biotite) and feldspar were observed to be present only in the sandy clay sediments in these two plains. The concentrations of Al_2O_3 (11.98 to 15.57 wt%) and Fe_2O_3 (4.40–5.53 wt%) in the Chianan plain sediments are lower compared to those from the Lanyang plain (18.57 to 19.12 wt% for Al_2O_3 and 7.01 to 7.33 wt% for Fe_2O_3) (Reza *et al.* 2009). These were measured in clay-rich sediments. The MnO content is <0.1 wt% in the Chianan plain and <0.17 wt% in the Lanyang plain throughout the Budai (150 m in depth) and Yichu drilled cores (200 m in depth) (Fig. 6.1). The fine-grained sediments enriched in phyllosilicates (such as mica) contain higher amounts of P_2O_5, indicating that the finer sediment concentrates more P in the Chianan and Lanyang plains. The P_2O_5 content correlates well with the CaO content, suggesting that P in the sediments is fixed as apatite in these two plains (Reza *et al.* 2009). XRF analysis also revealed that fine-grained sediments in these two plains contain a higher amount of trace elements because of their high surface area and adsorption capacity (Reza *et al.* 2009). The strong positive correlation between As, Fe_2O_3, and MnO in the Chianan plain sediments suggests a strong adsorption of As by Fe- and Mn-oxyhydroxides. Furthermore, a good positive and statistically significant correlation is found between As and organic carbon in the Chianan plain sediments ($R^2 = 0.77$, $p < 0.05$). However, As can be desorbed from Fe- and Mn-oxyhydroxides into the groundwater of the reducing aquifer in the west coast of Chianan plain. The shallow oxidizing aquifer of the Chianan plain is distributed in the groundwater recharge area in the east of the plain which is mainly composed of the mountainous coarser-grained materials (Lin, Y.-B. *et al.* 2006) (Fig. 6.5).

Chen (2001) revealed that organic matter in sediments can be depleted by bacteria and results in strong organic mineralization. This releases considerable amounts of dissolved fluorescent humic substances, phosphates, and alkalinity, as well as As and Fe, into the groundwater to form organometallic complexes, thereby causing high As, Fe, and fluorescent humic substances in the pore water. Lee (2008) analzyed a total of 149 geological core samples from 5 drilling wells located at mid- and distal-fan areas and showed a positive correlation of As and Fe contents in marine sequences and the adsorption of As onto/coprecipitated with non-crystalline Fe hydroxides.

6.4.3   *Mobilization and transport of arsenic*

6.4.3.1   *Arsenic speciation*
In the the confined aquifers of the BFD endemic area in the northern Chianan plain, the ratio of As(III)/As$_{tot}$ was much higher than that of As(V)/As$_{tot}$ (Table 6.4). The highest As(III)/As$_{tot}$ ratio was 98.6% (Table 6.4) at Yenshuei-4 (Fig. 6.1); in most places the ratios exceed 75% (e.g., Hsuechia-1, Liujiao-2, Yenshuei-1–4 and 5A, and Yichu-1A). Only at two sampling points (i.e., Jiangjiun-1B and Beimen-2A) (Fig. 6.1) the As(III)/As$_{tot}$ ratios were as low as 56.9% and 46.4%, respectively, revealing that the confined aquifer has strongly reducing conditions and As(III) is the predominant As species (>75%) (Table 6.4).

6.4.3.2   *Redox-mediated mobilization and transport of arsenic*
Arsenic can be mobilized and released into groundwater from sediments through redox-mediated processes. In the Chianan plain groundwater the ORP (oxidation reduction potential) values in 2006 were revealed to be negative (Table 6.3). Trace element analysis of groundwater samples from 11 sampling points in 2005 and 13 sampling points in 2006 (Table 6.3) revealed that As concentrations were higher than those of Fe and Mn. Moreover, the regression analysis demonstrated that As was not correlated with Fe and Mn in 2005 ($r = -0.342$, $p = 0.276$, $n = 12$, and $r = -0.554$, $p = 0.097$, $n = 12$, respectively) or in 2006 ($r = -0.180$, $p = 0.538$, $n = 14$, and $r = -0.467$, $p = 0.092$, $n = 14$, respectively). Arsenic was released into the groundwater from sediments much more than Fe and Mn, yielding low concentrations of Fe and Mn in the

Table 6.4.   Distributions of arsenic species in the Chianan plain groundwater (Yang 2006). The locations can be seen in Figure 6.1.

Sampling site	Well depth (m)	As$_{tot}$ (μg L^{-1})	As(III) (μg L^{-1})	As(V) (μg L^{-1})	Reduction ratio As(III) / As$_{tot}$ (%)	Oxidation ratio As(V) / As$_{tot}$ (%)	As(III)/ As(V)
Hsuechia-1	31	0.07	0.06	0.01	85.7	14.3	6.00
Beimen-2A	277	0.28	0.13	0.15	46.4	53.6	0.87
Jiangjiun-1B	336	0.58	0.33	0.25	56.9	43.1	1.32
Lucao-1B	10	0.02	Nd	Nd	Nd	Nd	Nd
Liujiao-1	13	0.04	Nd	Nd	Nd	Nd	Nd
Liujiao-2	67	0.16	0.14	0.02	87.5	12.5	7.00
Yenshuei-1	23	0.27	0.24	0.03	88.9	11.1	8.00
Yenshuei-2	23	0.49	0.46	0.02	93.9	4.08	23.0
Yenshuei-3	23	0.80	0.78	0.01	97.5	1.25	78.0
Yenshuei-4	23	0.71	0.70	0.01	98.6	1.41	70.0
Yenshuei-5A	233	0.06	0.05	0.002	83.3	3.33	25.0
Yichu-1A	20	0.04	0.03	0.003	75.0	7.50	10.0

Nd: not detected.

goundwater but high As concentrations. The experimental results of adsorption and desorption showed poor correlation between As, Fe and Mn in the Chianan plain groundwater, and a positive correlation between As and Fe but poor correlation between As and Mn in the Lanyang plain groundwater (Nath *et al.* 2008, Reza *et al.* 2009). It can be concluded that: (1) the presence of As at higher concentrations than Fe or Mn in the Chianan plain groundwater could be due to As desorption without dissolving Fe and Mn hydroxide minerals, or (2) Fe and Mn precipitated subsequently. Dissolved Fe may be re-precipitated as $FeCO_3$ (siderite) under reducing conditions (Nickson *et al.* 2000). However, the concentrations of Fe and Mn in the Lanyang plain groundwater were as high as those of As, suggesting that the As is desorbed with dissolving Fe and Mn hydroxide minerals. Arsenic(V) can be reduced to As(III). Fe(III) is reduced and solubilized as Fe(II), thus resulting in the desorption of Fe(II) and As from the sediments which is subsequently released into the groundwater. Thus, redox processes are principally responsible for the mobilization of As in groundwater.

Besides As, iron, manganese, and strontium are also present in the Chianan plain groundwater. The results reported by Nath *et al.* (2008) revealed that the Chianan plain groundwater in the Beimen area contained As: 292 μg L^{-1}, Fe: 5.8 μg L^{-1}, Mn: below detection limit, Na: 162,000 μg L^{-1}, Cl: 99,000 μg L^{-1}, SO$_4$: 7500 μg L^{-1}, NO$_2$: below detection limit, whereas the results reported by Reza *et al.* (2009) showed that the Lanyang plain groundwater at the Wujie sampling site contained As: 543 μg L^{-1}, Fe: 235 μg L^{-1}, Mn: 130 μg L^{-1}, Na: 194,040 μg L^{-1}, Cl: 9,760 μg L^{-1}, SO$_4$: 390 μg L^{-1}, NO$_2$: not detected.

As mentioned in section 6.2, the uppermost aquifer located in alluvium and the Tainan formation (loosely consolidated rocks, mainly composed of silts) (depth: 0–60 m) was mostly formed in the Holocene (Lau and Mink 1971, Wu 1978). The deposits in deep aquifers of the Liushuang and Erzong river formations (depth: 60–300 m) were classified as Pliocene to Pleistocene mudstones, occasionally intercalated with sandstone, which were formed diagenetically from sands and silts mixed with organic muds of the deep-sea to shallow-sea marine sequence (Wu 1978, Chou 1971). The minerals of organic muds are mainly chlorite, illite, and quartz (Chou 1971) with elevated concentrations of trace elements, which are associated with organic compounds in connate water (Wu 1978). These concentrated elements in organic mud and organic compounds in connate water are released into the groundwater under reducing conditions, resulting in anomalous groundwater quality and methane emission in the Chianan plain aquifer (Wu 1978). The groundwater quality analyses by Wu (1978) indicated that the deep aquifer in the Chianan plain contains high concentrations of As (Chen *et al.* 1962,

Tseng 1973), organochlorides, glucose, acetone, yeast, and trihalomethanes (e.g., chloroform and tribromomethane) as well as ergots (e.g., ergotamine tartrate, ergostetrine, ergosterol) and vitamin D2, which are the derivatives of minerals and organic compounds present in the clay (or mud) layers of the Liushuang and Erzong river formations (Lu *et al.* 1977). Slate and mudstone containing As concentrations up to 11–15 mg kg^{-1} (Wu 1978) along with these organic components may deteriorate the groundwater quality in the shallow aquifer of the Chianan plain (Wu 1978). The shallow aquifer is generally characterized by oxidized conditions, in which As may be bound with sulfide minerals and precipitated. In contrast to Fe/Mn hydroxide minerals, As(III) can be oxidized to As(V) and adsorbed onto iron (oxy)hydroxides in sediments. The deep aquifer is generally under reducing conditions, in which As(V) can be reduced to As(III) and sequentially desorbed from sediments and released into groundwater, deteriorating the groundwater quality in the BFD area of the Chianan plain.

6.4.3.3 *Microbe-mediated mobilization and transport of arsenic*

Another mechanism by which As can be mobilized from sediments and transported in groundwater is through microbial processes. Microbes use sulfate as an electron acceptor for anaerobic respiration. Sulfate is depleted, resulting in reducing conditions in groundwater. Thus, As(V) in sediments can be reduced to the dissolved phase As(III) and released into groundwater. However, the abiotic reduction of As(V) with sulfide is slow, so direct microbial reduction is more likely to play a role. Direct microbial reduction of sorbed As(V) to As(III) using arsenate as the electron acceptor by As(V)-reducing bacteria may also mobilize As.

In the Chianan and Lanyang plains, As mobilization into groundwater from sediments is mediated by bacteria via the oxidation of organic matter. The bacteria are proposed to play the same role in releasing As into the groundwater from sediments under reducing conditions in both plains. Alternately, As is originally present in the form of chemically reduced minerals, like arsenopyrite (FeAsS). These reduced minerals are oxidized by bacteria resulting in the oxidation of As(III) as well as iron and sulfide, coupled with the fixation of CO_2 into organic matter. The As(V) can subsequently be adsorbed onto oxidized mineral surfaces, like iron oxyhydroxide. The influx of substrate organic materials derived either from buried peat deposits or bacteria promote microbial respiration. Anaerobic bacteria then respire adsorbed As(V), resulting in the release of As(III) into the aqueous phase (Oremland and Stolz 2003). The Chianan plain aquifer in the southwestern coastal area has also been subjected to seawater intrusion. Thus, anaerobic bacteria release more As and humic substances into groundwater because seawater provides sulfate as an electron acceptor for bacterial growth.

The bacterial breakdown of organic matter produces weak organic acids such as ethanoic acid (CH_3COOH). Ethanoic acid dissolved in groundwater produces ethanoate anions (CH_3COO^-) and after reduing ethanoate anions with FeOOH combined with H_2CO_3 (dissolved CO_2), As and Fe^{2+} are released from sediments into groundwater according to the equation below:

$$CH_3COOH\ (aq) + 8FeOOH(s) + 14H_2CO_3(aq) \rightarrow 8Fe^{2+}\ (aq) + 16HCO_3^-\ (aq) + 12H_2O$$

Bacterial reductive dissolution of iron oxyhydroxides releases As from sediments into groundwater under reducing environment and through the oxidation of organic matter or humic substances. The degradation of organic matter in the sediment results in an overall reducing environment and facilitates the release of As in the aquifers (Bhattacharya *et al.* 2006a). Enhanced bacterial activity accelerates the diagenetic process, enhancing the mobilization of As from sediments with a high organic matter content (Akai *et al.* 2004).

6.5 ARSENIC IN MUD VOLCANOES AND HOT SPRINGS

Another source of the As found in the groundwater of the Chianan plain may be mud volcanoes. To date, mud volcanoes have been reported both onshore (Shih 1967, Wang *et al.* 1988)

Figure 6.18.   Wushanting mud volcano.

and offshore (Liu *et al.* 1997, Chow *et al.* 2001) of Taiwan. In total, 64 active mud volcanoes were described and classified according to their geomorphologial features and tectonic terrains in 17 land areas (Shih, 1967, Wang *et al.* 1988). Several mud volcanoes are distributed in the east of Chianan plain (Fig. 6.2). (e.g., Chunglun, Guanzihling, Wushanting (Fig. 6.18), Hsiaokunshuei, Kunshueiping) from which only Chunglun mud volcano at Choushueitan (23.369104°N, 120.568203°E) releases a considerable amount (>90%) of CO_2. The other mud volcanoes release large amounts (>70%) of methane. The erupted muds and fluids were analyzed and revealed to contain high concentrations (12.0–27.7 $\mu g/L^{-1}$) of As, which could be transported to the Chianan plain aquifer. The mud volcano fluids probably correspond to the pore water of marine sediments of the accretionary complex (Gieskes *et al.* 1992) which may mix with magmatic water and meteoric water and undergo further chemical changes due to water-rock interactions along its pathway to the earth's surface (Kopf and Deyhle 2002, Yeh *et al.* 2004). To date, the source and corresponding chemical evolution of the emanating fluids from the mud volcanoes are still uncertain (Liu, C.-C. *et al.* 2009).

The median concentrations of Na, Cl, SO_4, as well as alkalinity, salinity, and conductivity were lower in the fluids of the Wushangting mud volcano in October 2004, March 2005, and June 2005 than those in the Hsiaokunshuei mud volcano fluids (Table 6.5). These concentrations were 1989 *vs.* 6658 mg L^{-1} for Na, 2730 *vs.* 9856 mg L^{-1} for Cl, 121 *vs.* 269 mg L^{-1} for SO_4, 445.3 *vs.* 1146.8 mg L^{-1} for alkalinity, 3.8 *vs.* 10‰ for salinity, 7.2 *vs.* 24 mS cm^{-1} for conductivity. However, the median concentrations of SO_4 in the muds of the Wushangting mud volcano are higher than in the Hsiaokunshuei mud volcano muds, (970 *vs.* 916 mg kg^{-1} for SO_4), whereas the concentrations of Na and Cl in the muds of the Wushangting mud volcano are lower than in the muds of Hsiaokunshuei mud volcano, (1948 *vs.* 9886 mg kg^{-1} and 5642 *vs.* 17576 mg kg^{-1}, respectively) (Table 6.6). More SO_4 ions are adsorbed in the muds of Wushangting mud volcano than in the muds of Hsiaokunshuei mud volcano. However, less Na and Cl ions are in the fluids and muds of Wushangting mud volcano compared to the Hsiaokunshuei mud volcano, suggesting that the Hsiaokunshuei mud volcano contains more connate water than the Wushangting mud volcano.

Since the connate water has long been present in the deep aquifer of Chianan plain, the sulfate in connate water can be reduced to sulfides under reducing conditions (Wu 1978). The concentrations of chlorides and sulfates in connate water are positively correlated. Under the reducing conditions of the deep aquifer of Chianan plain, the groundwater has a chloride concentration exceeding 15 meq L^{-1}, but as the conditions become more oxidized, the chloride concentrations are lower than 15 meq L^{-1} (Wu 1978). A groundwater quality survey at Budai in Chianan plain (Well number K-01-T27, well depth: 152 m) by the Groundwater Engineer Division of the Taiwan Provincial Government in 1959 demonstrated that the well water contained very high concentrations of chloride (up to 20,500 mg L^{-1} or 578.6 meq L^{-1}; 1.1 times higher than that of seawater) and high sulfate concentrations (up to 93.7 mg L^{-1}) (Taiwan

Table 6.5. Chemical characteristics of the mud volcano fluids collected from the eruption center during different seasons (a) Wushanting (WST), and (b) Hsiaokunshuei (HKS) mud volcanoes (modified from Liu, C.-C. et al. 2009).

(a)

Time of sample collection	Na$^+$	K$^+$	Mg^{2+}	Ca^{2+}	Cl$^-$	NO$_3^-$	NO$_2^-$	SO$_4^{2-}$	Alkalinity as HCO$_3$ (mg L^{-1})	Temp. (°C)	pH	Salin. (‰)	ORP (mV)	Conduct. (mS cm^{-1})	As	Cu	Mn	Zn
	(mg L^{-1})														(µg L^{-1})			
October 2004	1989	50.8	10.2	37.7	2730	130.2	73.6	121.0	445.3	28	7.7	3.8	-49	7.2	21.7	108.0	15.9	15.0
March 2005	3589	70.4	29.2	48.1	5460	155	82.8	98.0	140.3	23	8.1	4.0	-30	7.3	27.7	1.2	14.8	20.3
June 2005	1655	50.8	17.8	39.7	2446	BDL	15.2	62.5	341.6	30	7.2	3.8	-50	7.3	12.0	2.1	31.9	19.0

(b)

Time of sample collection	Na$^+$	K$^+$	Mg^{2+}	Ca^{2+}	Cl$^-$	NO$_3^-$	NO$_2^-$	SO$_4^{2-}$	Alkalinity as HCO$_3$ (mg L^{-1})	Temp. (°C)	pH	Salin. (‰)	ORP (mV)	Conduct. (mS cm^{-1})	As	Cu	Mn	Zn
	(mg L^{-1})														(µg L^{-1})			
October 2004	6658	129.0	60.8	92.2	9856	BDL	4.60	269.0	1146.8	26	7.5	10	-50	24	NA	NA	NA	NA
March 2005	10040	156.4	94.8	88.2	15564	BDL	3.22	153.7	646.6	24	7.8	15	-32	22	NA	NA	NA	NA
June 2005	8729	152.5	85.1	88.2	13401	BDL	2.76	76.9	933.3	28	7.9	12	-48	21	NA	NA	NA	NA

Table 6.6. Chemical characteristics (exchangeable and/or extractable fractions) of the muds collected from the mud volcanoes during different seasons at (a) Wushanting (WST), and (b) Hsiaokunshuei (HKS) mud volcanoes (modified from Liu, C.-C. et al. 2009).

(a)

Time of sample collection	Na^+	NH_4^+	K^+	Mg^{2+}	Ca^{2+}	Cl^-	NO_3^-	NO_2^-	SO_4^{2-}	Alkalinity as HCO_3 ($mg\,kg^{-1}$)	Temp (°C)	pH	Salin. (‰)	ORP (mV)	Conduct. ($mS\,cm^{-1}$)	As	Cu	Mn	Zn
	($mg\,kg^{-1}$)															($\mu g\,kg^{-1}$)			
October 2004	1948	5409	7776	14268	2220.0	5642	713	N/A	970	N/A	N/A	N/A	N/A	N/A	N/A	2.3	0.63	27.5	2.6
March 2005	1892	159	9781	14858	112.6	3464	2631	N/A	1258	N/A	N/A	N/A	N/A	N/A	N/A	7.5	1.27	25.8	3.9
June 2005	1353	4156	9943	37173	76.5	6306	514	N/A	1144	N/A	N/A	N/A	N/A	N/A	N/A	0.7	0.25	26.4	0.5

(b)

Time of sample collection	Na^+	NH_4^+	K^+	Mg^{2+}	Ca^{2+}	Cl^-	NO_3^-	NO_2^-	SO_4^{2-}	Alkalinity as HCO_3 ($mg\,kg^{-1}$)	Temp (°C)	pH	Salin. (‰)	ORP (mV)	Conduct. ($mS\,cm^{-1}$)	As	Cu	Mn	Zn
	($mg\,kg^{-1}$)															($\mu g\,kg^{-1}$)			
October 2004	9886	1314	20059	3969	1421.	17576	451	N/A	916	N/A	N/A	N/A	N/A	N/A	N/A	1.50	0.63	24.7	1.31
March 2005	1431	5626	9600	8667	24.1	14332	3131	N/A	567	N/A	N/A	N/A	N/A	N/A	N/A	6.74	0.63	30.8	4.58
June 2005	1684	1228	12923	38565	65.3	877	3426	N/A	380	N/A	N/A	N/A	N/A	N/A	N/A	12.7	0.19	40.1	1.31

Table 6.7. The water quality comparison between Kunshueiping mud volcano fluids and the seawater of Taiwan Strait near Kaohsiung city (Figs. 6.1 and 6.2).

| Ions | Kunshueiping mud volcano fluids (mg L^{-1}) | | Seawater (mg L^{-1}) (Lau and Mink 1971) |
	1973 (Lau and Mink 1971)	2007 (Jean *et al.* unpublished)	
Ca^{2+}	140	8	400
Mg^{2+}	36	139	1272
Na$^+$	33235	4328	10561
K$^+$	1400	47	380
CO$_3^{2-}$	2766	–	–
HCO$_3^-$	3293	–	140
SO$_4^{2-}$	433	8	2649
Cl$^-$	45370	5020	18980

Provincial Government 1959). The connate water in the deep aquifer of Gutingkeng mudstone can also be collected from mud volcano fluids. The Kunshueiping mud volcano fluid samples were analyzed by Taiwan Sugar Company in 1973 (Table 6.7) (Lau and Mink 1971), revealing that the concentrations of chloride and sodium in the mud volcano fluids (possibly connate water) were as high as 45,370 mg L^{-1} (*cf.* Cl in seawater: 18,980 mg L^{-1}) and 33,235 mg L^{-1} (*cf.* sodium in seawater: 10,561 mg L^{-1}), respectively (Table 6.7). Thus, the Kunshueiping mud volcano fluids contain considerable amounts of connate water in which the concentrations of sodium and chloride are much higher than the seawater of Taiwan Strait near Kaohsiung city (Figs. 6.1 and 6.2). Hsu (1975) indicated that the seawater contained As at concentrations up to 100 µg L^{-1} (data not shown).

6.6 CONCLUDING REMARKS

In Taiwan, As in the groundwater can result from (1) As-affected aquifers (e.g., Chianan plain in SW Taiwan and Lanyang plain in NE Taiwan), (2) mining activity (e.g., Chinkuashi gold mine in northern Taiwan), and (3) geothermal waters (e.g., Geothemal Spring Valley in northern Taiwan and mud volcanoes in southern Taiwan). However, the geogenic sources of As in the Chianan plain and Lanyang plain aquifers are still uncertain, whereas the source in Guandu plain is certainly the Geothermal Spring Valley in Beitou district, Taipei city. The geothermal waters also caused the soil and groundwater contamination of Guandu plain. The seawater and groundwater in the vicinity of the Chinkuashi gold mine area were polluted with As from the acid mine drainage after mining activity.

Although no new cases of Blackfoot Disease (BFD) occurred in the Chianan plain in SW Taiwan after 1990, the etiological agents that really caused BFD still remain unclear. The mean concentrations of As as well as the mean concentrations and relative fluorescence intensity of humic substances in the Chianan plain groundwater were higher than in the Lanyang plain groundwater. Such high As concentrations and the high relative fluorescence intensity of humic substances may be responsible for causing the BFD in SW Taiwan before 1990. It is essential to investigate and compare the differences in hydrochemical factors between BFD and non-BFD areas in Taiwan.

The mobilization and transport mechanisms of As in the groundwater of Chianan plain and Lanyang plain are different. Arsenic is released into the Chianan plain groundwater from sediments to a greater extent than Fe and Mn, either due to direct desorption of As or precipitation of Fe and Mn as hydroxides. In the Lanyang plain, As is desorbed due to the dissolution of Fe and Mn hydroxide minerals in groundwater. High As contents were observed under reducing conditions, which may also suggest the reductive dissolution of As-bearing ferric (oxyhydro)oxides.

CHAPTER 7

Arsenic in soils and plants: accumulation and bioavailability

In chapter 6 we discussed the occurrence of high concentrations of arsenic (As) in groundwater of the deep reducing aquifers of Chianan and Lanyang plains, which are used at present for irrigation and aquaculture, and formerly also for drinking water supply causing severe health effects in the population. The common feature of these deep aquifers in Chianan and Lanyang plains is stagnant groundwater and reducing conditions, whereas in both areas the shallow wells which tap the phreatic aquifers (oxidizing conditions) deliver water with low As concentrations. The known sources of the groundwater As as of 2004 (Drury 2006) are as follows: As-contaminated aquifers, As released by mining activities and As release related to geothermal activities, the last explaining high As concentrations in the area of Guandu plain. In the present chapter, we will first overview the accumulation and behavior of As in the Taiwanese soil with emphasizing on the Guandu plain where the As concentration in the paddy soil greatly exceeded the 30 mg kg^{-1} threshold for concern under Taiwanese regulation. We will then discuss the bioaccumulation of As in plants and crops and compare the As concentrations in soil, rice, crops and vegetables in BFD and non-BFD areas as well as in the Guandu plain. Moreover, concentrations of total As (As$_{tot}$) and various As species in rice in Taiwan will be compared with those reported from different countries, to elucidate the potential health risk associated with ingestion of As-enriched rice.

7.1 ACCUMULATION AND BEHAVIOR OF ARSENIC IN SOIL

In Taiwan, As concentrations in soils are generally low to medium. The average As concentrations in 14 townships of Taiwan surveyed in 1973, at soil depths 0–12, 12–24 and 24–36 cm were 8 (range 1–44), 14 (range 2–34) and 19 mg kg^{-1} (range 3–177), respectively (Li et al. 1979a). Later, in 1976, the Taiwan Plant Protection Center surveyed the As concentrations in soils of 45 other townships. The average As concentrations in soil at depths 0–12, 12–24 and 24–36 cm were 7.86 (range 1–22), 10 (range 2–19) and 10 (range 6–19) mg kg^{-1}, respectively (Li et al. 1979a).

During 1982–2001, the Taiwan Environmental Protection Agency conducted a four-phase nationwide survey of the concentration of eight heavy metals and metalloids (As, Cd, Cu, Cr, Hg, Ni, Pb and Zn) in soils (Table 7.1). The results of the 2nd phase survey, in which one sample represented an area of 100 ha, indicated that the average As concentration was 7 mg kg^{-1} (range 3–13 mg kg^{-1}) (Fig. 7.1). In the 4th phase of the detailed survey, 100,000 ha were surveyed with one sample taken per ha. The results showed that only 4 samples exceeded

Table 7.1. National survey of heavy metals and metalloid contamination in soil by Taiwan Environmental Protection Agency (Taiwan EPA 2002).

Phase	Period	Survey density	Covering area
1	1982–1986	1 sample/80 ha	11.60×10^3 ha
2	1987–1990	1 sample/100 ha	400×10^3 ha
3	1991–1998	1 sample/25 ha	200×10^3 ha
4	1999–2001	1 sample/ha	100×10^3 ha

115

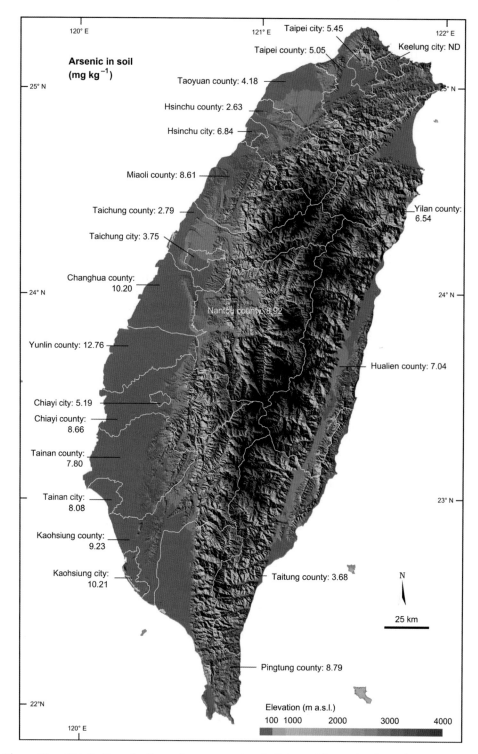

Figure 7.1.   Distribution of soil arsenic concentrations in Taiwan. Mean value for each county or city is given in mg kg⁻¹. Measuring density: 1 sample/100 ha. ND: not detected.

the 30 mg kg⁻¹ threshold for concern under Taiwanese regulations. Three of those samples were situated in Taipei city with an average As concentration of 69 mg kg⁻¹. One sample was in Hsinchu city with an average As concentration of 42 mg kg⁻¹. The contamination source of As in the soil of Taipei city is suspected from the natural hot spring and is classified as geogenic, whereas As in the Hsinchu city is mainly from the agrochemical manufacturing industry, an anthropogenic As source (CTCI Corporation 2001). Notably, the warning and control standards of As concentrations in soil, enforced by the Taiwan EPA, are 30 and 60 mg kg⁻¹, respectively (Taiwan EPA 2003).

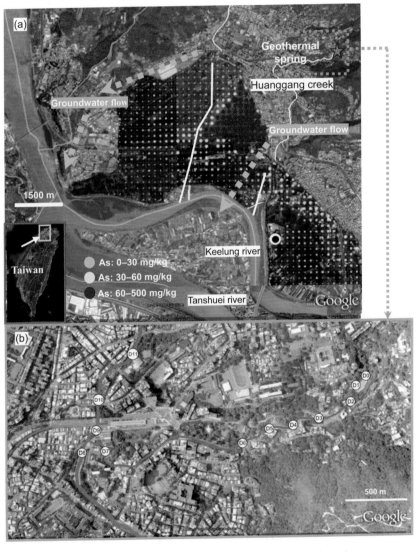

Figure 7.2.   Locations of the sites in the Guandu plain and hot springs where the water was sampled. Superimposed on the satellite images from Google Earth: (a) Surface and groundwater generally flow from the northeastern geothermal spring area to the downstream Keelung river. High As concentrations were accumulated in the surface soil of Guandu plain; (b) Sampling points for As analysis in the upstream geothermal spring area (modified from Taipei EPB 2006).

In 2006, a detailed soil survey of As was conducted in Guandu plain by the Taipei city government (Fig. 7.2a). The average As concentration in Guandu plain of Taipei city was 151 mg kg^{-1} (range 10–465 mg kg^{-1}) which greatly exceeded the 60 mg kg^{-1} control level (Shyu and Lin 2006). Shyu *et al.* (2009) carried out an intensive investigation on the distribution of As concentrations in Guandu plain. Soil samples were collected from 842 ha for each 1 ha unit (Fig. 7.2a). Mean As levels of topsoil (depth 0–15 cm) and subsoil (depth 15–30 cm) were 122 mg kg^{-1} (range 5–458 mg kg^{-1}) and 117 mg kg^{-1} (range 5–513 mg kg^{-1}), respectively. There were 143 ha with soil As concentration ranging from 30–60 mg kg^{-1}, and 128 ha with As concentrations exceeding 60 mg kg^{-1} at 0–15 cm depth. Over 61,500 kg of As present in the top- and in the subsoils are estimated to originate from external sources. This large quantity of As was mainly from natural sources due to the absence of any anthropogenic As sources in Guandu plain. The possible sources of As in Guandu plain are volcanic eruptions and hot spring water discharge (Shyu *et al.* 2009). Table 7.2 lists the As concentrations of hot spring water sampled along the creeks in the so-called "geothermal spring area". The sampling points themselves are shown in the map of Fig. 7.2b. As concentrations are high in the geothermal spring but decrease sharply downstream due to river water mixing and precipitation and accumulation in the bottom sediments. Local farmers have used this water with high concentrations of As for irrigation for hundreds of years. This has led to As accumulation in the paddy soil.

Although rice has been cultivated in the Guandu plain over the last 200 years, there has been no epidemiologic evidence showing As threat to public health through rice ingestion, suggesting low bioavailability of As in the Guandu soils (Su *et al.* 2007). To investigate the vertical distribution of As concentrations, soil samples were collected at different depths of As-containing paddy land and an adjacent dry-land (up to a depth of 140 cm) (Wu *et al.* 2009). Chemical extractions were applied to determine the concentrations of As$_{tot}$ and selected elements, such as Fe, Al, Pb, and Ba. As speciation in soils within those two profiles was also characterized using sequential extractions (Wenzel *et al.* 2001). The results showed that As was predominately associated with Fe oxides and was less recalcitrant in the surface soils of the rice-paddy land (Wu *et al.* 2009). Nonetheless, the vertical distributions of As showed no indication of enhanced mobility in the rice paddy soil compared with its dry-land counterpart. Thus, Wu *et al.* (2009) hypothesized that the high concentrations of Fe oxides or hydroxides in those soils may restrict As mobility and availability during flooding

Table 7.2.   Arsenic concentrations in the hot spring water sampled along two creeks in the geothermal spring area. Arsenic is predominately present as As(V), with exceptions of sampling points near the spring where As(III) is the predominant species, and no organic arsenic species were detected.

Sample number	As(III) (μg L^{-1})	As(V) (μg L^{-1})	As$_{tot}$ (μg L^{-1})
D0	ND	ND	ND
D1	4785.0	880.2	5665.2
D2	3917.3	1488.1	5405.4
D3	25.3	708.0	733.3
D4	30.9	953.4	984.2
D5	ND	919.0	918.8
D6	ND	1153.4	1153.2
D7	ND	717.6	717.4
D8	ND	348.2	348.0
D9	ND	2.6	2.4
D10	ND	1.8	1.8
D11	ND	0.8	0.8

ND: not detected.

seasons. A large portion of As was associated with the fine fractions (<0.05 mm) of the soils at any depth. As concentration was the highest in the coarsest fractions (>0.50 mm and 0.50–0.25 mm) of soils and decreased with decreasing particle size. Thus, the high As concentrations of the soils in the Guandu plain may result from weathering of As-bearing parent materials which is similar to the conclusions drawn by Shyu *et al.* (2009). The environmental impacts of high As concentrations in soil, its fate and transport in the Guandu plain require further study.

7.2 BIOACCUMULATION OF ARSENIC IN PLANTS AND CROPS

The bioavailability, uptake and accumulation of As are dependent on a number of factors that include source and concentration of the element, its chemical form, soil properties such as clay content, pH and redox conditions, other ions, type and amount of organic matter present, as well as numerous plant-dependent factors, including plant species. The chemical form of As was reported to be more important than the level of As_{tot} concentration in solution in determining phytotoxicity effect on plants (Marin *et al.*, 1992, Carbonell-Barrachina *et al.* 1999).

In Taiwan, rice is one of the most important crops. Although the daily intake rate of rice has decreased gradually from 366 g per capita in 1972 to 133 g per capita in 2005 (Taiwan COA 2005), the daily intake rate of rice still makes up 54.3% of total cereals consumed. Rice is the staple food of Taiwanese.

Irrigation water for rice paddies in Taiwan is mainly provided by local irrigation associations where 85% of water comes from surface river water, 10% from rain water and the remaining 5% from groundwater. The bioaccumulation of As in rice from irrigation water is considered unimportant. However, bioaccumulation of As in plants and crops has received great attention in the hyper-endemic areas of a blackfoot disease (BFD). Lo *et al.* (1983) conducted an extensive survey of As concentrations in rice and other crops. Several crops and vegetables including rice, corn, potato, water spinach and Chinese white cabbage were collected from BFD areas and analyzed for their As concentrations. Samples of the same crops and vegetables from non-BFD areas were also analyzed for comparison. Additionally, 35 and 45 soil samples from BFD and non-BFD areas, respectively, were collected and analyzed. The average As concentrations of different parts of rice plants including: grain, stem, leaf and husk were 1.5, 3.3, 4.6 and 1.6 mg kg^{-1} dry weight (dw), respectively, in BFD areas and 0.5, 0.8, 2.2 and 1.0 mg kg^{-1} (dw) in non-BFD areas. Moreover, the results showed that the average As concentrations in corn, corn leaf, potato, potato leaf, water spinach and Chinese white cabbage were 0.4, 0.9, 0.4, 0.4, 0.7 and 0.6 mg kg^{-1} (dw), respectively, in BFD areas, whereas the average As concentrations in the same plants were 0.2, 0.5, 0.5, 0.3, 0.5 and 0.2 mg kg^{-1} (dw), respectively, in the non-BFD areas. The average As concentrations in soils in the BFD and non-BFD area were 9.9 mg kg^{-1} (range 2.8–15.7) and 4.6 mg kg^{-1} (range 1.2–10.4), respectively; thus, the average As concentration in BFD soil was by 2.1-fold higher than that in non-BFD soil (Lo *et al.* 1983). The detailed results are listed in Table 7.3. Bioaccumulation of As in rice plants, crops and vegetables in the BFD areas were significantly higher than those in the non-BFD areas.

The first nationwide survey of As concentration in rice grains was conducted in 1975. Thereby, 328 samples of rice grains were collected from 86 townships in Taiwan. The results indicated that 95% of the samples contained detectable amounts of As, and the average As concentration was 0.30–0.53 mg kg^{-1} with a maximum value of 1.74 mg kg^{-1} (dw) (Li *et al.* 1979b). Moreover, the As concentrations in rice exceeding 1 mg kg^{-1} (dw) were found in the Pingtung, Kaoshiung, Chiayi, Yunlin and Taoyuan counties, and Chiayi had the maximum value of 1.74 mg kg^{-1} (dw) (Fig. 7.1). Since As concentration in rice plant was greatly concerned in the blackfoot disease hyper-endemic areas, rice plants were collected

Table 7.3.   Arsenic concentrations in soil, rice, crops and vegetables in BFD and non-BFD areas (mg kg^{-1}) (dw) (Lo *et al.* 1983).

	BFD area			Non-BFD area		
	Range	Mean	Sample #	Range	Mean	Sample #
Soil	2.76–15.7	9.91	30	1.22–10.4	4.60	45
Rice grain	0.74–2.03	1.54	14	0.09–2.33	0.55	28
Rice stem	0.62–6.80	3.32	14	0.08–3.78	0.84	28
Rice leaf	1.78–8.58	4.57	14	0.37–6.22	2.21	28
Rice husk	1.22–2.14	1.62	14	0.29–2.86	1.04	28
Corn	0.31–0.54	0.45	13	1.09–0.32	0.21	6
Corn leaf	0.38–1.46	0.87	16	0.28–0.94	0.54	12
Potato	0.31–0.57	0.44	2	0.20–1.14	0.47	7
Potato leaf	0.24–0.60	0.42	2	0.11–0.51	0.34	7
Water spinach	0.31–1.12	0.72	2	–	0.54	1
Chinese white cabbage	0.40–0.80	0.60	2	–	0.20	1

and analyzed in that area (Lo *et al.* 1983). As concentrations in some of the rice grain samples exceeded the Chinese statutory limit of 0.15 mg inorganic As kg^{-1} (dw) (GB2762-2005 China National Standard). Notably, no legal standards for As in food and plants are set in Taiwan.

Schoof *et al.* (1999) conducted a market basket survey of inorganic As (i-As) in food in Taiwan. They analyzed As$_{tot}$ and i-As concentrations in rice grains collected from Taiwan in 1993 and 1995. The As$_{tot}$ concentration and the percentage of i-As in As$_{tot}$ in the polished rice samples were 0.30 mg kg^{-1} (dw) and 59% in 1993 and 0.12 mg kg^{-1} (dw) and 72% in 1995, respectively. Lin, H.-T. *et al.* (2004) conducted a second nationwide survey and collected 137 samples from different rice storage houses and 280 samples from market baskets. The average As concentrations from rice storage houses and market baskets were 0.05 mg kg^{-1} (range 0.01–0.14 mg kg^{-1}) and 0.1 mg kg^{-1} (range 0.01–0.63 mg kg^{-1}). Williams *et al.* (2005) reported that As concentrations of three lots or batches of white long grain rice in Taiwan were 0.19, 0.20 and 0.76 mg kg^{-1}, respectively (all dw).

Table 7.4 shows a compilation of the As concentration in rice from surveys in various countries. Zavala and Duxbury (2008) estimated the normal (uncontaminated) levels of As$_{tot}$ in rice grain as 0.08–0.20 mg kg^{-1} (dw) based on the results of previous studies. If we adopt these data as a preliminary standard, then it appears that As concentrations in some of Taiwanese rice grain samples exceed the normal range.

Particularly, soil contamination in Guandu plain area has received much public attention. The grains of rice grown in As-contaminated soils can potentially accumulate high levels of As. Possible uptake of As by rice plants may play an important role in the transfer of this toxic element into food chains, resulting in potential threats to human health (Meharg and Rahman 2003). The food safety issues raised by consumption of crops produced on As-contaminated soil located at Guandu plain was of great concern to local residents and government. The preliminary results showed that various vegetables, fruits, plants and rice produced in Guandu plain were safe for consumers even though the As$_{tot}$ concentrations in soils were very high, in the range 100–500 mg As kg^{-1}, because the concentrations of bio-available As in soils were extremely low (Shyu and Lin 2006). Yau (2008) reported that the average As concentration in rice grown in the As-contaminated Guandu paddy zone was 0.23 ± 0.08 mg kg^{-1} (dw) (mean ± SD; range 0.11–0.49 mg kg^{-1}). Arsenic concentrations in rice grains only slightly exceeded the normal range, which confirms low bioavailability of As in the Guandu soils. Su and Chen (2008) investigated As fractionation in soils and its relationships with As concentration in rice grain. As$_{tot}$ concentrations in 13 soil samples

Table 7.4. Survey of arsenic concentration in the rice grain produced in different countries.

| Country | As_{tot} in grain in mg kg^{-1} (dry weight) | | Sample number | Reference |
	Min–Max	Mean ± SD		
Taiwan	0.10–0.63	0.10 ± 0.08	280	Lin, H.-T. *et al.* 2004
	0.19–0.22	0.20		Schoof *et al.* 1998
	0.19–0.76	0.38		Williams *et al.* 2005
Thailand	0.06–0.14	0.10 ± 0.01	15	Williams *et al.* 2006
Philippines	0.00–0.25	0.07 ± 0.02	22	Williams *et al.* 2006
Australia	0.02–0.04	0.03 ± 0.00	5	Williams *et al.* 2006
India (basmati rice)	0.03–0.07	0.05 ± 0.00	10	Williams *et al.* 2006
China (Beijing)	0.07–0.19	0.12 ± 0.01	32	Williams *et al.* 2006
Bangladesh	0.03–0.28	0.11	17	Williams *et al.* 2005
West Bengal (India)	0.11–0.40	0.26	7	Roychowdhury *et al.* 2002
Vietnam	0.03–0.47	0.21	31	Phuong *et al.* 1999
United States	0.11–0.40	0.26	7	Williams *et al.* 2005

Table 7.5. Arsenic concentrations in soil and crops in the Guandu plain (mg kg^{-1}) (Su *et al.* 2009).

| Sample point | Arsenic concentration in soil (mg kg^{-1}) | Garland | | Taros | | |
		Surface part	Subsurface part	Surface part	Stem	Root
1	6.33	0.36	40.7	–	–	–
2	68.8	1.30	154.4	–	–	–
3	72.8	0.90	92.1	–	–	–
4	9.81	1.37	109	–	–	–
5	111	0.78	92.7	–	–	–
6	221	4.81	ND	–	–	–
7	224	–	–	0.38	0.16	15.3

ranged widely from 12–535 mg kg^{-1}, but the As concentrations in brown rice were all below 0.35 mg kg^{-1} (dw) and no relationships were found between them. The concentrations of non-specifically-bound As in soil, which represented bioavailable As, were very low (less than 1% of As_{tot}) which may explain the facts that As concentrations in rice grain cultivated in seriously As-contaminated soils were not significantly enhanced, and no adverse effects were shown on rice growth. Pot experiments conducted by Su *et al.* (2009) gave similar results, and showed that carrot cultivated in highly As-contaminated soil was safe for consumers, however, the As concentrations in garland *chrysanthemum* increased with increasing As_{tot} concentrations in soil (Table 7.5). Arsenic concentrations in field-collected white mushroom were elevated, which is inconsistent with the results published by Environmental Protection Bureau of Taipei City Government (Taipei EPB 2006). Therefore, it should be concluded that the crops produced in Guandu plain may pose potential risks for human health and establishing a regular monitoring program of crops is suggested. In addition, the agronomic practices should be carefully considered, since the application of lime materials or phosphorous fertilizers may enhance the bioavailability of As in soil (Wenzel *et al.* 2001). A significant linear relationship was found between specifically-bound As in soil and As concentration in garland *chrysanthemum* ($r^2 = 0.830$, $p < 0.001$). This suggests that specifically-bound As in soil may be used as an indicator for assessing the food safety of vegetable crops cultivated in As-contaminated soil of the Guandu plain (Su *et al.* 2009).

Table 7.6.   Average concentrations of arsenic species in Taiwanese rice (referred to dry weight).

Sample	As_{tot} ($\mu g\ g^{-1}$)	MMA ($\mu g\ g^{-1}$)	DMA ($\mu g\ g^{-1}$)	Sum of As_{org} species (% of As_{tot})	Sum of i-As species ($\mu g\ g^{-1}$)	Sum of i-As species (% of As_{tot})
Rice (grain) (n = 1)[1]	0.76	0.058	0.046	17	0.506	83
Rice (polished) (n = 2)[1]	0.20	0.016	0.030	28	0.118	72
Rice (polished) (n = 5)[1]	0.15	0.013	0.013	19	0.11	81
Rice (polished)[2]	0.19	0.03	0.015	27	0.12	73
Rice (polished)[2]	0.20	0.03	0.02	31	0.11	69
Rice (polished)[2]	0.76	0.05	0.06	18	0.51	82

[1]Schoof *et al.* 1998, [2]Williams *et al.* 2005.

The data on As speciation in the rice grain are limited in Taiwan. Williams *et al.* (2006) and Schoof *et al.* (1998) have collected samples and determined the As speciation in rice grain of Taiwan. The results are listed in the Table 7.6. Inorganic As made up 69–83% of As_{tot} in rice grain, whereas organic As (As_{org}) amounted 17–31% of As_{tot}, indicating a low methylation capability of As in rice grain. Williams *et al.* (2005) reported various amounts of As(V) and As(III) in rice collected from various sites in the world, and the percentage of i-As(V) was 64% for European, 80% for Bangladeshi and 81% for Indian rice. In contrast, DMA (dimethylarsenic acid) was the predominant species of As in rice from USA with only 42% being i-As.

The average As_{tot} concentrations in rice grain in Taiwan were in the range 0.30–0.53 mg kg^{-1} (dw) as surveyed by Li *et al.* (1979b) and 0.11–0.51 mg kg^{-1} (dw) as surveyed by Williams *et al.* (2005) and Schoof *et al.* (1998). In most cases, i-As exceeded the concentration of 0.15 mg kg^{-1} which is the China standard for i-As (GB2762-2005 China National Standard).

Assessment of the health risk caused by the presence of As in rice has largely been based on i-As concentrations because these species have generally been proved more toxic than As_{org} (Fitz and Wenzel 2002). The mean contribution of i-As to As_{tot} in grains was reported by Schoof *et al.* (1998) and Williams *et al.* (2005) to be 70.7%. These authors had surveyed As concentrations in the rice grain in Taiwan. Assuming that i-As makes up 75% of As_{tot} in the rice grain in Taiwan, and that As_{tot} concentrations in rice grains are in the range of 0.30–0.53 mg kg^{-1}, the i-As health effects of the concentrations in the rice grains will be assessed at the concentration level of 0.23–0.40 mg kg^{-1}. This exceeds the standard in China of i-As (0.15 mg kg^{-1}). However, the Taiwanese government has not set up a similar standard yet. Limit data set available in Taiwan indicated there were rice samples exceeded food standards set overseas. It will require a more comprehensive survey of arsenic in rice in Taiwan in order to have a better understanding the extent of potential risk from arsenic-contaminated rice.

CHAPTER 8

Potential threat of the use of arsenic-contaminated water in aquaculture

8.1 INTRODUCTION

Arsenic (As) has been well-documented to be a major risk factor for blackfoot disease (BFD). BFD is an endemic peripheral vascular disease that is frequently observed among inhabitants in a small coastal area of Chianan plain in the southwest of Taiwan, where water in deep wells contains high concentration of As. Extensive epidemiological evidence has proven that drinking groundwater with a high As content is strongly associated with the occurrence of BFD (Ch'i and Blackwell 1968). Chen *et al.* (1994) reported total As levels of well water ranging 470–897 μg L^{-1} from the hyperendemic BFD area in southwestern Taiwan. Ninety-five percent of As compound in the groundwater is inorganic (As(III) and As(V)), and the predominant As species is As(III).

Groundwater has been used abundantly as an alternative to surface water in the coastal region of Chianan plain, Taiwan, where surface water resources are severely deficient because of the high demand of water for domestic, irrigated, aquacultural, and industrial uses. Nowadays, most inhabitants in this area do not drink well water directly. However, very large quantities of groundwater are used for crop irrigation, fish and shellfish aquaculture. Aquatic animals farmed in this region may bioaccumulate large quantities of As from ponding water. Several As species are present in plants and aquatic organisms; these include arsenobetaine (AsB), monomethylarsonic acid [MMA(V)], dimethylarsinic acid [DMA(V)], arsenite [As(III)] and arsenate [As(V)]. Inorganic As species are generally more toxic than methylated As species but monomethylarsonous acid [MMA(III)] is more toxic than As(III) (Oremland and Stotlz 2003).

Arseniasis-endemic areas are referred to the areas situated in southwestern of BFD-endemic area of Chianan plain and northeastern of Lanyang plain of Taiwan (Hsueh *et al.* 2003, Tsai *et al.* 2003, Yang *et al.* 2003, Chen *et al.* 2004) (Fig. 8.1). Budai, Yichu, Hsuechia and Beimen are four townships located in the flat region of Chiayi and Tainan counties in the BFD hyperendemic areas of southwestern Taiwan (Fig. 8.1). The area of four townships is approximately 260.9 km^2.

The pathway of As exposure to residents at BFD-endemic area is mainly through the consumption of farmed fishes mainly including tilapia (*Oreochromis mossambicus*), milkfish (*Chanos chanos*), clam (*Meretrix lusoria*) and mullet (*Mugil cephalus*), whereas in the Lanyang plain, the routes of As exposure include farmed smelt fish (ayu, *Plecoglossus altirelis*) and grasp shrimp (*Penaeus monodon*) consumption and groundwater ingestion. Taiwanese Fishery Annual Report in 2004 (FACOA 2004) indicated the percentages of total area of fishponds in which tilapia, milkfish, mullet and clams are farmed in Budai, Yichu, Hsuechia and Beimen townships amounting to 72.4, 47.4, 69.7 and 82.2%, respectively (Fig. 8.2). The ratios of aquacultural areas of the four townships to the total aquacultural areas in Taiwan of the four species are 14.4% for tilapia, 17.9% for milkfish, 2% for mullet and 13.2% for hardclam, respectively (Taiwan COA 2007). Moreover, the amount of oyster produced in the Budai township consists 35.6% of the total amount produced in Taiwan. Table 8.1 shows the domestic production and export of fish and seafood in Taiwan (Taiwan COA 2004).

Arsenic can be easily uptaken by aquatic organisms and As can also accumulate in animals and human food chains (Cullen and Reimer 1989, Edmonds *et al.* 1997). Humans are exposed

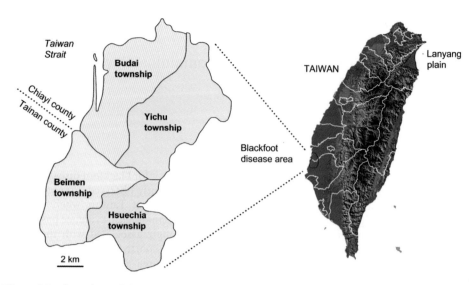

Figure 8.1.   Locations of the BFD area and Lanyang plain.

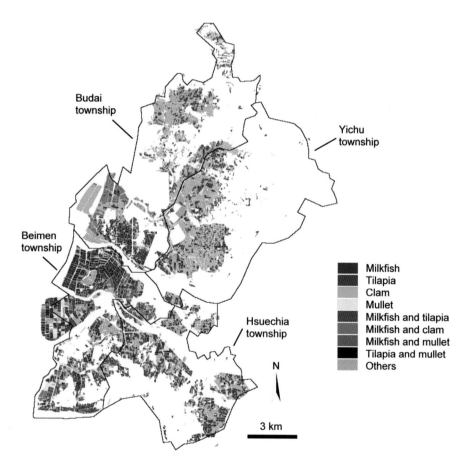

Figure 8.2.   Locations of fish ponds in the BFD areas.

Table 8.1.   Domestic production and export of fish and seafood in Taiwan (Taiwan COA 2004).

Fish and seafood	Domestic production of fish and seafood (M.T.)	Export of fish and seafood (M.T.)
Tilapia	89302	40570
Milkfish	56852	8164
Mullet	3557	96
Hardclam	26223	10

M.T.: million metric tons.

to As from various sources, such as food, water, air and soil; food is a significant source of As to which humans are exposed. The US Food and Drug Agency (US FDA 1993) indicated that approximately 90% of the As in US diet comes from seafood.

Suhendrayatna *et al.* (2002a, b) found that freshwater organisms in Japan accumulated and transformed As in their bodies. Approximately 90% of accumulated As was depurated to water. Arsenical toxicity, determined from the prevalence of carcinogenesis and vascular disorders in an earlier As endemic areas of southwest coast of Taiwan, follows the order of MMA(III) > As(III) > As(V) > MMA(V) ≈ DMA(V) (Lin *et al.* 1998).

8.2   ARSENIC IN AQUACULTURAL ORGANISMS

Five major fish and shellfish are farmed in the BFD area including tilapia, milkfish, mullet, clam and oyster. These species are farmed and harvested manually by hand or fishnet. Large volumes of groundwater with high As concentration are pumped to supply fishpond needs, especially in a dry season of fall and winter (Figs. 8.3 and 8.4). Aquatic species farmed in this area may thus bioaccumulate large quality of As. However, no regulation for As concentration in aquacutural product has been set in Taiwan to date.

8.2.1   *Tilapia*

Farming tilapia (*Oreochromis mossambicus*) is a popular industry in the BFD area because of its high market value. The fishes are fed with commercial feed, which does not contain As. These fishes are maintained in the ponds for at least 6 months (from April to October) before harvest. At present, data on the actual effects of As to tilapia are limited. Generally, the accumulation of metals in aquatic organisms has been linked to decreased survival and reduced reproductive ability. If As levels in pond water are excessively high there may be severe effects on cultured fish, reducing their market price and leading to the closure of fish farms. Suhendrayatna *et al.* (2002b) suggested that tilapia could be used as a bioindicator for studying the bioaccumulation and transformation of As in freshwater organisms.

Arsenic in groundwater indirectly enters the food chain via various paths and bioaccumulates in human bodies. Han *et al.* (1998) found the total As concentrations of 0.13–1.45 µg g^{-1} (dry wt) bioaccumulated in tilapia purchased from Taipei fish market. Huang *et al.* (2003) surveyed As levels in farmed tilapia and pondwater in BFD area. The mean As concentration in pondwater was 48.9 ± 18.4 µg L^{-1} (n = 21). Moreover, total As levels of tilapia obtained from dorsal muscle ranged 0.1–1.4 mg g^{-1} (dry wt) (n = 68), and the inorganic As level was measured to be 7.4% of total As in farmed tilapia from BFD-endemic area. This is the first study conducted in a BFD hyperendemic area to determine the As species levels in tilapia and in water from culture ponds. The As levels of fish significantly increased with those of pond

Figure 8.3.　Five major fish/shellfish, fishpond and farming activities in BFD area (a) tilapia,
(b) milkfish, (c) mullet, (d) clam, (e) oyster and (f) farmed oyster by hand.

water. It appears concentration of inorganic As in water is positively related to the concentration of As species in tilapia (Huang *et al.* 2003).

Liao *et al.* (2003) conducted a field bioaccumulation investigation in tilapia farms in BFD-endemic area and reported that the mean As pond water concentrations ranged from 26.3 ± 16 (mean $\pm$ SD) to 251.7 ± 12.2 µg L^{-1}, whereas the mean As concentrations in tilapia tissues were 29.3, 10.9, 5.37, 5.04, and 3.55 µg g^{-1} dry wt in intestine, stomach, liver, gill, and muscle, respectively. The highest bioconcentration factor (BCF) was found in the intestine and the order of BCFs was: intestine > stomach > liver = gill > muscle (Liao *et al.* 2003). Liu *et al.* (2005) indicated that the inorganic As concentrations in tilapia, summation of the concentrations of As(III) and As(V), ranged from 0.032 to 0.095 µg g^{-1} with a geometric mean of 0.053 µg g^{-1}. Moreover, Chen and Liao (2004) measured the uptake and depuration rate constants of As by tilapia with only water phases in the laboratory and found that BCF was 3.2. Sediment phase was not considered in their experiment. Huang *et al.* (2003), however, estimated an average BCF of 41.8 ± 31.4 based on the As concentration in the pond water and tilapia collected from 21 farmed ponds which contained both water and sediment phases

Figure 8.4.   Fish/shellfish farming activities in BFD area: (a) farmed oyster by rack, (b) harvest fish by hand, (c) by fishnet and (d) extraction of groundwater for fishpond needs.

in the BFD area. To resolve the discrepancy of BCF of tilapia, Wang, S.-W. *et al.* (2007) proposed a first order two-compartment model to describe the uptake and transformation of total As by tilapia in the fish pond involving both water and sediment phases. They measured the total As concentrations in pond water, sediment and tilapia which were 4.4–39.4 µg L^{-1}, 5.7–25 µg g^{-1} and 357.3–1047.9 µg g^{-1}, respectively, in four fish ponds in the BFD area. The As concentrations in the sediment were several orders higher than those in the tilapia. Using the first order two-compartment model including water and sediment phases, Wang, S.-W. *et al.* (2007) were able to resolve the discrepancy and accurately estimated the As concentration of farmed tilapia in the BFD area.

8.2.2   *Milkfish*

Milkfish (*Chanos chanos*) is one of the most commercially important aquacultural species in Taiwan. Most milkfish ponds are located in the southwest coasts of Taiwan, where most part of the area is situated in the BFD area. Milkfish culture needs a large amount (38,000–49,000 tons ha^{-1}) of freshwater. Groundwater is used for aquaculture because river water in this area is polluted. The bioaccumulation of As in milkfish and the potential hazards of As in the aquacultural environment were assessed (Lin, M.-C. *et al.* 2004).

Since milkfish is a common seafood in Taiwan, ingestion of contaminated fish could result in As accumulation in human bodies and lead to adverse health effects. Lin, M.-C. *et al.* (2004) collected samples of farmed milkfish and ambient water from aquacultural ponds in BFD-endemic area and reported that the averaged As concentration in the pond water was 65.5 ± 7.6 µg L^{-1} (mean ± SD), whereas averaged As level in juvenile and adult milkfish tissue were 15.2 ± 5.1 and 0.7 ± 0.7 (dry wt) µg g^{-1}, respectively. The values of BCF of the juvenile and adult milkfish were 556.16 ± 187.98 (n = 63) and 11.6 ± 4.4 (n = 36), respectively, indicating that the former has a higher bioaccumulation effect of As. Thus, ingestion of juvenile milkfish may cause a higher risk to health. Lin and Liao (2008) further analyzed the total and inorganic As levels in milkfish ranging from 0.21 ± 0.02 to 3.35 ± 0.32 and 0.11 ± 0.03 to 1.69 ± 0.74 mg kg^{-1}, respectively. The ratio of inorganic As to the total As was $44.1 \pm 10.2\%$ in the milkfish and is much higher than other fish and shellfish. Donohue and Abernathy (1999) reported that the amount of inorganic As in general seafood ranged from <3–7% of the total As.

8.2.3   Mullet

Mullet (*Mugil cephalus*) is a detritivorous fish and has several edible tissues with high economic values. The ovary of mature female mullet produces valuable mullet roe. Many wild mullets swim through the ocean around Taiwan from the middle of November to early February. Aquacultural mullet grows in inland ponds all year round. Farming of large-scale mullet is a promising aquaculture in the BFD area in Taiwan because of high market value. Fishermen collect juvenile mullet from the sea, and keep the fish in ponds. These fish are maintained in the ponds for 8 months (from March to October) before they are sold in the market. According to statistical data in the Taiwanese Fishery Annual Report in 2002 (http://www.fa.gov.tw/), the total area of fishponds farmed mullet is 613 hectares (Lin 2004).

   Arsenic can be accumulated in fish tissues, therefore the As accumulation in the fish from the ambient water is necessary to be assessed to predict the risk to human health. Lin *et al.* (2001) have sampled ambient water from aquacultural ponds in Budai and Yichu located in BFD endemic area indicated that the mean concentrations of As in the culture ponds was 49 µg L^{-1} (range: 13.0–169.7 µg L^{-1}). Liu *et al.* (2006b) surveyed the contents of As species in edible tissues of mullet in the region of BFD area in Taiwan. The mean total As contents of the testis, ovary, stomach and muscle of mullet were 12.94, 9.17, 7.35 and 3.28 µg g^{-1} (dry wt), respectively. These results were similar to those of Kirby and Maher (2002) and Maher *et al.* (1999), who investigated mullet in the seas around Australia. The BCF of the aquacultured mullet is about 122 based on the As concentration of 49 µg L^{-1} in aquaculture water. Mullet is a benthic feeding fish. Sediment is ingested when mullets eat. The total As concentrations in pond sediment which ranged 20.2–44.9 µg g^{-1} in the BFD areas markedly exceeded those in sea water that is unpolluted by As (Wang 2003). The average inorganic As content to the total As in mullet was 1.6%, indicating a high ability of methylation of inorganic As species. Maher *et al.* (1999) reported that inorganic As level accounts for 3.4% of the total As in mullet in Australia.

8.2.4   Clam

Clams live at the interface between sandy sediment and water. Sediment in farmed ponds is full of organic material and can adsorb large quantities of heavy metals and metalloids (Baudrimont *et al.* 2005), including As. Furthermore, clams are sediment-dwelling animal and are cultivated in a brackish aquatic environment. Seawater and groundwater are primary water resources in the farmed ponds with a mixture ratio of 1:2. When farmers pump the groundwater to cultivate clams, the pond water and sediment become As-polluted. Hence, clams may bioaccumulate As from sediment and pond water. Many studies have indicated that the inorganic As level in clams exceeds those in other bivalves, such as oysters, scallops and others (Edmonds and Francesconi 1993, Lai *et al.* 1999, Muñoz *et al.* 2000). Clams and oysters in Taiwan represented 76.4% of the shellfish offered for sale in 2003. From 1994 to 2003, the number of clams as a percentage of the number of shellfish offered for sale increased from 32.6% to 43.9% and the increases were mostly from eastern Taiwan where no elevated As concentration was found.

   Liu *et al.* (2007) surveyed the total As and As species contents in clams (*Meretrix lusoria*) farmed in areas of BFD in southwestern Taiwan. Total As content in clams varies 0.21–3.25 µg g^{-1} (wet wt). This range is wider than that the normal As content in blue mussels obtained through the monitoring program in Norway (1.3–2.8 µg g^{-1} (wet wt)) (Airas *et al.* 2004) and in Mid-Atlantic (USA) (0.93–1.53 µg g^{-1} (wet wt)) (Greene and Crecelius 2006). The average total As contents in medium-sized and small clams were 7.62 and 10.71 µg g^{-1} (dry wt), respectively. The average ratios of inorganic As contents to total As contents in clams ranged 12.3–14.0% which are much higher than that found in the farmed oyster (*C. gigas*), indicating that humans may expose to larger quantities of inorganic As by ingesting the same amount of clams as oysters.

The As bioaccumulation in farmed clams is not linearly related to the exposure environment. The BCF from sediment ranges 0.74–0.35 and the BCF from pond water ranges 363.8–164.5. The BCF at an equilibrium, of internal biota concentration to exposure concentration, describes partitioning between the exposure medium (water and sediment) and clams. Few BCF measurements are available and the mechanisms that involve adsorption/desorption in sediment are complex. Accordingly, the inter-relationship of As content in clams to that in sediment and pond water is unable to be determined. McGeer *et al.* (2003) found an inverse relationship between BCF and the concentrations of metals in the aquatic environment. Additionally, the metal BCFs derived from sediment studies tend to be orders of magnitude lower than those from aquatic environments (McGeer *et al.* 2003). The ratios of BCF derived from pond water and sediment of two ponds by Liu *et al.* (2007) were 492.3 and 467.5, respectively and exposure to an environment with high As results in low BCF and *vice versa*. The adoption of BCF as an indicator of the long-term hazard potential was due to the sparse data on the chronic toxicity of As for clam. As such, BCF is of most value when limited data are available (OECD 2001).

8.2.5   Oyster

Oysters are the most popular shellfish in Taiwan and the major farmed areas are off the coast of southwest Taiwan, so the As content in oysters has received much attention. Han *et al.* (1997, 1998) collected oyster samples from the different coasts of Taiwan during the period from August 1991 to July 1998 and measured the total As concentrations of oysters in Budai (BFD area), obtaining values of 12.3–21.4 µg g^{-1} in 1997 with a geometric mean of 18.7 µg g^{-1}, and 3.15–7 µg g^{-1} in 1998 with a geometric mean of 4.86 µg g^{-1}. Han *et al.* (1998) used a assumed 10% inorganic to total As content to assess the potential health risks from consuming oyster and reported a high cancer risk of 5.14×10^{-4}. Guo (2002) critically examined the work of Han *et al.* (1998) and concluded that insufficient and improper data and assumptions were responsible for the unrealistic results.

Liu *et al.* (2008) comprehensively surveyed the total As and As species contents of oysters in four townships in the BFD area over a 14-month period, indicating that the highest and lowest total As contents were 22.90 ± 7.14 and 4.22 ± 0.38 µg g^{-1} (mean ± SD). The average inorganic to total As contents is 1.5%. The spatial concentration distributions (mean ± SD) of total As concentration in oysters were 7.90 ± 1.80, 10.68 ± 4.57, 8.71 ± 1.30 and 9.78 ± 1.90 µg g^{-1} in Anping (Tainan city), Budai (Chiayi county), Dongshih (Chiayi county) and Wangkung (Changhua county), respectively (Fig. 7.1). Budai had the highest mean total As concentration whereas Anping had the lowest one. The total As concentration of oyster in Budai ranged 3.11–33.37 µg g^{-1} with a geometric mean of 9.99 µg g^{-1}, suggesting that an annual temporal effect may apply. However, the seasonal variation of the total As concentration in oysters was statistically insignificant. Notably, the BCF of the oyster farmed in Anping, Budai, Dongshih and Wangkung were 5550, 7790, 6390 and 6740, respectively, based on the As concentration 1.5 µg L^{-1} in seawater. Sanchez-Rodas *et al.* (2002) analyzed the same species of oyster (*Crassostrea gigas*) farmed off the Atlantic coast of Spain and showed that the total As concentration (mean ± SD) in oyster was 17.24 ± 0.25 µg g^{-1}. Vilano and Rubio (2001) also analyzed the same species of oyster farmed in northwest Spain. They found a total As concentration (mean ± SD) in oyster of 9.74 ± 0.37 µg g^{-1}. Kohlmeyer *et al.* (2002) examined the same species of oyster farmed in the Arcachon Bay of France and found a total As concentration (mean ± SD) of 26.7 ± 0.5 µg g^{-1}.

Edmonds and Francesconi (1993) reported that 1.4% of the total As in oysters (*C. gigas*) from Japan was inorganic As. Kohlmeyer *et al.* (2002) showed 3% of the As in oysters (*C. gigas*) from the northwest of Spain was inorganic. 1.5% reported for oysters from the BFD area (Lin 2004) is between these values for Japan and Spain.

8.2.6   *Arsenic levels in groundwater and farmed fish/shrimp in Lanyang plain*

Because of the abundance of underground water in the area, residents in Lanyang plain have been using groundwater from shallow wells since the late 1940 in that As levels in groundwater ranged from undetectable (<0.15 µg L^{-1}) to 3590 µg L^{-1} (Chiou *et al.* 2001). The variation of As level in groundwater at Lanyang plain was much more striking than that of the BFD-endemic area (Fig. 8.5). They further pointed out that the main exposure to inorganic As of local residents in Lanyang plain was through groundwater ingestion. Lee *et al.* (2008a) reported that the mean As level in groundwater was 110 µg L^{-1} ranging from undetectable (<0.5 µg L^{-1}) to 1010 µg L^{-1} of 40 monitoring wells and the average inorganic As level exceeded 90% of the total As in groundwater.

Although the implementation of a tapwater system started in Lanyang plain in the 1990s, some residents around 50% still drank the As-contaminated groundwater or used it for

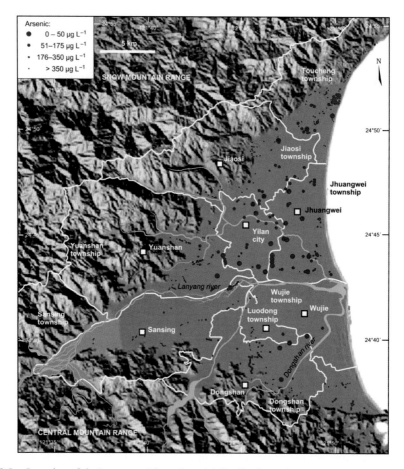

Figure 8.5.   Location of the Lanyang plain and spatial distributions of measured arsenic concentrations in groundwater.

aquaculture (Lin and Chiang 2002). A field investigation was conducted to analyze the As levels in groundwater located at Dongshan and Wujie in Lanyang plain, indicating that there were 64.4% of 180 groundwater samples were higher than the Taiwan drinking water standard (10 µg L^{-1}), and the maximum concentration of As was 1145 µg L^{-1}. Lin and Chiang (2002) collected samples of smeltfish (ayu *Plecoglossus altirelis*) and grass shrimp (*Penaeus monodon*) from seven culture ponds in Lanyang plain and reported that the levels of As in smeltfish and grass shrimp were 25.6 and 16.65 µg g^{-1} (dry wt), respectively. Smeltfish is an aqucultural product with high economic value in Taiwan. The yield of smeltfish in the Lanyang plain is as high as 92% of total yield of smeltfish in Taiwan, and this proportion is increasing annually (Taiwan FACOA 2005). Most smeltfish is consumed domestically and the amount consumed increases annually (Taiwan FACOA 2005). Smeltfish is farmed in freshwater ponds fed by groundwater and spring water. However, groundwater has become the major water source for ponds because the water temperature is stable and it is extracted easily (Han 2003). Since the available information is scarce, Ling and Liao (2007) adopted the data from Japan and assumed that inorganic As is also 5% of the total As in smeltfish. For shrimp, on the other hand, Larsen *et al.* (1997) reported that inorganic As was estimated to be 1.6% of the total As in shrimp.

8.3   ARSENIC METHYLATION CAPABILITY

After inorganic As exposure, As(V) is readily reduced to As(III) in red blood cells and subsequently methylated to MMA and DMA in human bodies (ATSDR 2007). Metabolism in fish and shellfish seems similar to that in human bodies (Maher *et al.* 1999). Most As species in pond water in inorganic forms are As(III) and As(V). We used the ratio of inorganic As to total As as an indicator to illustrate the percentage of As species, which had been subjected to methylation process in fish and shellfish (Table 8.2). The inorganic As to total As follows the order: milkfish > clam > tilapia > mullet > oyster. Percentage of inorganic As in milkfish was much higher than that in other fish and shellfish, when they were exposed to As in the same aquatic environment. Milkfish has high fatty tissue in the belly which may accumulate inorganic As species and prevent As from methylation, resulting in significant higher inorganic As content. Clam has the second highest inorganic to total As ratio indicating that clam might have a lower ability to convert the inorganic As into organic forms. Inorganic As methylation capabilities of the clams were lower than those of tilapia, mullet and oyster. Mason *et al.* (2000) reported that low-trophic biota accumulated more mercury in the body and has less methylmercury percentage. Low-trophic biota accumulated toxic As from the aquatic environment because they cannot methylate inorganic As to MMA or DMA efficiently. In general, methylation capabilities of fishes are considered higher than those of clams when they are cultivated in similar aquatic environment except for milkfish.

Table 8.2.   Statistics concerning inorganic arsenic in fish and shellfish in the BFD area.

Fish and shellfish	Sample number	Inorganic As content[1] (µg g^{-1})	Inorganic As/total As (%)
Tilapia[2]	68	LN (0.03, 2.94)[7]	7.31 ± 7.18
Milkfish[3]	36	LN (0.21, 2.43)	44.1 ± 1.02
Mullet[4]	24	LN (0.03, 1.81)	1.56 ± 0.99
Oyster[5]	112	LN (0.12, 2.11)	1.51 ± 0.92
Clam[6]	21	LN (0.83, 2.33)	13.50 ± 6.94

[1] inorganic As is the summation of As(III) and As(V); Values were taken from: [2] Huang *et al.* (2003), [3] Lin and Liao (2008), [4] Liu *et al.* (2006b), [5] Liu *et al.* (2008), [6] Liu *et al.* (2007). [7] LN(μ_g, σ_g) denotes a log-normal distribution with a geometric average of μ_g and a geometric standard deviation of σ_g.

Total As levels in oysters were higher than those in fishes and clams, but the percentages of inorganic As were lower than those in the other organisms. Total As levels in clams and oysters are all within the FDA safety level in regulations and guidance (US FDA 2001). The percentages of As species different in the various species of fish/shellfish sampled and determining the percentage of As species especially for toxic inorganic As in aquacultural species provides a more rational way to assess As-induced human health effects through seafood consumption.

8.4   HEALTH RISK ASSESSMENT

Potential human health risks associated with inorganic As uptake from various kinds of seafood have been evaluated by Han *et al.* (1998) and Liu *et al.* (2005, 2006a, b, 2007, 2008) for Taiwan. The methodology for estimation of target risk (TR) of cancer used was provided in "US EPA Region III Risk-Based Concentration Table" (US EPA 1998). It is expressed as excess probability of contracting cancer over a lifetime (70-year). According to the report of US EPA (1998), the dose calculations were made using the standard assumption for an integrated US EPA risk analysis, including exposure over an entire 70-year lifetime and to a 65-kg body weight for an average Taiwanese adult. In addition, it was assumed in accordance with the US EPA guidance (US EPA 1989) that the ingested dose is equal to the absorbed contaminant dose and that cooking has no effect on the contaminants (Cooper *et al.* 1991).

An integrated health risk assessment through ingestion of the five fish and shellfish in the BFD area was conducted herein. According to the geographical information system (GIS) database obtained from Taiwan Fishery Agency (2003), approximately 30% of land uses in the BFD townships are aquacultural ponds. For the inland aquaculture, 37.8, 18.4, 2.5 and 7.6% of fish ponds farm milkfish, tilapia, mullet and clam, respectively. The ratio of aquacultural areas in the BFD region to the total aquacultural areas in Taiwan for the aforementioned species are 22.8% for oyster, 17.9% for milkfish, 14.4% for tilapia, 13.2% for clam and 2.0% for mullet (Taiwan COA 2007). Moreover, the amount of oysters produced in the Budai township counts for 35.6% of the total amount produced in Taiwan.

Inorganic As contents measured in tilapia, milkfish, mullet, oyster and clam were obtained from Huang *et al.* (2003), Lin and Liao (2008), Liu *et al.* (2006a, b), Liu *et al.* (2008) and Liu *et al.* (2007), respectively. Table 8.2 provides the statistics of inorganic As contents to total As in the tilapia, milkfish, mullet, oyster and clam farmed in As-affected groundwater areas. Clam has the highest inorganic As content (0.83 µg g^{-1}), while mullet has the lowest one (0.03 µg g^{-1}). Moreover, for average ratios of inorganic As contents to total As contents, milkfish presents the highest ratio of 44.1%, whereas oyster is the lowest one with 1.51%.

Inorganic As contents in various fish and shellfish were probabilistically treated. Daily food consumption data of aquacultural species for the general population of Taiwan were adopted from the Taiwanese Food Supply and Demand Annual Report of 2006 (Taiwan COA 2006). The daily intake of inorganic As from each fish and shellfish species was calculated by multiplying the individual concentration in each species with the average amount of fish and shellfish consumed by public. Total intake of inorganic As was obtained by summing the products for five aquacultural species.

Other parameters including the daily ingestion rate (IR) of five fish and shellfish per person, the ratio of the edible weight to the total weight of fish and shellfish (α), the average water contents in fish and shellfish (ω) and the ratio of farmed fish and shellfish in the BFD area to the total farmed area in Taiwan were listed in Table 8.3.

Kolmogorov-Smirnov (K-S) test was first utilized to determine the best fitting probability distributions of inorganic As contents. Subsequently, the Monte Carlo technique was employed to characterize uncertainty based on the determined best-fitting distribution of inorganic As contents in fish and shellfish. Probabilistic analyses of the daily intake by consumer corresponding to the 50th and 95th percentiles were conducted.

Table 8.3.   Parameters used for cancer risk estimation.

Fish and shellfish	Tilapia[5]	Milkfish[6]	Mullet[7]	Oyster[8]	Clam[9]
IR[1] (g day^{-1} wet wt)	4.07	3.49	0.19	2.48	1.22
α^2	0.32	0.32	0.32	0.2	0.2
ω^3	0.77	0.77	0.76	0.85	0.83
A[4]	0.641	0.41	0.43	0.49	0.17

[1] IR is the daily ingestion rate, Taiwan COA (2007); [2] α is the ratio of the edible weight to the total weight of fish and shellfish; [3] ω is the ratio of water content in fish and shellfish; [4] A is the ratio of the individual production of tilapia, milkfish, mullet, oyster and clam in Chiayi and Tainan counties to total production of those from Taiwan area; Values were taken from: [5] Huang *et al.* (2003), [6] Lin and Liao (2008), [7] Liu *et al.* (2006a), [8] Liu *et al.* (2008), [9] Liu *et al.* (2007).

Table 8.4.   Estimated target risk (1×10^{-6}) for various aquacultural species with different percentils of TR in the black foot disease area.

| Species | Percentiles | | | | |
	95th	75th	50th	25th	5th
Tilapia	6.55	2.28	1.08	0.50	0.18
Milkfish	30.57	12.79	6.87	3.68	1.56
Mullet	0.19	0.10	0.07	0.05	0.03
Oyster	10.3	4.94	2.93	1.75	0.85
Clam	40.09	17.49	9.67	5.35	2.35

The estimated TRs for different aquacultural species with various level percentiles were calculated (Table 8.4). The TRs of clam and milkfish all exceed 10^{-6}. TR values of clam are significantly higher than other aquacultural species at all percentiles. Milkfish has the second highest TRs for all percentiles. On the other hand, mullet has the lowest TR at the all percentiles and all less than one millionth. The values of TR show that consumption of farmed aquacultural species of As-affected groundwater area might cause an overexposure of inorganic As and pose potential cancer risks to human health.

Three key factors influence upon the results of the health risk assessment: (1) the bioaccumulation of inorganic As contents in fish and shellfish, (2) As concentrations in groundwater and (3) ingestion rates. The bioaccumulation of inorganic As contents in fish and shellfish is associated with BCF and ratios of inorganic As contents to total As contents (β). Clam has the highest BCF and the second highest β, and milkfish has the highest β, causing high potential risks to human health upon ingesting clam and milkfish. Although the geometric means of BCF and β in tilapia are in the middle range among the species, tilapia has high geometric standard deviations of BCF and β and the highest ingestion rates, indicating that high risks and wide risk ranges occurred at the 95th percentile of TR. Mullet with the lowest ingestion rates and β leads to a lowest risk to human health. High As concentrations in groundwater are mainly distributed in the southwestern and western coastal regions and the central regions, which enhances risks to human health. Particularly, clam is mostly farmed in the western coastal regions and milkfish is mainly cultivated in Beimen (BFD area) (Fig. 8.1), the southwestern and western coastal regions of Taiwan. Low As concentrations in groundwater are mainly situated in the southern and northern regions, where tilapia is principally farmed, and hence reduce risks to human health.

In northeastern Taiwan, Lee *et al.* (2008b) spatially analyzed potential carcinogenic risks associated with ingesting As from farmed smeltfish ayu (*Plecoglossus altirelis*) from the Lanyang plain. Sequential indicator simulation (SIS) was adopted to reproduce As exposure distributions in groundwater based on their three-dimensional variability. A target risk (TR) of cancer associated with ingesting As in aquacultural smeltfish was employed to evaluate the potential risk to human health. Safe and hazardous aquacultural regions were mapped to elucidate the safety of groundwater use. The *TR* values determined from the risks at the 75th percentiles exceed one millionth, indicating that ingesting smeltfish that are farmed in the highly As-affected regions represents a potential cancer threats to human health. The 75th percentile of *TR* is considered in formulating a strategy for the aquacultural use of ground-water in the preliminary stage.

This investigation focuses on the use of groundwater with various levels of As exposure which induces the potential risk of cancer associated with the consumption of aquacultural smelt-fish. Since TR values exceed one millionth, the use of groundwater in aquaculture should be reduced or the use of spring water considered. Groundwater is a major water source that meets the needs of aquaculture in the Lanayng plain. Regarding the development of aquacultural smeltfish businesses, Lee *et al.* (2008b) suggested that smeltfish aquaculture should be relocated to other safe regions, or shallow groundwater or spring water should be used.

Lin, H.-T. *et al.* (2008) analyzed total As and As species content in cephalopod, small and large shape fish purchased from the market in Taiwan. Notably, these types of fish and shellfish are not farmed in the BFD area. The predominant As compound found in samples was AsB. Arsenic(V) was detected in low concentrations (range in 0.02–0.34 mg kg^{-1}, wet wt), whereas DMA, MMA and As(III) contents were undetectable. Average AsB content in cephalopods, small fish and large fish were 5.42, 1.57 and 1.54 mg kg^{-1} As (wet wt), respectively. The weekly intake of As was calculated based on the consumption of fish by Taiwanese residents. The calculated results demonstrated that the intake is 30.6 µg kg^{-1} (total As) body weight/week, higher than the acceptable weekly intake of 15 µg kg^{-1} body weight/week for inorganic As that was suggested by WHO. However, around 87% of As in fish muscle was AsB is non-toxic and non-carcinogenic to human, and is rapidly excreted after ingestion. Lin, H.-T. *et al.* (2008) suggested that the intake of fish muscle from non-BFD area is low risk in Taiwan based on the market basket survey.

The benefits of fish and shellfish consumption are on a wide range of public health endpoints. If farmed fish and shellfish are not contaminated by As, they are the healthy food with valuable nutrients, such as omega-3 polyunsaturated fatty acid and muscle proteins, that are well known to have certain benefits to human health. However, fish and shellfish consumption involves potential risks and benefits, both risk and benefit information should provide to the public. To accurately assess the risk/benefit ratios from consumption of farmed fish and shellfish are complicated, cautious interpretation of present data may substantially reduce the likelihood in dealing with uncertainty and risk management. The first case in which theoretical human health risks for consuming As-contaminated farmed fish and shellfish and groundwater in the arseniasis-endemic areas are alarming under a conservative condition based on a probabilistic risk assessment framework. The probabilitstic integrated assessment (PIA) framework—probabilistic physiologically based pharmacokinetic/pharmacodynamic (PBPK/PD) model together with risk diagrams—is an effective representation of state-of-the-art results of scientific assessments for human As exposure through consumption of contaminated farmed fish/shellfish and groundwater. Despite the great uncertainty in many aspects of integrated assessment, the As toxicity, its concentration in farmed species or groundwater, and daily ingestion rates that may modify the outcomes of risk estimate, cautious interpretation of observations obtained from current epidemiological data can substantially reduce this likelihood. Although the suitability and effectiveness of techniques for presenting uncertain results is context-dependent, the probabilistic framework and methods can be taken seriously because they produce general conclusions that are more robust than estimates made with a limited set of scenarios or without probabilistic presentations of outcomes. Besides, the predictive risk modeling technique also offers a risk-management framework for future discussion in deriving risk thresholds for human As exposure.

CHAPTER 9

Current solutions to arsenic-contaminated water*

9.1 INTRODUCTION

In aqueous environments, the speciation of arsenic (As) is complicated and depends on the water chemistry. There are four stable valence states for As (+5, +3, 0, and −3). In natural water systems, the pentavalent As(V) and trivalent As(III) are most common. Two inorganic salts, arsenate and arsenite, and two organic acids, monomethylarsinic acid and dimethyl-arsinic acid, were reported to be the major species in natural groundwater; however, the two organic acids are normally less than 1 µg L^{-1} (Chen et al. 1994, Edwards 1994, Kondo et al. 1999).

The speciation of As may strongly affect its removal from water. For example, arsenate is generally easier to remove than arsenite in many treatment processes, such as coagulation/ sedimentation and adsorption. Under natural pH conditions, arsenate is present in a nega- tively charged ionic form and arsenite is in a non-ionic form (Xu et al. 1991). Therefore, the anionic species, arsenate, has stronger interaction (specific binding) with the metal oxides and anion exchange resins commonly used in treatment processes. Since the type of As depends upon the local water chemistry, the ratios of As(V) to As(III) may vary over a wide range depending upon the geological conditions. For example, in Budai township, a blackfoot dis- ease epidemic area in southwestern Taiwan, the As(III) to As(V) ratio was in the range of 1.1 to 5.2 (2.6 on average) (Chen et al. 1994), while that in Yilan county in northeastern Taiwan and Chiayi county in southwestern Taiwan was 0.4–2.6. The ratio was between 0.068 and 2.5 in the Fukuoka area in Japan (Kondo et al. 1999) and was 0.5 on average in the USA (Edwards 1994).

The particulate and dissolved types of As are also an important factor in the As removal process. In the USA, an average of 39% of the As was the particulate type (>0.45 µm) (Edwards 1994), while in Budai only 3% was particulate (Chen et al. 1994).

Arsenic problems in drinking water can be managed in three ways; change of water source, removal of As at the central system, and the application of point of use/entry (POU/E) systems. In this chapter, the three approaches are introduced.

9.2 CHANGE OF WATER SOURCE

Alternate sources of supply may offer lower cost and simple solutions to resolve As problems in drinking water. These may include locating surface water sources and drilling new wells. The water may also be blended to the existing water sources (Rubel 2003). Since surface water generally contains much less As, replacement of As-laden well water with a public water supply, especially with surface water as source water, is feasible in many areas. For example, much of the groundwater in both northeastern and southwestern coastal areas of Taiwan contains a high concentration of As (Lin 1999). In the southwestern coastal area, As-laden groundwater was discovered to be responsible for blackfoot disease in the 1950s. Reservoirs were built in the area and people were provided with a public water supply prior to 1970.

---

* This chapter was prepared in collaboration with Yi-Fong Pan, PhD student, Department of Environ- mental Engineering, National Cheng Kung University, Tainan City, Taiwan

Later, in the 1990s, the groundwater in a few townships in the northeastern area of Taiwan was also found to have high concentrations of As. Within 2 years, all the people in the area were supplied with public water (Chen 2004). The change of water supply was much faster (only 2 years) as compared to the previous case (>10 years), primarily due to better economic condition in the 1990s, compared to the 1960s.

Although the replacement of groundwater with surface water is one of the quickest ways to mitigate As in drinking water, providing a public water supply and/or switching to a surface water source may not be feasible in many countries. It may be limited by financial issues or a lack of reliable surface water sources. For example, groundwater with a lower As content, typically 20–39 μg L^{-1} (though as high as 300 μg L^{-1}) in a part of southwestern Taiwan is still used in a few small water treatment plants (Lin 1999). In addition, As-laden groundwater is still used in some remote areas for drinking. Therefore, other approaches, including blending of different well water sources to achieve a lower As concentration and the treatment of As-laden water, are still needed.

9.3   WATER TREATMENT PROCESSES FOR CENTRALIZED SYSTEMS

Conventional water treatment processes may not be able to remove As to reach the desired level, which is 10 μg L^{-1} according to Taiwan Drinking Water Standards (Taiwan EPA 1998), WHO guidelines (WHO 2008), and Maximum Contaminant Levels (MCLs) of the US EPA (US EPA 2009). Lin (1999) surveyed the As concentrations and removal efficiency in 10 water treatment plants (WTPs) using groundwater in southwestern and northeastern Taiwan. In all the WTPs, conventional treatment processes, mostly aeration, pre-chlorination, sedimentation, filtration, and chlorination are used. Field samples showed that 0 to 70% of the As (typically less than 10–20%) was removed in the plants. The As concentrations in two of the WTPs were higher than 100 μg L^{-1}, and three were around 20 μg L^{-1}. Hence, the finished water would not be able to comply with the 10 μg L^{-1} standard. Therefore, more rigorous methods should be used to reduce the As concentration in the finished water.

The processes commonly used for the removal of As can be categorized into three groups: coagulation/precipitation, adsorption/ion exchange, and membrane technology. The three methods are discussed in the following sections 9.3.1 to 9.3.3.

9.3.1   *Precipitation methods*

In precipitation-based methods, aluminum sulfate (alum), iron-based coagulants, and lime are three groups of chemicals commonly used for the removal of As. In applying alum and ferric chloride, coprecipitation with the floc and adsorption onto metal oxides are both important mechanisms for removing As (Edwards 1994). Ferric chloride often outperforms alum, primarily due to its stronger floc formation and better adsorption affinity for As (Cheng *et al.* 1994). In the (co)precipitation process, pH is an important factor governing the removal efficiency. The pH needs to be below 7.0 for alum coagulation, while a wider range of 5.0–7.8 is acceptable for ferric chloride (Edwards 1994, Hering *et al.* 1996). Under most pH conditions, the removal of As(V) is better than that of As(III), as the anionic form of As(V) has a higher affinity for the positively charged surfaces of metal oxides. Therefore, pre-oxidation using chlorine, ozone, or permanganate may improve the removal if As(III) is a major portion of the total dissolved As in the water. McNeill and Edwards (1997a) proposed a general approach for improving As removal using coagulation/precipitation processes. The first step is to oxidize As(III) to As(V) if As(III) is a major fraction. Then, the operation can be optimized by changing the pH or alum dosage. For alum, changing the pH from 7.3 to 6.8 may improve the efficiency by up to 100% (As removal increases from 21 to 43%). Finally, if the finished water cannot meet the standard, replacement of alum with ferric chloride may be needed.

The removal of As using lime softening also involves adsorption, coprecipitation, and precipitation (McNeill and Edwards 1997b). The efficiency of the method is often between 20 and 100%. Compared to other precipitation-based methods, lime softening cannot obtain very low As concentrations by itself (US EPA 2005). To obtain better removal, the pH needs to be over 10.5 (US EPA 2005). At this pH, coprecipitation with calcium carbonate may remove about 40% of the As. If a higher removal efficiency is required, the pH may need to be increased to 11.0–11.5, at which point magnesium cations in the water would form magnesium hydroxide and remove with As(V) by coprecipitation.

Since iron oxide is a stronger adsorbent of As, it may enhance the removal of As in many instances. For example, in many groundwaters, iron may be present at high concentrations. The removal of iron may also result in As removal (Sorg 2007). In addition, during lime softening, the addition of ferric salts may improve the removal efficiency (McNeill and Edwards 1997b). US EPA (2005) proposed a rule-of-thumb guidance for the simultaneous removal of As and iron from source water. For water with a high iron concentration (>3 mg L^{-1}) and a high Fe/As ratio (>20), As can be removed from the source water during the iron removal process.

9.3.2   *Adsorption and ion exchange methods*

Among the treatment processes appropriate for the removal of As, adsorption and ion exchange are considered to be less expensive than membrane separation. Many granular adsorbents have been shown to be effective for the removal of inorganic As species. These adsorbents include activated alumina (AA) (Clifford 1990, Lin and Wu 2001), iron-oxide-modified AA (Westerhoff *et al.* 2008), titania-based adsorbents, including MetsorbG and Adsorbsia (Westerhoff *et al.* 2008), and many iron (hydro)oxide-based adsorbents, including commercial media like Aqua-Bind MP, Bayoxide E33 ferric oxide, Granular Ferric Hydroxide (GFH), and MEDIA G2 (Driehaus *et al.* 1998, Lin, T.-F. *et al.* 2006, Petrusevski *et al.* 2007, Westerhoff *et al.* 2008). In addition, there are many low-cost materials under development, including iron-based sand (Joshi and Chaudhuir 1996), iron-based diatomite (Jang *et al.* 2006), and many other materials based on agricultural and industrial by-products (Malik *et al.* 2009). The uptakes of As onto the low-cost iron-based adsorbents are generally low, often <<1 mg As/g^{-1} adsorbent. Low As capacity shortens the service life of the materials, limiting the usage of these techniques in the field. However, some adsorbents were shown to have a high As capacity, at a level similar to those of commercial products. For example, iron-oxide-coated diatomite and goethite were shown to have capacities of around 10 mg As/g^{-1} adsorbent, which is in the range of the capacities of E33 (Lin, T.-F. *et al.* 2006) and GFH (Driehaus *et al.* 1998). High As uptake allows for a long service life for the treatment systems. For example, in a few WTPs in Germany, the treated volumes of water were as great as 30,000 bed volumes (BV) (Driehaus *et al.* 1998). This is certainly much longer than that for iron-oxide-coated sand, which is in the order of 200–800 BV (Joshi and Chauduri 1996, Viraraghvan *et al.* 1999).

Although the materials are called adsorbents, the interaction is a form of ion exchange. Therefore, the effectiveness of As removal is strongly influenced by the pH and competing ions. Lin and Wu (2001) and Lin, T.-F. *et al.* (2006) observed that the arsenate uptake decreases with increasing pH of the water for AA and E33. Since arsenate in the aqueous phase is present as the anionic forms, $H_2AsO_4^-$ and $HAsO_4^{2-}$, over the pH range tested (3.5–12.0), the surface charge of the aluminum-oxide-based adsorbent, AA, and the iron-hydroxide-based adsorbent, E33, changed with pH due to protonation and deprotonation of the surface. As the pH increases, the portion of positively charged surface sites on the two media decreases, causing the reduction in As uptake.

In addition to the effect of pH, the presence of other anions, including nitrate, sulfate, phosphate, and silica may reduce the uptake of As by the adsorbents (Meng *et al.* 2000, 2002, Lin, T.-F. *et al.* 2006, Petrusevski *et al.* 2007). Lin, T.-F. *et al.* (2006) studied the influence of nitrate on the equilibrium uptake of As onto E33. They observed a 35 to 45% reduction in As uptake at various pH levels for E33 in the presence of 0.01 N nitrate. Meng *et al.* (2000, 2002) conducted a few batch studies for As removal using ferric oxides. They observed that the presence of silica, phosphate, and bicarbonate may reduce the adsorption capacity. The results suggested the combined effects of phosphate, silica, and bicarbonate caused the high mobility of As in the Bangladesh water they studied. Thus, the selectivity for common anions (NO_3^-, HCO_3^-, SO_4^{2-}, etc.), in comparison to different arsenate ions, should be considered before the application of adsorbents.

Anion exchange resin is a promising media for the removal of As species from water sources (Horng and Clifford 1997). Ion exchange does not remove As(III) because it is present in a molecular form in natural water. Therefore, pre-oxidation is normally required. Lin, T.-F. *et al.* (2006) examined the uptake of As onto an anion exchange resin from Purolite, USA, and found that the uptake was in the order of 10–40 mg g^{-1}, which is similar to E33 and GFH. However, unlike the metal-oxide-based adsorbents, the uptake increases with increasing pH. This may be attributed to the difference of the surface interaction between the resin and arsenate anions. The two arsenate anions present in the aqueous phase change with pH. As the pK_1 and pK_2 are 2.4 and 7.3, respectively, for arsenate, the concentration of the monovalent anion $H_2AsO_4^-$ would decrease as the pH increases and the divalent anion $HAsO_4^{2-}$ would increase. For a typical anion exchange resin, the affinity between the resin surface and the anion is dependent on the valency of the anion. The higher valent arsenate species will be preferred by the resin (Horng and Clifford 1997). Therefore, as the pH increases, the ratio of $HAsO_4^{2-}/H_2AsO_4^-$ increases, causing the increase in arsenate uptake.

In the last few years, a group of newly developed hybrid sorbents, such as ArsenX^{np} from Purolite, USA, FO 36 from Lanxess, Germany, and READ-As from Japan Seawater, Japan, have been used in the field for As removal. The hybrid sorbents are spherical resin/polymer beads either containing dispersed nanoparticles of hydrated ferric oxide (HFO) (for ArsenX^{np} and FO 36) or hydrated cerium oxide (for READ-As), enabling the adsorption of As from water. For example, ArsenX^{np} has been used in the remote villages of West Bengal, India (Sarkar *et al.* 2007) and in POU/E devices in North America (Möller *et al.* 2009) (see 9.4 for the POU/E case). In the West Bengal case (Sarkar *et al.* 2007), each As removal device includes either one column with 100 kg of ArsenX^{np}, or two columns, one with 50 kg of AA for preliminary As removal and another one with 50 kg of ArsenX^{np} for polishing. The device is attached to a hand-pump driven well and may provide As-safe water to 1000 villagers. The device runs for more than 20,000 BV on average before reaching a breakthrough of 50 µg L^{-1} of As (the MCL for drinking water in India).

The selection of adsorptive media is strongly dependent upon the water chemistry and media properties. Therefore, they are selected mostly based on the long-term operation results of pilot studies (Westerhoff *et al.* 2008). A manual for the design of adsorptive media for As removal has been published by US EPA (Rubel 2003). Details about the preliminary design, final design, capital cost, and operation can be found in the manual. To provide a lower cost alternative for full-scale system design, Westerhoff *et al.* (2008) applied the rapid small-scale column test (RSSCT) method, which was originally used for evaluating the performance of granular activated carbon, for the estimation of media performance. They found that only three to four weeks of testing was required to predict the performance of full-scale systems.

9.3.3   *Membrane technology*

Common membrane processes used for As removal include microfiltration (MF), ultrafiltration (UF), nanofiltration (NF), and reverse osmosis (RO). The feasibility of As removal for MF and UF strongly depends on the sizes of As-laden particles in the source water. As most As in groundwater is in the dissolved phase (Edwards 1994, Chen *et al.* 1994, Petrusevski *et al.* 2007), the two methods cannot be used as stand-alone processes for As removal. However, they may be combined with other pre-treatment methods, such as coagulation, to achieve high efficiency. For example, Brandhuber and Amy (1998) studied the removal of As from groundwater in Phoenix, Arizona, USA, using coagulation and MF. In their study, ferric chloride was used as a coagulant, and MF with 3 pore sizes (0.1, 0.2, and 1.2 μm) was tested. They found that a coagulant dosage of 7.0 mg L^{-1} and MF of 0.2 μm was adequate to remove 39–89% of As. The same process was adapted in a demonstration WTP operated by the Albuquerque Bernalillo County Water Utility Authority, New Mexico, US (ABCWUA 2009). The plant, with a capacity of about 20,000 m^3 day^{-1}, was completed in July 2007. Currently, the plant is the largest facility of its kind in the world.

Unlike MF and UF, NF and RO are able to remove 90 to >95% of As from water (Brandhuber and Amy 1998). They can be used as stand-alone methods for As removal. Brandhuber and Amy (1998) examined three NF and four RO systems for the removal of As. For As(V) at 25.5 μg L^{-1}, >95% removal was achieved for all the NFs and ROs, while for As(III) at an initial concentration of 18.5 μg L^{-1}, the removal was much less. Arsenic(III) rejection was only 40% for the NFs, and 74% for the ROs. Xia *et al.* (2007) studied the removal of As(V) and As(III) using NF. They found that As(V) was almost completely removed, while only 5% of As(III) was removed. While RO and NF (with pre-oxidation of As(III)) are considered promising technologies for the removal of As, their cost is higher than those of other membrane and conventional methods. In addition, the waste streams from the two systems may be up to 80% of the influent water, which also limits their application in areas of water scarcity.

9.4   POINT-OF-USE AND POINT-OF-ENTRY DEVICES

Point-of-use and point-of-entry devices (POE/E) devices have been extensively used in different continents, including North America (Möller *et al.* 2009) and Asia (Hussam and Munir 2007, Petrusevski *et al.* 2008). The common processes employed in these devices include membrane-based technologies, such as RO and NF (Lin *et al.* 2002), and adsorbent media, including AA (Thomson *et al.* 2003), iron-based (Ngai *et al.* 2007), and hybrid sorbent media (Möller *et al.* 2009).

Groundwater in both the northeastern and southwestern coastal areas of Taiwan contains high concentrations of As. Since no central water supply system is available in some of those areas, POU water purification devices are considered as an option for providing safe drinking water. Lin *et al.* (2002) studied the removal of As using two types of POU purification device, RO systems and distillers. Three commercially available RO systems and two distillers were selected to test their As removal efficiency from synthetic and real groundwater. Each RO system tested comprised of three pre-treatment units, including a 5 μm particle filter, a granular activated carbon (GAC) filter, and a 1 μm particle filter or pressed activated carbon filter. An RO membrane unit equipped with a booster pump was connected to the end of the pre-treatment filters. The three RO membranes were all made of thin film composite but from different manufacturers. For a typical household distillation system, water is added to a stainless steel boiling chamber and is then boiled. The steam passes through a cooling coil and condenses. The condensate is collected in the collection tank as distilled water. The volume of the boiling chamber of the tested distillers was 4 L, and the time for water production was 8 hours.

Experiments using the three RO systems, namely Systems A, B, and C, using synthetic groundwater at 100 and 1000 μg L^{-1} of As(III) showed that only System C successfully reduced As(III) to a concentration level of lower than that required in the Drinking Water Standards of Taiwan (10 μg L^{-1}). In subsequent experiments using real groundwater with 700 μg L^{-1} of As, only the treated water of System C was able to meet the standard after producing nearly 1000 L of water. In some instances, System A produced water with As concentrations 10 times higher than the standard. The experimental results of the two distillers using synthetic groundwater at 700 μg L^{-1} As(III) and real groundwater at around 700 μg L^{-1}

of total dissolved As showed that the finished water concentrations were both well below the new standard, with removal efficiencies higher than 99%. The total dissolved solids (TDS) reductions for the tested RO systems were all above 90% on average. Although the POU system with the higher As removal seemed to have a higher TDS reduction efficiency, no correlation was found between As removal and TDS reduction efficiency. Using TDS as a surrogate indicator for As concentration in the treated water may not be appropriate.

Geucke *et al.* (2009) examined a commercially available marine RO desalinator with three different membrane modules for As removal. Tapwater spiked with up to 2500 μg L^{-1} of As(III) and As(V) was used as feed water. Similar to other studies (Lin 1999, Xia *et al.* 2007, Möller *et al.* 2009), the As(V) rejection was better than As(III). Two of the membranes were able to comply with the generic As standards (10 μg L^{-1}), even at a feed concentration of 2400 μg L^{-1}. However, for As(III), the treated water can meet the standard only at feed concentrations below 350 μg L^{-1}.

Thomson *et al.* (2003) designed and constructed a POU system using commercially available AA to remove As from the chlorinated tapwater of the City of Albuquerque. Assuming an average As concentration of 23 μg L^{-1}, 1 L of adsorbent would provide water for direct consumption by a family of four for 435 days. A monthly cost of about \$10 was estimated to purchase, install, and operate this POU system.

In a recent study, Möller *et al.* (2009) monitored 275 POU/E As removal systems in eastern New England and New Jersey in the USA. The hybrid hydrous iron oxide/polymer media (ArsenX^{np}) was used for the adsorption of As from the water. The POE systems consist of two columns operating in a lead-lag configuration, while the POU devices contain 4 L of media and are designed to treat the water from faucets. Although the lead column is expected to last 1.5 years in most cases, the actual life of the column depends on the As concentration, the water volume treated, and the water quality. For the case with the largest operation capacity (>44,500 BV), the inlet As concentration was relatively low, between 15 and 68 μg L^{-1}, with a low pH (6.8). In one extreme case with higher As concentration (100–150 μg L^{-1}), only 2580 BV were obtained before As bleed was detected in the treated water. Although no direct evidence was linked to the lower operation capacity, competition from other anionic species, including phosphate and silica, the presence of arsenite (As(III)), and higher pH may be the reasons that caused the lower bed life.

In addition to the membrane-based and relatively expensive adsorbent-based POU/POE systems, low cost POU units have also been tested/used extensively in many countries, such as Bangladesh (Hussam and Munir 2007, Petrusevski *et al.* 2008), Nepal (Ngai *et al.* 2007), and Vietnam (Berg *et al.* 2006). Berg *et al.* (2006) tested 43 household sand filters for As removal in rural areas of the Red River delta in Vietnam. An average of 80% As removal was obtained for the groundwaters containing 10–380 μg L^{-1} As, <0.1–48 mg L^{-1} Fe, <0.01–3.7 mg L^{-1} P, and 0.05–3.3 mg L^{-1} Mn. Only 40% of the tested filters could achieve 10 μg L^{-1}, which is the suggested drinking water standard. They concluded that dissolved iron was the key factor for As removal. A Fe/As ratio of 250 was required to ensure As removal to a level below 10 μg L^{-1}.

Hussam and Munir (2007) developed a low cost iron-based filter, called SONO, and applied it in Bangladesh. The major component of the filter, called composite iron matrix, is comprised of 92–94% iron, 4–5% carbon, and 1–2% SiO$_2$, manganese, sulfur, and phosphorus. The filter was tested in six tubewells with 6 different families. All the filters removed As to <10 μg L^{-1} from an input range of 32 to 2400 μg L^{-1} of As, with a total water yield of 68,000 to 128,000 L. The filter costs about US\$40/5 years and produces 20–30 L hour^{-1} of drinking water. About 30,000 SONO filters were deployed in Bangladesh by 2007.

Petrusevski *et al.* (2008) reported 30 month performance results for As removal by 11 family filters in rural Bangladesh. The filter materials were made of iron-oxide-coated sand, a by-product of iron removal plants. The arsenic level in the filtrate reached 10 μg L^{-1} after 50 days of operation at one test site, and after 18 months at 3 other test sites. For the other 7 sites, As remained below 10 μg L^{-1} until the end of the study.

Ngai *et al.* (2007) reported a two-year technical and social evaluation of over 1000 *Kanchan* Arsenic Filters (KAFs) deployed in rural villages of Nepal. The filter (KAF), with a flow rate of 10–15 L hour^{-1}, is comprised of two units: As adsorption on ferric hydroxide and microbial removal by a slow sand filtration process. The KAF typically removes 85–90% of the arsenic, 90–95% of the iron, 80–95% of the turbidity, and 85–99% of the total coliforms. The initial cost of the filter is about 15–25 US$, and operation cost is about 2–5 US$ year^{-1}. By January 2007, over 5000 filters have been installed in Nepal.

9.5   CASE STUDY IN SOUTHWESTERN TAIWAN

As mentioned earlier, the groundwater in southwestern Taiwan contains relatively high concentrations of As (Lin 1999). Although most of the wells with As problems have been abandoned, some of them are still used in a few small water treatment plants in the area. For example, in a coastal town of Yunlin county (Fig. 7.1), groundwater is still the sole source for a few WTPs, primarily due to a shortage of reliable surface water. One of the WTPs discussed here, called WTP A1, produces 12,500 m^3 day^{-1} of water from 12 wells, and was constructed in 1968. Although some minor changes have been made, the processes include only pre-chlorination and rapid sand filtration.

The water quality of the nine source water wells for WTP A1 is shown in Figure 9.1. According to their As concentrations, the wells can be categorized into three groups. Four wells contained relatively low As concentrations, usually from <0.01 to 20 µg L^{-1} (Fig. 9.1a), three wells were in the medium range, mostly below 50 µg L^{-1} (Fig. 9.1b), and four wells had a high As content, often >500 µg L^{-1} (Fig. 9.1c). Figure 9.2a shows the raw and finished water quality of the WTP since 2004. The variation of As concentrations in the raw water depended upon the mixing ratio of the water pumping from all the wells. As shown in the figure, the As concentrations were mostly similar for raw and finished water before 2008, suggesting almost no As removal in WTP A1. The process removes only very small portion of As, similar to the ratio (10–20%) suggested by Lin (1999) for that area. With this low removal ratio, the As content in the finished water occasionally exceeded the 10 µg L^{-1} standard of Taiwan (see Fig. 9.2a). Since no reliable surface water source is available in the area, groundwater is still used as the sole water source. To mitigate the As issue in the area, two approaches were employed by the local water authority, including the construction of a new plant with As removal processes and the management of water wells.

Although the upgrade of WTP A1 may reduce As concentrations in the finished water, no land is available to add new processes. Therefore, a new WTP, called A2, was constructed near A1 in October 2008, for the removal of As. The processes in A2 include pre-chlorination, ferric chloride coagulation, sedimentation, and rapid sand filtration, with a production rate of 3000 m^3 day^{-1}. Well water with high As content, wells A11, 19, and 20 shown in Figure 9.1c, was directed to WTP A2 for better As removal. Figure 9.2b shows the As concentrations in the raw and finished water of WTP A2. The removal efficiency for As was between 45 and 97%, with an average of 79%. As indicated in Figure 9.2b, this improved removal efficiency keeps the finished water concentration well below the required limit.

Water from the three wells with high As content was diverted to WTP A2. The remaining wells, containing lower As levels, were still treated by WTP A1. After October 2008, WTP A1 treated only 75% of its original capacity, 9500 m^3 day^{-1}, allowing some flexibility for well management. The water quality at each well is regularly monitored for As. Arsenic in the raw water of WTP A1 is from all the wells pumped into the WTP. Therefore, to meet the As requirement for the finished water, the As concentration in the mixed raw water is always considered in the selection of the production rate for each well. As shown in Figure 9.2a, after October 2008, by properly choosing production rates for low As-content wells, the finished water was able to meet the As standard.

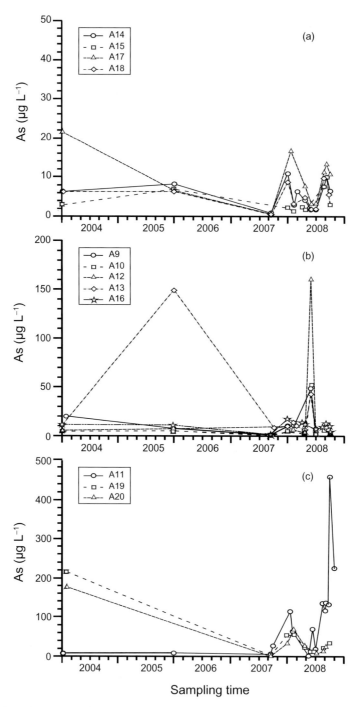

Figure 9.1.   Arsenic concentrations in 12 source water wells for a WTP in southwestern Taiwan (data courtesy of Taiwan Water Supply Corp.).

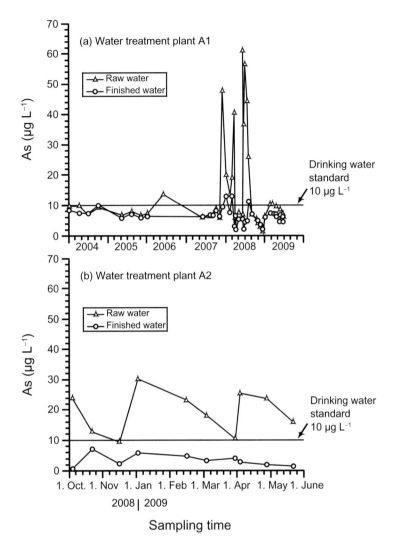

Figure 9.2.   Arsenic concentrations in the raw and finished water in WTP A1 and A2 in southwestern
Taiwan (data courtesy of Taiwan Water Supply Corp.).

9.6   RECOMMENDATIONS

Many drinking water systems around the world have As problems. There are many potential
alternatives to mitigate the issue. In this chapter, three types of approaches were introduced,
namely change of water source, treatment in centralized systems, and POU/E devices.

Table 9.1 summarizes the advantages and limitations of the methods discussed in this
chapter. In brief, a change of source water to surface water or blending with well water with
a lower As concentration is a simple way to resolve the issue, provided that new water sources
are available or the development of source water is not financially limited. For the treatment
of groundwater, if treatment processes such as coagulation are present, optimization of the
process and/or change of coagulant type may often be a less expensive way to reduce the
As concentrations in the finished water. However, careful operation is needed to maintain
stable water quality. In addition, pre-oxidation is often needed if the As(III) concentration

Table 9.1.   Summary of the three approaches for arsenic removal.

Approach	Advantages	Limitations
Change of water source		
	Rapid solution; Usually low As concentration	Low As water sources are required; Cost may be expensive for source water development
Treatment processes for centralized systems		
Precipitation-based	If processes are present, optimization/change of coagulation may be enough; Usually low-cost	Careful operation is needed; Pre-oxidation may be needed for As(III); If no existing processes, land space and construction costs need to be considered
Adsorption/ion exchange	Usually high removal efficiency of As	Medium range cost; Efficiency may be affected by other anions; Lower efficiency for As(III)
Membrane	Very high removal efficiency of As	High capital and running costs; Low recovery of water; Lower efficiency for As(III) with nano-filtration
Point of use/entry devices		
	Good for remote areas; Usually high removal efficiency of As	Removal efficiency may depend on the processes/ systems employed; Service time depends on the chemistry of local water; Lower efficiency for As(III) with nano-filtration

in the water is significant. Adsorption and ion exchange can also be used for As mitigation. Although the cost is often higher compared to that for the precipitation-based method, it provides a high removal efficiency and the cost is often cheaper than membrane-based methods. The drawback of the methods is that some anions present in groundwater may compete for the adsorption/exchange sites; the removal efficiency of As(III) is also usually lower. For membrane-based methods, both RO and NF may provide high removal efficiencies for As. However, the capital and running costs are usually high, which would limit the application in centralized systems. In addition, the recovery may be as low as 20%. This may also limit the application if water scarcity is an issue. POU/POE water purification devices are another alternative for areas where centralized drinking water systems are not available. In particular, many low cost POU systems, mostly iron (oxide) based, have been successfully tested in developing countries with As issues. The removal efficiency of As depends on the removal processes employed. Even applying the same processes, the efficiency also depends upon the systems used. In addition, the service life strongly depends upon the local water quality. Therefore, more *in-situ* assessments are needed before POU/POE devices are used in a specific location.

CHAPTER 10

Future areas of study and tasks for the Taiwan arsenic problem

The Taiwan arsenic issue has attracted scientists from Taiwan and the rest of the world, and has been the subject of numerous studies over the last five decades. Despite these intensive studies, there remain many mysteries which need further investigation. For example, in Chianan and Lanyang (Yilan) plains, which are the principal areas affected by groundwater As, the sources of arsenic (As) remain unclear, though it is evident that the arsenic is geogenic in both of these areas. The principal triggers for As mobilization also remain unclear. Possible controls include the redox conditions, microbiological processes, the presence of humic acids (which can form soluble complexes with As and so maintain it in solution) and ions competing for As adsorption sites. Although the occurrence of blackfoot disease (BDF) is well-studied in Taiwan, the precise causes of this endemic disease still remains a mystery today. We only know that blackfoot disease is a product of the co-exposure to As and at least one other water component, probably humic acids. This would explain why blackfoot disease occurs in Chianan plain, where the concentration and relative fluorescence intensity of humic substances in the groundwater are high, but not in Lanyang plain where the As concentrations in the groundwater can be as high as those in Chianan plain in some places; however, the median As concentration in the Chianan plain groundwater is 402 μg L^{-1} (range: 4.9–704 μg L^{-1}, n = 5), which is higher than that in the Lanyang plain groundwater (range: 2.5–543 μg L^{-1}, median: 20.6 μg L^{-1}, n = 6) (Reza *et al.* 2009), but where the concentration and relative fluorescence intensity of humic substances are relatively low.

If we look at the present-day situation regarding human exposure to As in Taiwan, we can see that even today, not all of the population receives drinking water with As concentrations below the national regulatory limit of 10 μg L^{-1}. The use of groundwater with elevated As concentrations in southwestern and northeastern parts of Taiwan requires water treatment plants to remove As. In many cases this is not effective, mostly due to technical and economic constraints, so the treated water still has As concentrations exceeding the regulatory limit. Another small part of the population has no access to tapwater and still depends on As-contaminated groundwater as their only available drinking water resource. Further improvement to the As removal process and the establishment of new water treatment plants equipped with state-of-the-art technology is required to rectify this situation.

Exposure of humans to As through the food chain can also be important. In areas of Taiwan affected by high concentrations of As in the groundwater, the use of this water for aquaculture potentially introduces As into the human food chain. This highlights the demand for further studies to provide alternative water sources or to implement well-head As removal techniques.

In the Guandu plain (Taipei area), where As of geothermal origin resulted in a high As content in soils, further studies are required to estimate potential triggers which may in the future increase the mobility of this soil As. Under the present soil conditions the mobility fortunately is low, so that the rice harvested from that area does not seem to be a threat to consumers.

10.1 SOURCES OF ARSENIC AND MOBILIZATION IN GROUNDWATER

Although there have been a number of studies investigating the source(s) of groundwater As in Taiwan (Chianan plain, Lanyang plain, and Guandu plain) as discussed in section 6.3.1 of chapter 6, there are still many uncertainties related to the origin of As and its mobilization in the groundwater. In both Chianan and Lanyang plains, the source of As, though geogenic, has not yet been established. In Guandu plain, the As occurrences can clearly be related to high-As geothermal water.

To date, the temporal and vertical spatial distributions of arsenic contents in groundwater and sediments of the Chianan and Lanyang plains are not very clear. With regard to this, in August 2007 the National Science Council of Taiwan allocated a three-year integrated project (August 1, 2007 to July 31, 2010) to the Hydrogeology Laboratory of the Earth Sciences Department of National Cheng Kung University, Tainan City, Taiwan to carry out adsorption and desorption experiments which were conducted by Professor Jiin-Shuh Jean and his research team. Sediment and water samples were taken from two boreholes (150–200 m deep), each with ten piezometers in the BFD endemic areas of Chianan plain. Further research is being conducted as follows:

- to identify BFD factors in Chianan plain;
- prospective studies in Lanyang and Guandu plains, where high arsenic exists without BFD;
- comparison of the chemical components and biochemical structure of causative agents of BFD in Chianan plain;
- understanding the different competing models for arsenic release including geomicrobiological study;
- fundamental chemistry and complexation behavior of arsenic species interacting with humic substances and related organic compounds.

The experimental results of Reza *et al.* (2009) indicated that the concentrations of As in the Chianan plain groundwater were much higher than those of Fe and Mn. This could be because the As was desorbed without dissolving oxide minerals, or it could be that Fe and Mn have precipitated subsequently. However, the concentrations of Fe and Mn in the Lanyang plain groundwater were as high as As, suggesting that the As was desorbed along with the dissolution of oxide minerals. This should be investigated further by increasing the number of samples from different sampling locations.

The redox condition is one of the principal factors controlling arsenic mobility in the aquifers of Chianan and Lanyang plains. The experimental results in Nath *et al.* (2008) and Reza *et al.* (2009) demonstrated that, due to the abundance of organic matter, the Chianan plain aquifer is more reduced in nature compared to that of the Lanyang plain. This results in the release of more As to the groundwater in Chianan plain, where it is predominantly present as As(III). To date, spatial and temporal variations in Eh values in the groundwater of Chianan and Lanyang plains have not been well documented. Nath *et al.* (2008) revealed that the groundwater Eh values in the Chianan plain were much lower in 2006 (decreasing trend of redox potential) as compared with 2005. Several years of Eh data are required to confirm whether the redox potential is decreasing with time. Further work should be performed on temporal chemical variations in the aquifers of Chianan and Lanyang plains. Though long-term groundwater monitoring well networks have been established in Chianan and Lanyang plains by the Taiwan Water Resources Bureau, no results have yet been published.

Potential electron acceptors, such as NO_3^- and SO_4^{2-}, may be derived from seawater intrusion into the aquifers to allow further oxidation of organic matter, which is widely believed to be an important contributor to the generation of anoxic conditions. Microbial processes in the aquifer sediments rich in organic matter create a favorable reducing environment, facilitating mobilization of As in the Chianan plain aquifer. However, the addition of SO_4^{2-} from seawater may even lower dissolved As concentrations by facilitating the precipitation of sulfide minerals. Whether As is released to groundwater or precipitated in sulfide minerals in the groundwater of BFD area of Chianan plain must be investigated.

To date, few studies have been conducted to investigate the geochemical and microbiological processes that influence the speciation and mobility of As (and Fe and Mn) in the aquifers of Chianan and Lanyang plains. Microbially-mediated reductive dissolution of amorphous/

crystalline Fe-/Mn-oxyhydroxides is thought to be the primary mechanism for releasing As from sediments to the groundwater of Chianan and Lanyang plains (Reza *et al.* 2009). However, the Chianan plain groundwater contains much higher concentrations of organic matter than the Lanyang plain groundwater. Organic matter can play a role in creating a favorable reducing environment to mobilize As in the Chianan plain groundwater. Bacterial reduction of Fe-/Mn-oxyhydroxides would release ferrous Fe/Mn, As, and bicarbonate species to the groundwater (Nath *et al.* 2008, Reza *et al.* 2009). Reduced Fe and Mn may be precipitated as $FeCO_3$ and $MnCO_3$ solids and subsequently react with bicarbonate species in groundwater (Nickson *et al.* 2000). Sulfate-reducing bacteria can reduce sulfates to sulfides and then react with desorbed As to form precipitated As-bearing sulfide minerals. Reducing bacteria (e.g., Fe, Mn, As, etc.) should be isolated to confirm which types of microbes mediate mobilization/immobilization of arsenic in the aquifers of Chianan and Lanyang plains as well as Guandu plain and the Chinkuashi gold mine area.

Kulp and Jean (2009) demonstrated the potential for reductive desorption and mobilization of As from sediments to groundwater by the endogenous population of arsenate-reducing bacteria in the Chianan plain aquifer. The reduced humic substances in the aquifer can serve as electron donors for biological As(V) reduction. Further work is needed to clarify if direct microbial reduction is an important mechanism for mobilizing adsorbed As from the aquifer sediments of Chianan and Lanyang plains as well as Guandu plain and the Chinkuashi gold mine area.

10.2   HUMAN IMPACT THROUGH THE FOOD CHAIN

Understanding As exposure to humans through the food-chain, primarily crops and fish, is important in assessing the As-related health effects in Taiwan and other countries. The following suggestions are presented based on Taiwanese observations, and deserve further study:

- Low bioavailability of As to rice has been found in the Guandu paddy. The cause is attributed to the low concentration of non-specifically-bound As in soil or to the formation of iron plaque in the rhizosphere.
- A long-term monitoring program must be established to assess the fate and transport of arsenic in the Guandu plain. Future studies should investigate various factors including land-use type, irrigation pattern, fertilizing and farming habits that may affect the As mobilization and its bioavailibility.
- The estimation of spatial distribution of contaminant groundwater quality is important in the health risk assessment. Geo- and multivariate statistical methods are widely used for the estimation of spatial variability. In future studies, the construction of risk maps of the BFD farmed area, Lanyang and Guandu plains are welcomed (Liang *et al.* 2009). Moreover, advanced techniques and models such as multi-scale (individual level, county level) hierarchical Bayesian model (HBM) could be applied to investigate the pathways of human exposure to As.
- Recent studies have provided the data for human exposed to inorganic As through raw seafood. However, the method was unable to provide a real intake, since most of these foods are cooked before consumption, and the results may not reflect the changes in concentration of the various arsenic species during the cooking treatment. In modern dietary habits, the various cooking treatments, such as grilling, roasting, baking, stewing, boiling, steaming, or microwaving have to be considered. Therefore, future work may be focused on the As concentration and speciation in different seafoods after various cooking treatments.
- Marine fishes have normally much less inorganic As compared to freshwater fish. Worldwide very few studies have been performed on As in freshwater fish compared to marine seafood.

- Human health risks were estimated based on the consumption of contaminated fish and shellfish due to the use of aquacultural water containing high concentrations of As. Thus, many cultured stocks, such as eel, carp and shrimp, from the As-affected groundwater area should be included in the assessment. The other source of human exposure to As is from drinking water.
- Information concerning the dietary intake of As by children is rather limited as most studies focus on the general population. Exposure to As during growth and development can result in acute long-term effects on the health of children. Ingestion of contaminated seafood and freshwater fish may present an important route of exposure to arsenic hazards. As childrens bodies are developing, they generally consume more seafood on a body weight basis than adults. Also metabolism in children is different from adults. Children are at particular risk of illness from exposure to As hazards in seafood.
- Regulatory agencies including EPA, Council of Agriculture and Department of Health should jointly establish the health standard of As concentrations (inorganic) in food to protect human health from dietary intake of As through the food chain (Liu, C.-W. *et al.* 2009).

10.3   HEALTH EFFECTS OF ARSENIC IN DRINKING WATER, TREATMENT, RISK ASSESSMENT AND PREVENTION

Following the delivery of surface water to supply drinking water to the Chianan and Lanyang plains, the residents living in arseniasis-endemic areas in Taiwan were no longer exposed to high levels of As in drinking water. However, the health effects of As in drinking water seem to exist for a long period after the exposure. The exposed population in the endemic areas still had an increased risk of skin and internal cancers, circulatory diseases, hypertension, diabetes, cataract, pterygium, neurological disorders, erectile dysfunction, and developmental retardation for a long time after the cessation of the intake of As-contaminated well water. Recently, a reduction in mortality from internal cancers, circulatory disease and diabetes was observed in the arseniasis-endemic area in southwestern Taiwan (Chen *et al.* 2009). However, the mortality rates of these diseases were still higher than the general population in the unexposed areas in Taiwan. Further follow-up studies are needed to elucidate whether the exposure to arsenic in drinking water will have a life-long health risk and whether the offspring of the exposed population are also affected.

Since only a proportion of As-exposed residents were affected with various diseases in the endemic areas, both genetic and acquired susceptibility deserve further investigation to identify the characteristics responsible for the individual differences in the occurrence of arsenic-induced health hazards. Understanding this individual variation in susceptibility is important to the prevention and treatment of As-induced diseases as well as to the risk assessment of arsenic exposure. Biomarkers of nutritional intake, humoral and cellular immunity, hormones, and genetic factors may be used to explore the individual susceptibility to arseniasis (Chen *et al.* 2005). In the post-genomic era, the impact of gene-environment interaction on As-induced health hazards needs intensive evaluation. The genes involved in As methylation, DNA repair, cell cycle regulation, oxidative stress, carcinogenesis and atherogenesis are especially important in the delineation of genetic susceptibility to As-induced health hazards. The epigenetic effects of ingested arsenic should also be elucidated for better understanding of the toxicological mechanisms of ingested arsenic.

The risk assessment of As in drinking water has been primarily based on the excess risk of internal cancers, especially lung cancer and urothelial cancer, among residents in areas where the As levels in drinking water exceeded 100 µg L^{-1}. The assessment of human cancer risk associated with As levels less than 100 µg L^{-1} is needed for further validation of the current Maximal Contamination Level for As in drinking water (10 µg L^{-1}). Long-term follow-up studies on a large number of study participants may provide the best evidence for such risk assessments. In addition to the residents living in Lanyang (Yilan) plain who have been followed-up since 1994, residents in the other countries with As problems indicated in chapter 1 may also be enrolled to examine their cancer risk due to exposure to low levels of As in drinking water. Early biological markers, such as micronuclei, sister chromatid exchanges, and chromosomal aberrations of exfoliated urothelial cells combined with cystoscopic examination may be used for the early detection of urothelial carcinoma in these studies.

The health risk assessment based on the incidence of cardiovascular diseases, hypertension or diabetes is also important for the comprehensive evaluation of pleiotropic health effects of ingested As (Chen *et al.* 2007, Wang, C.-H. *et al.* 2007). Given that diabetes and hypertension are major risk predictors of cardiovascular diseases associated with ingested As, the characterization of the associations between ingested As and these two prevalent diseases may improve the risk assessment of cardiovascular diseases associated with As in drinking water. The human health hazards associated with *in utero* exposure to As in drinking water should also be compared with those associated with postnatal exposure to As in drinking water. This is especially important in arseniasis-endemic areas where residents are still using high-As well water. Risk assessment of developmental retardation and pediatric disorders is urgently needed in these areas.

10.4 FUTURE TREATMENT DEMANDS, INCLUDING NANOTECHNOLOGY

To date, numerous treatment processes have been developed for the treatment of As-rich water worldwide. These include precipitation-based, adsorbent/ion exchange-based, and membrane-based methods. Precipitation-based methods are often the cheapest water treatment process for As removal. This process is appropriate for centralized systems, but is difficult to employ in remote areas. In addition, large amounts of space are needed for the installation of the technology. A few of the water treatment plants (WTPs) in southwestern Taiwan adopted a precipitation process based upon the addition of ferric chloride for As treatment, and 45 to 97% (average: 79%) As removal was achieved. However, the process requires careful operation and the As removal efficiency is strongly influenced by the water chemistry. Therefore, a better understanding of the chemistry and kinetics involved in the removal processes are needed to support the reliable operation of these plants and provide a stable treated water quality. Simple models and operational protocols developed from fundamental theory would be useful for the operators of WTPs.

Adsorbent- and ion-exchange-based methods can either be used in centralized systems or at well heads. Many adsorbents or resins based upon the oxides of iron, titanium, and cerium have been successfully developed for As removal from drinking water (see chapter 9). In Taiwan, a few adsorbents, including AA, E33, and GFH, have been tested in the field for As removal (Lin, T.-F. *et al.* 2002, 2006). However, the studies were only limited to small scale trials. Adsorption has not been employed in any of the WTPs in Taiwan. The uptake and removal efficiencies of As(V) for the adsorbents/resins are generally high, resulting in a long service life. To achieve high As removal and stable water quality, installation of adsorbent-based As removal technologies is one of the choices for WTPs in the coastal areas of southwestern Taiwan. In addition, the process needs less land space compared to precipitation-based methods. This is also an advantage for the area because of limited space and the high price of land. In applying the process, the competitive effect of other anions and the low removal efficiency of As(III) need to be taken into account. Although some new adsorbents were shown to have better removal of As(III) than earlier media, pre-oxidation is still often needed in many applications. Another disadvantage of adsorption is its high cost. The cost is generally higher than that of precipitation-based methods used in conventional WTPs, which limits the applications. In addition, the cost of POE/POU or well head treatment devices using adsorbents or resins may still be costly for people living in rural areas or in developing countries. Therefore, the development of low-cost, high-efficiency adsorbents with low waste generation is needed.

Membrane-based methods have been applied in POE/POU devices as well as in WTPs in different parts of the world. These methods, ROs and NFs in particular, generally provide a very high rejection of As(V) as well as many other organic and inorganic compounds. At present, the method is still expensive compared with other techniques. From an economical consideration, POE/POU devices may be used in remote areas of Taiwan where a centralized system is not available. However, due to their high cost, membrane methods may be difficult to employ in WTPs. However, the method may be used in some specific conditions. For example, As may co-exist with other contaminants, such as nitrate and organic pollutants, in groundwater. In this case, the method can remove most of the chemicals simultaneously, and the cost may be lower compared to that of using several treatment processes together. Since water recovery may be as low as 20–30% for the process, this will also limit its application.

In the last few years, nanotechnologies have drawn much attention in the water treatment field. The reactivity of nano-sized particles, including reaction rate and capacity, has been greatly improved, primarily due to the large accessible surface area. In drinking water treatment, the direct application of nanoparticles is not easy since their recovery from the water is difficult. Therefore, techniques to immobilize reactive nanoparticles into granular or fabric media have been developed. For example, typical hybrid ion exchange (HIX) adsorbents are spherical resin beads containing dispersed nanoparticles of hydrated ferric oxide (HFO) (Sarkar *et al.* 2007). In addition, nano-sized iron oxide has also been impregnated into granular activated carbon, activated alumina and other media to increase the As adsorption capacity (Gu *et al.* 2005, Hristovski *et al.* 2009). While these hybrid sorbents possess a high capacity for As, the price is still relatively high for many WTPs as well as for POE/POU users in developing countries. Reducing the cost of these As removal media is certainly an important goal to pursue. In addition to the granular type adsorbents, magnetism can also be used to separate As-laden particles from treated water. The particles may be made of a single material that can adsorb As and is magnetic, such as magnetite and maghemite (Yavuz *et al.* 2006, Tuutijarvi *et al.* 2009), or the particles can be a core-shell type of material, with a magnetic core and a nano-sized reactive shell (Lim *et al.* 2009). This magnetic property enables the separation of particles from water using a magnetic field. Although the application of nanoparticles in drinking water looks promising for the removal of As, the impact on human health and the environment is not clear. More studies should be performed in this area.

The removal of As from source water is highly dependent on the specific site and the water chemistry. Therefore, a simple way to characterize parameters for the design of a field device/process is very important. The RSSCT methodology has been proposed in the design of adsorbent-based processes (Westerhoff *et al.* 2008). However, no simple and rapid assessment method is available for other As removal technologies. Therefore, it will be very useful if a simple protocol or approach can be developed for determining the design parameters for field devices/processes. In addition, the water quality is always very important no matter which process is employed in water treatment. Rapid monitoring methods for As and other anions that may influence As removal will greatly improve the ability to predict the effectiveness of a process and to control the water quality of the finished water. Therefore, further development of these methods is also needed.

References

ABCWUA: Water quality report 2008. Albuquerque Bernalillo County Water Utility Authority (ABCWUA), Albuquerque, NM, 2009.

Abernathy, C.O., Chappell, W.R, Meek, M.E., Gibb, H. and Guo, H.-R.: Is inorganic arsenic a "threshold" carcinogen? *Fund Appl. Toxicol.* 29 (1996), pp. 168–175.

Acharyya, S.K., Chakraborty, P., Lahiri, S., Raymahashay, B.C., Guha, S. and Bhowmik, A.: Arsenic poisoning in the Ganges delta. *Nature* (1999), pp. 401–545.

Adriano, D.: *Trace elements in terrestrial environments: biogeochemistry, bioavailability, and risks of metals*: Chapter 7: *Arsenic.* 2nd ed, Springer, Berlin, Germany, 2001, pp. 220–256.

Ahamed, S., Sengupta, M.K., Mukherjee, A., Hossain, M.A., Das, B., Nayak, B., Pal, A., Mukherjee, S.C., Pati, S., Dutta, R.N., Chatterjee, G., Mukherjee, A., Srivastava, R. and Chakraborti, D.: Arsenic groundwater contamination and its health effects in the state of Uttar Pradesh (UP) in upper and middle Ganga plain, India: A severe danger. *Sci. Total Environ.* 370 (2006), pp. 310–322.

Ahmed, K.M., Bhattacharya, P., Hasan, M.A., Akhter, S.H., Alam, S.M.M., Bhuyia, M.A.H., Imam, M.B., Khan, A.A. and Sracek, O.: Arsenic enrichment in groundwater of the alluvial aquifers in Bangladesh, an overview. *Appl. Geochem.* 19: 2 (2004), pp. 181–200.

Airas, S., Duinker, A. and Julashamn, K.: Copper, zinc, arsenic, cadmium, mercury, and lead in blue mussels (*Mytilus edulis*) in the Bergen harbor area, western Norway. *Bull. Environ. Contam. Toxicol.* 73 (2004), pp. 276–284.

Akai, J., Izumi, K., Fukuhara, H., Masuda, H., Nakano, S., Yoshimura, T., Ohfuji, H., Anawar, H.M. and Akai, K.: Mineralogical and geomicrobiological investigations on groundwater arsenic enrichment in Bangladesh. *Appl. Geochem.* 19 (2004), pp. 215–230.

Akai, J., Kanekiyo, A., Hishida, N., Ogawa, M., Nagamura, T., Fukuhara, H. and Anawar, H.N.: Biogeochemical characterization of bacterial assemblages in relation to release of arsenic from South East Asia (Bangladesh) sediments. *Appl. Geochem.* 23 (2008), pp. 3177–3186.

Albores, A., Brambila, C., Calderón, A., Del Razo, L., Quintanilla, V. and Manno, M.: Stress proteins induced by arsenic. *Toxicol. Appl. Pharmacol.* 177 (2001), pp. 132–148.

Alsina, M., Saratovsky, I., Gaillard, J.-F., Pasten and P.A.: Arsenic speciation in solid phases of geothermal fields. In: M.O. Barnett and D.B. Kent (eds): *Adsorption of metals by geomedia* II: *Variables, mechanisms, and model applications.* Elsevier, Amsterdam, The Netherlands, 2007, pp. 417–440.

Altamirano Espinoza, M. and Bundschuh, J.: Natural arsenic groundwater contamination of the sedimentary aquifers of southwestern Sébaco valley, Nicaragua. In: J. Bundschuh, M.A. Armienta, P. Birkle, P. Bhattacharya, J. Matschullat and A.B. Mukherjee (eds): Geogenic arsenic in groundwater of Latin America. In: J. Bundschuh and P. Bhattacharya (series eds): *Arsenic in the environment*, Volume 1. CRC Press/Balkema Publisher, Leiden, The Netherlands, 2009, pp. 109–122.

Anawar, H.M., Akai, J., Yoshioka, T., Konohira, E., Lee, J.Y., Fukuhara, H., Alam, M.T.K. and Garcia-Sanchez, A.: Mobilization of arsenic in groundwater of Bangladesh: evidence from an incubation study. *Environ. Geochem. Health* 28 (2006), 553–565.

Andreae, M.O.: Arsenic in rain and the atmospheric mass balance of arsenic. *J. Geophys. Res.* 85 (1980), pp. 4512–4518.

Appelo, C.A.J. and Postma, D.: *Geochemistry, groundwater and pollution.* 2nd ed, AA Balkema, Rotterdam, The Netherlands, 2006.

Appelo, C.A.J., van der Weiden, M.J.J., Tournassat, C. and Charlet, L.: Surface complexation of ferrous iron and carbonate on ferrihydrite and the mobilization of arsenic. *Environ. Sci. Technol.* 36 (2002), pp. 3096–3103.

Arai, Y., Elzinga, E.J. and Sparks, D.L.: X-ray absorption spectroscopic investigation of arsenite and arsenate adsorption at the aluminum oxide–water interface. *J. Colloid Interface Sci.* 235 (2001), pp. 80–88.

Arai, Y., Sparks, D.L. and Davis, J.A.: Effects of dissolved carbonate on arsenate adsorption and surface speciation at the hematite–water interface. *Environ. Sci. Technol.* 38 (2004), pp. 817–824.

Arellano, V.M., García, A., Barragán, R.M., Izquierdo, G., Aragón, A. and Nieva, D.: An updated conceptual model of the Los Humeros geothermal reservoir (Mexico). *J. Volcanol. Geotherm. Res.* 124:1:2 (2003), pp. 67–88.

Arkesteyn, G.J.M.W.: Pyrite oxidation in acid sulphate soils: The role of microorganisms. *Land and Soil* 54:1 (1980), pp. 119–134.

Armienta, M.A. and Segovia, N.: Arsenic and fluoride in the groundwater of Mexico. *Environ. Geochem. Health* 30 (2008), pp. 345–353.

Armienta, M.A., Amat, P.D., Larios, T. and López, D.L.: América Central y México. In: J. Bundschuh, A. Pérez-Carrera and M.I. Litter (eds): *Distribución del arsénico en las regiones Ibérica e Iberoamericana*. Editorial Programa Iberoamericano de Ciencia y Tecnologia para el Desarrollo, Buenos Aires, Argentina, 2008, pp. 187–210. Available at: http://www.cnea.gov.ar/xxi/ambiental/iberoarsen/

Astolfi, E.A.N., Maccagno, A., García-Fernández, J.C., Vaccaro, R. and Stimola, R.: Relation between arsenic in drinking water and skin cancer. *Biolog. Trace Elements Res.* 3 (1981), pp. 133–143.

Astolfi, E.A.N., Besuschio, S.C., García-Fernández, J.C., Guerra, C. and Maccagno, A.: *Hidroarsenicismo Crónico Regional Endémico*. Edit. Coop. Gral. Belgrano, Buenos Aires, Argentina, 1982.

ATSDR: Toxicological profile for arsenic. US Department of Health and Human Services, Agency for Toxic Substances and Disease Registry, 2007.

Avena, M.J. and Koopal, L.K.: Desorption of humic acids from an iron oxide surface. *Environ. Sci. Technol.* 32 (1998), pp. 2572–2577.

Ayerza, A.: Arsenicismo regional endémico (keratodermia y melanodermia combinadas). *Bol. Acad. Medicina* 2–3 (1917a), pp. 11–24.

Ayerza, A.: Arsenicismo regional endémico (keratodermia y melanodermia combinadas) (continuación). *Bol. Acad. Medicina* 2–3 (1917b), pp. 41–55.

Ayerza, A.: Arsenicismo regional endémico (keratodermia y melanodermia combinadas) (continuación). *Bol. Acad. Medicina* (1918), pp. 1–24.

Ayotte, J.D., Nielson, M.G. and Robinson, G.R.: Relation of arsenic concentrations in ground water to bedrock lithology in eastern New England. *Geol. Soc. Amer. Annual Meet. Abstracts with Programs*, St. Louis, MO, A-58, 1989.

Ball, J.W., Nordstrom, D.K., Jenne, E.A. and Vivit, D.V.: Chemical analyses of hot springs, pools, geysers, and surface waters from Yellowstone National Park, Wyoming, and Vicinity, 1974–1975. USGS Open-File Rep. 98–182, 1998.

Ballantyne, J.M. and Moore, J.N.: Arsenic geochemistry in geothermal systems. *Geochim. Cosmochim. Acta* 52:2 (1988): pp. 475–483.

Barragne, P.: Contribución al estudio de cinco zonas contaminadas naturalmente por arsénico en Nicaragua. UNICEF, Managua, Nicaragua, 2004.

Barrow, N.J.: The effect of time on the competition between anions for sorption. *J. Soil Sci.* 43 (1992), pp. 421–428.

Bates, M.N., Smith, A.H. and Hopenhayn-Rich, C.: Arsenic ingestion and internal cancers: A review. *Am. J. Epidemiol.* 135 (1992), pp. 462–475.

Bates, M., Marshall, G. and Smith, A.: Arsenic-related cancer mortality in northern Chile, 1989–98. *Epidemiology* 15:4 (2004), pp. S108–S108.

Baudrimont, M., Schäfer, J., Marie, V., Maury-Brachet, R., Bossy, C., Boudou, A. and Blanc, G.: Geochemical survey and metal bioaccumulation of three bivalve species (Crassostrea gigas, Cerastoderma edule and Ruditapes philippinarum) in the Nord Médoc salt marshes (Gironde estuary, France). *Sci. Tot* Süßwassermolasse Bayerns. In: U. Kleeberger, H. Frisch and G. Heinrichs (eds): *Arsen im Grund- und Trinkwasser Bayerns*. Sven von Loga, Köln, Germany, 1998, pp. 43–66.

Baza, D., Iturre de Aguirre, L. and Aguirre, S.: Hidroarsenisimo crónico. *Piel* 14 (2001), pp. 4–8.

Beak, D.G., Wilkin, R.T., Ford, R.G. and Kelly, S.D.: Examination of arsenic speciation in sulfidic solutions using X-ray absorption spectroscopy. *Environ. Sci. Technol.* 42 (2008), pp. 1643–1650.

Bauer, M. and Blodau, C.: Mobilization of arsenic by dissolved organic matter from iron oxides, soils and sediments. *Sci. Tot. Environ.* 354:2–3 (2006), pp. 179–190.

Baur, W.H. and Onishi, B.M.H.: Arsenic. In: K.H. Wedepohl (ed): *Handbook of geochemistry*. Springer, Berlin, Germany, 1969, pp. 33-A-133-0-5.

Bayer, M. and Henken-Mellies, W.-U.: Untersuchungen arsenführender Sedimente in der Oberen Süßwassermolasse Bayerns. In: U. Kleeberger, H. Frisch and G. Heinrichs (eds): *Arsen im Grund- und Trinkwasser Bayerns*. Sven von Loga, Köln, Germany, 1998, pp. 43–66.

Belkin, H.E., Zheng, B. and Finkelman, R.B.: Human health effects of domestic combustion of coal in rural China: a causal factor for arsenic and fluorine poisoning. *The Second World Chinese Conference on Geological Sciences, Extended Abstracts with Programs*, Stanford University, Stanford, CA, 2000, pp. 522–524.

Belkova, N.L., Zakharova, J.R., Tazaki, K., Okrugin, V.M., and Parfenova, V.V.: Fe-Si biominerals in the Vilyuchinskie hot springs, Kamchatka Peninsula, Russia. *Int. Microbiol.* 7:3 (2004), pp. 193–198.

Berg, M., Tran, H.C., Nguyen, T.C., Pham, H.V., Schertenleib, R., Giger, W.: Arsenic contamination of groundwater and drinking water in Vietnam: a human health treat. *Environ. Sci. Technol.* 35 (2001): pp. 2621–2626.

Berg, M., Luzi, S., Trang, P.T.K, Viet, P.H., Giger, W. and Stueben, D.: Arsenic removal from groundwater by household sand filters: Comparative field study, model calculations, and health Benefits. *Env. Sci. Technol.* 40 (2006), pp. 5567–5573.

Berg, M., Stengel, C., Trang, P.K.T, Viet, P.H., Sampson, M.L., Leng, M., Samreth, S. and Fredericks, D.: Magnitude of arsenic pollution in the Mekong and Red River Deltas-Cambodia and Vietnam. *Sci. Total Environ.* 372 (2007), pp. 413–425.

Berg, M., Trang, P.T.K., Stengel, C., Buschmann, J., Viet, P.H., Dan, N.V., Giger, W. and Stüben, D.: Hydrological and sedimentary controls leading to arsenic contamination of groundwater in the Hanoi Area, Vietnam: The impact of iron-arsenic ratios, peat, river bank deposits, and excessive groundwater abstraction. *Chem. Geology* 249 (2008), pp. 91–112.

BGS [British Geological Survey] and DPHE [Department of Public Health Engineering, Bangladesh]: Arsenic contamination of groundwater in Bangladesh. In: D.G. Kinniburg and P.L. Smedley (eds): BGS Technical Report WC/00/19, Keyworth, UK, 2001.

Bhattacharjee, Y.: A sluggish response to humanity's biggest mass poisoning. *Science* 315 (2007), pp. 1659–1661.

Bhattacharya, P.: Arsenic contaminated groundwater from the sedimentary aquifers of south-east Asia. In: E. Bocanegra, D. Martinez and H. Massone (eds): *Groundwater and Human Development*, Proceedings of the XXXII IAH and VI ALHSUD Congress, Mar del Plata, Argentina, 21–25 October 2002, pp. 357–363.

Bhattacharya, P., Chatterjee, D. and Jacks, G.: Occurrence of As-contaminated groundwater in alluvial aquifers from the Delta Plains, eastern India: option for safe drinking water supply. *Int J. Water Res. Dev.* 13 (1997), pp. 79–92.

Bhattacharya, P., Frisbie, S.H., Smith, E., Naidu, R., Jacks, G. and Sarkar, B.: Arsenic in the environment: a global perspective. In: B. Sarkar (ed): *Hand book of heavy metals in the environment*. Marcel Dekker, New York, NY, 2002a, pp. 145–215.

Bhattacharya, P., Jacks, G., Ahmed, K.M., Khan, A.A. and Routh, J.: Arsenic in groundwater of the Bengal Delta Plain aquifers in Bangladesh. *Bull. Environ. Contamin. Toxicol.* 69 (2002b), pp. 538–545.

Bhattacharya, P., Mukherjee, A.B., Jacks, G. and Nordqvist, S.: Metal contamination at a wood preservation site: Characterization and experimental studies on remediation. *Sci. Tot. Environ.* 290:1–3 (2002c), pp. 168–180.

Bhattacharya, P., Ahmed, K.M., Broms, S., Fogelström, J., Jacks, G., Sracek, O., von Brömssen, M., Routh, J.: Mobility of arsenic in groundwater in a part of Brahmanbaria district, NE Bangladesh. In: R. Naidu, E. Smith, G. Owens, P. Bhattacharya and P. Nadebaum (ed): *Managing arsenic in the environment: From soil to human health.* CSIRO Publishing, Melbourne, Australia, 2006a, pp. 95–115.

Bhattacharya, P., Claesson, M., Bundschuh, J., Sracek, O., Fagerberg, J., Jacks, G., Martin, R.A., Storniolo, A. and Thir, J.M.: Distribution and mobility of arsenic in the Rio Dulce alluvial aquifers in Santiago del Estero Province, Argentina. *Sci. Tot. Environ.* 358 (2006b), pp. 97–120.

Bhattacharya, P., Welch, A.H., Stollenwerk, K.G, McLaughlin, M., Bundschuh, J. and Panaullah, G. (eds): Arsenic in the environment: Biology and chemistry. Special Issue, *Sci. Total Environ.* 379:2–3 (2007a), pp. 109–265.

Bhattacharya, P., Welch, A.H., Stollenwerk, K.G, McLaughlin, M., Bundschuh, J. and Panaullah, G.: Arsenic in the environment: Biology and chemistry. In: P. Bhattacharya, A.H. Welch, K.G. Stollenwerk, M. McLaughlin, J. Bundschuh, G. Panaullah: Arsenic in the environment: Biology and chemistry. Special Issue, *Sci. Total Environ.* 379:2–3 (2007b), pp. 109–120.

Bhumbla, D.K. and Keefer, R.F.: Arsenic mobilization and bioavailability in soils. In: J.O. Nriagu (ed): *Arsenic in the environment*, Part I: *Cycling and characterization*. John Wiley & Sons, New York, NY, 1994, pp. 51–82.

Biagini, R.E., Salvador, M.A., Qüerio, R.S. de, Torres-Soruco, C.A., Biagini, M.M. and Diez-Barrantes, A.: Hidroarsenicismo crónico. Comentario de casos diagnosticados en el periodo 1972–1993. *Arch. Arg. Dermatol.* 45 (1995), pp. 47–52.

Bianchelli, T.: Arsenic removal from drinking water. Nova Science Pub. Inc., 2004.

Birkle, P.: Herkunft und Umweltauswirkungen der Geothermalwässer von Los Azufres. *Wiss. Mitt.* 6, Institute for Geology, Technical University Freiberg, Germany, 1998.

Birkle, P.: Estudio isotópico y químico para la definición del origen de los acuíferos profundos del Activo Pol-Chuc. Final report, Instituto de Investigaciones Eléctricas, Cuernavaca, Mexico, IIE/11/12093/I02/F, 2003.

Birkle, P.: Caracterización química e isotópica de los acuíferos profundos del campo Jujo-Tecominoacán. Final report, Instituto de Investigaciones Eléctricas, Cuernavaca, Mexico, IIE/11/2473/F, 2004.

Birkle, P. and Angulo, M.: Conceptual hydrochemical model of Late Pleistocene aquifers at the Samaria-Sitio-Grande petroleum reservoir, Gulf of Mexico, Mexico. *Appl. Geochem.* 20:6 (2005), pp. 1077–1098.

Birkle, P. and Bundschuh, J.: The abundance of natural arsenic in deep thermal fluids of geothermal and petroleum reservoirs in Mexico. In: J. Bundschuh, M.A. Armienta, P. Birkle, P. Bhattacharya, J. Matschullat and A.B. Mukherjee (eds): Geogenic arsenic in groundwater of Latin America. In: J. Bundschuh and P. Bhattacharya (series eds): *Arsenic in the environment*, Volume 1. CRC Press/Balkema Publisher, Leiden, The Netherlands, 2009, pp. 145–153.

Birkle, P. and Maruri, R.A.: Isotopic indications for the origin of formation water at the Activo Samaria-Sitio-Grande oil field, Mexico. *J. Geochem. Explor.* 78–79 (2003), pp. 453–458.

Birkle, P. and Portugal, E.: Caracterización química e isotópica de los acuíferos profundos de los campos petroleros Cactus, Níspero, Río Nuevo y Sitio grande en Chiapas: origen y dinámica de la migración. Final report 11-11840, Instituto de Investigaciones Eléctricas, Cuernavaca, Mexico, 2001.

Birkle, P., Portugal, E., Rosillo, J.J. and Fong, J.L.: Evolution and origin of deep reservoir water at the Activo Luna oil field, Gulf of Mexico, Mexico. *AAPG Bull.* 86:3 (2002), pp. 457–484.

Birkle, P., Martínez-García, B. and Milland-Padrón, C.M.: Origin and evolution of formation water at the Jujo–Tecominoacán oil reservoir, Gulf of Mexico. Part 1: Chemical evolution and water-rock interaction. *Appl. Geochem.* 24:4 (2009a), pp. 543–554.

Birkle, P., Martínez-García, B., Milland Padrón, C.M. and Eglington, B.M.: Origin and evolution of formation water at the Jujo-Tecominoacán oil reservoir, Gulf of Mexico. Part 2: Isotopic and field-production evidence for fluid connectivity. *Appl. Geochem.* 24:4 (2009b), pp. 555–573.

Blackwell, R.Q.: Estimated total arsenic ingested by residents in the endemic blackfoot disease area. *J. Formos. Med. Assoc.* 60 (1961), pp. 1143–1144.

Borgoño, J.M. and Greiber, R.: Epidemiological study of arsenism in the city of Antofagasta. *Trace Subst. Environ. Health* 5 (1972), pp. 13–24.

Borgoño, J.M., Vicent, P., Venturino, H. and Infante, A.: Arsenic in the drinking water of the city of Antofagasta: epidemiological and clínica: study before and after the installation of a treatment plant. *Environ. Health Perspect.* 19 (1977), pp. 103–105.

Borja, A., Calderón, J., Díaz Barriga, F., Goleen, A., Jiménez, C., Navarro, M., Rodríguez, L. and Santos, D.: Exposure to arsenic and lead and neuropsychological development in Mexican children. *Environ. Res. Sect.* A85 (2001), pp. 69–76.

Bostick, B.C., Fendorf, S. and Manning, B.A.: Arsenite adsorption on galena (PbS) and sphalerite (ZnS). *Geochim. Cosmochim. Acta* 67 (2003), pp. 895–907.

Bostick, B.C., Fendorf, S. and Brown, Jr., G.E.: *In situ* analysis of thioarsenite complexes in neutral to alkaline sulphide solutions. *Miner. Mag.* 69 (2005), pp. 781–795.

Boudette, E.L., Canney, F.C., Cotton, J.E., Davis, R.I., Ficklin, W.H. and Motooka, J.M.: High levels of arsenic in the groundwater of southeastern New Hampshire: A geochemical reconnaissance. USGS Open-File Report 85-202, 1985.

Bowell, R.J.: Sorption of arsenic by iron oxides and oxyhydroxides in soils. *Appl. Geochem.* 9 (1994), pp. 279–286.

Boyle, R.W.: Geology, geochemistry, and origin of the lead-zinc-silver deposits of the Keno Hill-Galena Hill area, Yukon territory. *Geol. Surv. Canada Bull.* 111, 1965a.

Boyle, R.W.: The geochemistry of arsenic, Keno Hill-Galena Hill area, Yukon, Canada. D.N. Wadia Commem. Vol., *Mining, Geol. Met. Inst. India* 757, 1965b.

Boyle, R.W. and Jonasson, I.R.: The geochemistry of arsenic and its use as an indicator element in geochemical prospecting. *J. Geochem. Explor.* 2 (1973), pp. 251–296.

Brandhuber, P. and Amy, G.: Alternative methods for membrane filtration of arsenic from drinking water. *Desalination* 117 (1998), pp. 1–10.

Brannon, J.M. and Patrick, W.H.: Fixation, transformation and mobilization of arsenic in sediments. *Environ. Sci. Technol.* 21 (1987), pp. 450–459.

Brennan, E.W. and Lindsay, W.L.: The role of pyrite in controlling metal ion activities in highly reduced soils. *Geochim. Cosmochim. Acta* 60:19 (1996), pp. 3609–3618.

Brown, K.G. and Chen, C.-J.: Significance of exposure assessment to analysis of cancer risk from inorganic arsenic in drinking water in Taiwan. *Risk Anal.* 15 (1995), pp. 475–484.

Brown, K.G., Boyle, K.E., Chen, C.-W. and Gibb, H.J.: A dose-response analysis of skin cancer from inorganic arsenic in drinking water. *Risk Anal.* 9 (1989), pp. 519–528.

Brown, K.L. and Simmons, S.F.: Precious metals in high-temperature geothermal systems in New Zealand. *Geothermics* 32:4 (2003), pp. 619–625.

Buchet, J.P., Lauwerys, R. and Roels, H.: Urinary excretion of inorhanic arsenic and its metabolites after reported ingestion of sodium metaarsenite by volunteers. *Int. Arch. Occup. Environ. Health* 48 (1981), pp. 111–118.

Bundschuh, J. and García M.E.: Rural Latin America—A forgotten part of the global groundwater arsenic problem? In: P. Bhattacharya, Al. Ramanathan, A.B. Mukherjee, J. Bundschuh, D. Chandrasekharam and Keshari, A.K.: *Groundwater for sustainable development: Problems, perspectives and challenges.* Balkema Publisher, The Netherlands, 2008, pp. 311–321.

Bundschuh, J. and Litter, M.I.: Situación de América Latina con relación al problema global del arsénico. In: M.I. Litter and J. Bundschuh (eds): *Situación del arsénico en la Región Ibérica e Iberoamericana. Posibles acciones articuladas e integradas para el abatimiento del arsénico en zonas aisladas.* Editorial Programa Iberoamericano de Ciencia y Tecnologia para el Desarrollo, Buenos Aires, Argentina, 2010, ISBN: 978-84-96023-73-4.

Buschmann, J., Berg, M., Stengel, C., Winkel, L., Sampson, M.L., Trang, P.T.K. and Viet, P.H.: Contamination of drinking water resources in the Mekong delta floodplains: Arsenic and other trace metals pose serious health risks to population. *Environ. Int.* 34:6 (2008), pp. 756–764.

Bundschuh, J., Bonorino, G., Viero, A.P., Albouy, R. and Fuertes, A.: Arsenic and other trace elements in sedimentary aquifers in the Chaco-Pampean Plain, Argentina: origin, distribution, speciation, social and economic consequences. In: P. Bhattacharya and A.H. Welch (eds): Arsenic in groundwater of sedimentary aquifers; *Pre-Congress Workshop, International Geological Congress,* Rio de Janeiro, Brazil, 2000, pp. 27–32; http://www.lwr.kth.se/Personal/personer/bhattacharya_prosun/KTH_DU_Special_Publication.pdf (accessed June 2009).

Bundschuh, J., Farías, B., Martin, R., Storniolo, A., Bhattacharya, P., Cortes, J., Bonorino, G. and Albouy, R.: Groundwater arsenic in the Chaco-Pampean Plain, Argentina: case study from Robles County, Santiago del Estero Province. *Appl. Geochem.* 19 (2004), pp. 231–243.

Bundschuh, J., García, M.E. and Bhattacharya, P.: Arsenic in groundwater of Latin America—A challenge of the 21st century. Geological Society of America Annual Meeting, Philadelphia, 22–25 Oct. 2006, *Geological Society of America Abstracts with Programs* 38:7, 2006, p. 320.

Bundschuh, J., Altamirano Espinoza, M. and Cumbal, L.: Geogenic arsenic in sedimentary aquifers of southwestern Sébaco valley, Nicaragua. Geological Society of América Annual Meeting, Denver, 28–31Oct. 2007, Session No. 192: Arsenic: From Nature to Human I. *Geological Society of America Abstracts with Programs* 39:6 (2007a), p. 518.

Bundschuh, J., Bhattacharya, P., von Brömssen, M., Jakariya, M., Litter, M.I. and García, M.E.: Targeting arsenic-safe aquifers as socially acceptable innovative method of remediation—What lessons can rural Latin America learn from Bangladesh experiences? *Proceedings International Conference on Water and Wastewater Treatment,* Shah Jalal University of Science & Technology, Sylhet, Bangladesh, 1–4 April 2007b.

Bundschuh, J., Pérez-Carrera, A. and Litter, M.I. (eds): *Distribución del arsénico en las regiones Ibérica e Iberoamericana.* Editorial Programa Iberoamericano de Ciencia y Tecnologia para el Desarrollo, Buenos Aires, Argentina, 2008a. Available at: http://www.cnea.gov.ar/xxi/ambiental/iberoarsen/

Bundschuh, J., Giménez-Forcada, E., Guérèquiz, R., Pérez-Carrera, A., Garcia, M.E., Mello, J. and Deschamps, E.: Fuentes geogénicas de arsénico y su liberación al medio ambiente. In: J. Bundschuh, A. Pérez-Carrera, and M.I. Litter (eds): *Distribución del arsénico en las regiones Ibérica e Iberoamericana.* Editorial Programa Iberoamericano de Ciencia y Tecnologia para el Desarrollo, Buenos Aires, Argentina, 2008b, pp. 33–47. Available at: http://www.cnea.gov.ar/xxi/ambiental/iberoarsen/

Bundschuh, J., Nicolli, H.B., Blanco, M. del C., Blarasin, M., Farías, S.S., Cumbal, L., Cornejo, L., Acarapi, J., Lienqueo, H., Arenas, M., Guérèquiz, R., Bhattacharya, P., García, M.E., Quintanilla, J.,

Deschamps, E., Viola, Z., Castro de Esparza, M.L., Rodríguez, J., Pérez-Carrera, A. and Fernández Cirelli, A.: Distribución de arsénico en la región sudamericana. In: J. Bundschuh, A. Pérez-Carrera and M.I. Litter (eds): *Distribución del arsénico en las regiones Ibérica e Iberoamericana*. Editorial Programa Iberoamericano de Ciencia y Tecnologia para el Desarrollo, Buenos Aires, Argentina, 2008c, pp. 137–186. Available at: http://www.cnea.gov.ar/xxi/ambiental/iberoarsen/

Bundschuh, J., Armienta, M.A., Birkle, P., Bhattacharya, P., Matschullat, J. and Mukherjee, A.B. (eds): Natural arsenic in groundwater of Latin America. In: J. Bundschuh and P. Bhattacharya (series eds): *Arsenic in the environment*, Volume 1. CRC Press/Balkema Publisher, Leiden, The Netherlands, 2009a.

Bundschuh, J., García, M.E., Birkle, P., Cumbal, L.H., Bhattacharya, P. and Matschullat, J.: Occurrence, health effects and remediation of arsenic in groundwaters of Latin America. In: J. Bundschuh, M.A. Armienta, P. Birkle, P. Bhattacharya, J. Matschullat and A.B. Mukherjee (eds): Geogenic arsenic in groundwater of Latin America. In: J. Bundschuh and P. Bhattacharya (series eds): *Arsenic in the environment*, Volume 1. CRC Press/Balkema Publisher, Leiden, The Netherlands, 2009b, pp. 3–15.

Bundschuh, J., Bhattacharya, P., von Brömssen, M., Jakariya, M., Jacks, G., Thunvik, R. and Litter, M.I.: Arsenic-safe aquifers as a socially acceptable source of safe drinking water—What can rural Latin America learn from Bangladesh experiences? In: J. Bundschuh, M.A. Armienta, P. Birkle, P. Bhattacharya, J. Matschullat and A.B. Mukherjee (eds): Natural arsenic in groundwater of Latin America. In: J. Bundschuh and P. Bhattacharya (series eds): *Arsenic in the environment*, Volume 1. CRC Press/Balkema Publisher, Leiden, The Netherlands, 2009c, pp. 687–697.

Byrd, D.M., Roegner, M.L., Griffiths, J.C., Lamm, S.H., Grumski, K.S., Wilson, R. and Lai, Shenghan, L.: Carcinogenic risks of inorganic arsenic in perspective. *Int. Arch Occupat. Environ. Health* 68 (1996), pp. 484–494.

CAA (Código Alimentario Argentino): Modification of Articles 982 and 983, May 22, 2007. Buenos Aires, Argentina, 2007, http://www.anmat.gov.ar/normativa/normativa/Alimentos/Resolucion_Conj_68-2007_96-2007.pdf

Cáceres, A.: Arsénico, normativas y efectos en la salud. *Actas del XIII Congreso de Ingeniería Sanitaria y Ambiental* (AIDIS). Santiago de Chile, Chile, 1999, pp. 1–13.

Cáceres, D.D., Pino, P., Montesinos, N., Atalah, E., Amigo, H. and Loomis, D.: Exposure to inorganic arsenic in drinking water and total urinary arsenic concentration in a Chilean population. *Environ. Res.* 98:2 (2005), pp. 151–159.

Cama, J., Rovira, M., Ávila, P., Pereira, M.R., Asta, M.P., Grandia, F., Martínez-Lladó, X. and Álvarez-Ayuso, E.: Distribución de arsénico en la región Ibérica. In: J. Bundschuh, A. Pérez-Carrera and M.I. Litter (eds): *Distribución del arsénico en las regiones Ibérica e Iberoamericana*. Editorial Programa Iberoamericano de Ciencia y Tecnologia para el Desarrollo, Buenos Aires, Argentina, 2008, pp. 95–136. Available at: http://www.cnea.gov.ar/xxi/ambiental/iberoarsen/

Carbonell-Barrachina, A.A., Burló, F., Valero, D., López, E., Matínez-Romero, D. and Matínez-Sánchez, F.: Arsenic toxicity and accumulation in turnip as affected by arsenic chemical speciation. *J. Agric. Food Chem.* 47 (1999), pp. 2288–2294.

Carter, J.M., Sando, S.K., Hayes, T.S. and Hammond, R.H.: Source, occurrence, and extent of arsenic contamination in the Grass Mountain area of the Rosebud Indian reservation, South Dakota. USGS Water Resources Investigation Report 97-4286, 1998.

Castro de Esparza, M.L.: The presence of arsenic in drinking water in Latin America and its effect on public health. In: J. Bundschuh, M.A. Armienta, P. Birkle, P. Bhattacharya, J. Matschullat and A.B. Mukherjee (eds): Natural arsenic in groundwater of Latin America. In: J. Bundschuh and P. Bhattacharya (series eds): *Arsenic in the environment*, Volume 1. CRC Press/Balkema Publisher, Leiden, The Netherlands, 2009, pp. 17–29.

Cebrián, M.E., Albores, A., Aquilar, M. and Blakely, E.: Chronic arsenic poisoning in the north of Mexico. *Hum. Toxicol.* 2 (1983), pp. 121–133.

Cebrián, M.E., Albores, A., García-Vergas, G. and Del Razo, L.M.: Chronic arsenic poisoning in humans: the case of Mexico. In: J.O. Nriagu (ed): *Arsenic in the environment*, Part II. John Wiley & Sons, New York, 1994, pp. 93–107.

Celico, P., Dall'Aglio, M., Ghiara, M.R., Stanzione, D., Brondi, M. and Prosperi, M.: Geochemical monitoring of the thermal fluids in the phlegraean fields from 1970 to 1990. *Boll. Soc. Geol. It.* 111 (1992), pp. 409–422.

Chakraborti, A.K. and Saha, K.C.: Arsenic dermatosis from tubewell water in West Bengal. *Indian J. Medical Res.* 85 (1987), pp. 326–334.

Chakraborti, D., Rahman, M.M., Pau, K., Chowdhury, U.K., Sengupta, M.K., Lodh, D., Chanda, C.R., Saha, K.C. and Mukherjee, S.C.: Arsenic calamity in the Indian subcontinent: What lessons have been learned? *Talanta* 58 (2002), pp. 3–22.

Chakraborti, D., Sengupta, M.K., Rrahaman, M.M., Ahamed, S., Chowdhury, U.K., Hossain, Md A., Mukherjee, S.C., Pati, S., Saha, K.C., Dutta, R.N. and Quamruzzaman, Q.: Groundwater arsenic contamination and its health effects in the Ganga-Meghna-Brahmaputra plain. *J. Environ. Monit.* 6:6 (2004), pp. 74N–83N.

Chandrasekharam, D. and Bundschuh, J. (eds): *Geothermal energy resources for developing countries.* A.A. Balkema, Lisse, The Netherlands, 2002.

Chang, C.-C., Ho, S.-C., Tsai, S.S. and Yang, C.-Y.: Ischemic heart disease mortality reduction in an arseniasis-endemic area in southwestern Taiwan after a switch in the tap-water supply system. *J. Toxicol. Environ. Health* A 67 (2004), pp. 1353–1361.

Chang, T.-C., Hong, M.-C. and Chen, C.-J.: Higher prevalence of goiter in endemic area of blackfoot disease of Taiwan. *J. Formosan Med. Assoc.* 90 (1991), pp. 941–946.

Chang, T.-K., Shyu, G.-S. and Lin, Y.-P.: Geostatistical analysis of soil arsenic content in Taiwan. *Environ. Sci. Health* A: *Toxic Hazard. Subst. Environ. Eng.* 34 (1999), pp. 1485–1501.

Chang, T.-K., Shyu, G.-S., Huang, C.-H. and Lin, C.-S.: Project of investigation and verification of heavy metals and arsenic in the municipal agricultural land soil of Taipei. Environmental Protection Agency of Taipei City Government, DEP-95-056, Taipei, Taiwan, 2007 (in Chinese).

Chappell, W.R., Abernathy, C.O. and Cothern, C. (eds): Arsenic exposure and health. *Sci. Technol. Lett.* (1994), Northwood, pp. 109–117.

Chappell, W.R., Abernathy, C.O. and Calderón, R.L. (eds): *Arsenic exposure and health effects.* Elsevier, Amsterdam, The Netherlands, 2001.

Chen, B.-C. and Liao, C.-M.: Farmed tilapia *Oreochromis mossambicus* involved in transport and biouptake of arsenic in aquacultural ecosystems. *Aquaculture* 242:1–4 (2004), pp. 365–380.

Chen, C.-J.: Environmental health issues in public health. In: S. Detel, R. Beaglehole, M.A. Lansang and M. Gulliford (eds): *Oxford textbook of public health.* 5th ed, Oxford University Press, New York, NY, 2009.

Chen, C.-J. and Wang, C.-J.: Ecological correlation between arsenic level in well water and age-adjusted mortality from malignant neoplasms. *Cancer Res.* 50 (1990), pp. 5470–5474.

Chen, C.-J., Chung, Y.-C., Lin, T.-M. and Wu, H.-Y.: Malignant neoplasms among residents of a black-foot disease-endemic area in Taiwan: High-arsenic artesian well water and cancers. *J. Cancer Res.* 45 (1985), pp. 5895–5899.

Chen, C.-J., Chuang, Y.-C., You, S.-L., Lin, T.-M. and Wu, H.-Y.: A retrospective study on malignant neoplasms of bladder, lung and liver in blackfoot disease endemic area in Taiwan. *Br. J. Cancer* 53 (1986), pp. 399–405.

Chen, C.-J., Kuo, T.-L. and Wu, M.-M.: Arsenic and cancers. *Lancet* 2 (1988a), pp. 414–415.

Chen, C.-J., Wu, M.-M., Lee, S.-S., Wang, J.-D., Cheng, S.-H. and Wu, H.-Y.: Atherogenicity and carcinogenicity of high-arsenic artesian well water. Multiple risk factors and related malignant neoplasms of blackfoot disease. *Arteriosclerosis* 8 (1988b), pp. 452–460.

Chen, C.-J., Chen, C.-W., Wu, M.-M. and Kuo, T.-L.: Cancer potential in liver, lung, bladder and kidney due to ingested inorganic arsenic in drinking water. *Br. J. Cancer* 66 (1992), pp. 888–892.

Chen, C.-J., Hsueh, Y.-M., Lai, M.-S., Hsu, M.-P., Wu, M.-M. and Tai, T.-Y.: Increased prevalence of hypertension and long-term arsenic exposure. *Hypertension* 25 (1995), pp. 53–60.

Chen, C.-J., Chiou, H.-Y. and Chiang, M.-H.: Dose-response relationship between ischemic heart disease mortality and long-term arsenic exposure. *Arterioscl. Throm. Vas. Biol.* 16 (1996), pp. 504–510.

Chen, C.-J., Hsu, L.-I., Shih, W.-L., Hsu, Y.-H., Tseng, M.-P., Lin, Y.-C., Chou, W.-L., Chen, C.-Y., Lee, C.-Y., Wang, L.-H., Cheng, Y.-C., Chen, C.-L., Chen, S.-Y., Wang, I.-H., Hsueh, Y.-M., Chiou, H.-Y. and Wu, M.-M.: Biomarkers of exposure, effect and susceptibility of arsenic-induced health hazards in Taiwan. *Toxicol. Appl. Pharmacol.* 206 (2005), pp. 198–206.

Chen, C.-J., Wang, S.-L., Chiou, J.-M., Tseng, C.-H., Chiou, H.-Y., Hsueh, Y.-M., Chen, S.-Y., Wu, M.-M. and Lai, M.-S.: Arsenic and diabetes and hypertension in human populations: a review. *Toxicol. Appl. Pharmacol.* 222 (2007), pp. 298–304.

Chen, C.-H., Chiou, H.-Y., Hsueh, Y.-M., Chen, C.-J., Yu, H.-J. and Pu, Y.-S.: Clinicopathological characteristics and survival outcome of arsenic related bladder cancer in Taiwan. *J. Urol.* 181 (2009), pp. 547–552.

Chen, C.-L., Hsu, L.-I., Chiou, H.-Y., Hsueh, Y.-M., Chen, S.-Y., Wu, M.-M., Chen, C.-J. and Blackfoot Disease Study Group: Ingested arsenic, cigarette smoking, and lung cancer risk: a follow-up study in arseniasis-endemic areas in Taiwan. *J. Am. Med. Assoc.* 292 (2004). pp. 2984–2990.

Chen, I.-J.: *Geochemical characteristics of porewater and sediments from Chung-hsing, Wu-jie and Long-de of I-Lan plain, Taiwan.* MSc Thesis, Department of Geosciences, National Taiwan University, Taipei, Taiwan, 2001 (in Chinese).

Chen, J.-R.: Taiwan's experience of arsenic and it's application in the world. Scientific American (Chinese Edition) 31 (2004), pp. 79–81.

Chen, K.-P.: Epidemiological study on blackfoot disease. Research report in 1974–1975. Blackfoot Disease Research Report 3, Taiwan Provincial Department of Health, Taichung, 1976, pp. 1–29.

Chen, K.-P. and Wu, H.-Y.: Epidemiologic studies on blackfoot disease: 2. A study of source of drinking water in relation to the disease. *J. Formos. Med. Assoc.* 61 (1962), pp. 611–617.

Chen, K.-P., Wu, H.-Y. and Wu, T.-C.: Epidemiologic studies on blackfoot disease in Taiwan: III. Phystaochemical characteristics of drinking water in endemic blackfoot disease areas. *Mem. College Med. Natl. Taiwan Univ.* 8 (1962), pp. 115–129.

Chen, K.-Y. and Liu, T.-K.: Major factors controlling arsenic occurrence in the groundwater and sediments of the Chianan coastal plain, SW Taiwan. *Terr. Atmos. Ocean. Sci.* 18:5 (2007), pp. 975–994.

Chen, S.-L., Dzeng, S.-R., Yang, M.-H., Chiu, K.-H., Shieh, G.-M. and Wai, C.-M.: Arsenic species in groundwaters of the blackfoot disease area, Taiwan. *Environ. Sci. Technol.* 28 (1994), pp. 877–881.

Chen, Y.-C., Guo, Y.-L.L., Su, H.-J.J., Hsueh, Y.-M., Smith, T.J., Ryan, L.M., Lee, M.-S., Chao, S.-C., Lee, J.Y.-Y. and Christiani, D.C.: Arsenic methylation and skin cancer risk in southwestern Taiwan. *J. Occup. Environ. Med.* 45 (2003a), pp. 241–248.

Chen, Y.-C., Su, H.-J.J., Guo, Y.-L.L., Hsueh, Y.-M., Smith, T.J., Ryan, L.M., Lee, M.-S. and Christiani, D.C.: Arsenic methylation and bladder cancer risk in Taiwan. *Cancer Cause. Control* 14 (2003b), pp. 303–310.

Cheng, R.-C., Liang, S., Wang, H.-C. and Beuhler, M.D.: Enhanced coagulation for arsenic removal. *J. AWWA* 86:9 (1994), pp. 79–90.

Cherry, J.A., Shaikh, A.U., Tallman, D.E. and Nicholson, R.V.: Arsenic species as an indicator of redox conditions in groundwater. *J. Hydrol.* 43 (1979), pp. 373–392.

Ch'i, I.-C. and Blackwell, R.Q.: A controlled retrospective study of blackfoot disease, an endemic peripheral gangrene diseases in Taiwan. *Am. J. Epidemiol.* 88 (1968), pp. 7–24.

Chiou, H.-Y., Hsueh, Y.-M., Liaw, K.-F., Horng, S.-F., Chiang, M.-H., Pu, Y.-S., Lin, J.-S., Huang, C.-H. and Chen, C.-J.: Incidence of internal cancers and ingested inorganic arsenic: a seven-year follow-up study in Taiwan. *Cancer Res.* 55 (1995), pp. 1296–1300.

Chiou, H.-Y., Huang, W.-I., Su, C.-L., Chang, S.-F., Hsu, Y.-H. and Chen, C.-J.: Dose-response relationship between prevalence of cerebrovascular disease and ingested inorganic arsenic. *Stroke* 28 (1997), pp. 1717–1723.

Chiou, H.-Y., Chiou, S.-T., Hsu., Y.-H., Chou, Y.-L., Tseng, C.-H., Wei, M.-L. and Chen, C.-J.: Incidence of transitional cell carcinoma and arsenic in drinking water: a follow-up study of 8,102 residents in an arseniasis-endemic area in northeastern Taiwan. *Am. J. Epidemiol.* 153 (2001), pp. 411–418.

Chiu, H.-F. and Yang, C.-Y.: Decreasing trend in renal disease mortality after cessation from arsenic exposure in a previous arseniasis-endemic area in southwestern Taiwan. *J. Toxicol. Environ. Health* A 68 (2005), pp. 319–327.

Chiu, H.-F., Ho, S.-C. and Yang, C.-Y.: Lung cancer mortality reduction after installation of tap-water supply system in an arseniasis-endemic area in southwestern Taiwan. *Lung Cancer* 46 (2004a), pp. 265–270.

Chiu, H.-F., Ho, S.-C., Wang, L.-Y., Wu, T.-N. and Yang, C.-Y.: Does arsenic exposure increase the risk for liver cancer? *J. Toxicol. Environ. Health* A 67 (2004b), pp. 1491–1500.

Choprapawon, C. and Rodcline, A.: Chronic arsenic poisoning in Ronpibool Nakhon Sri Thammarat, the southern province of Thailand. In: C.O. Abernathy, R.L. Calderon and W.R. Chppel (eds): *Arsenic: Exposure and health effects.* Chapman and Hall, London, UK, 1997, pp. 69–77.

Chow, J., Lee, J.-S., Liu, C.-S., Lee, B.-D. and Watkins, J.S.: A submarine canyon as the cause of a mud volcano: Liuchieuyu Island in Taiwan. *Mar. Geol.* 176 (2001), pp. 55–63.

Chou, J.-T.: A preliminary study of the stratigraphy and sedimentation of the mudstone formations in the Taiwan area, Southern Taiwan. *Petroleum Geology Taiwan* 8 (1971), pp. 187–219.

Chow, N.-H., Guo, Y.-L., Lin, J.-S., Su, H.-J, Tzai, T.-S., Guo, H.-R. and Su, I.-J.: Clinicopathological features of bladder cancer associated with chronic exposure to arsenic. *Br. J. Cancer* 75 (1997), pp. 1708–1710.

Chowdhury, U.K., Biswas, B.K., Roychowdhuri, T., Samanta, G., Mandal, B.K., Basu, G.K., Chandra, C.R., Lodh, D., Saha, K.C., Mukherjee, S.K., Roy, S., Kabir, S., Quamruzzaman, Q. and Chakraborti, D.: Groundwater arsenic contamination in Bangladesh and West Bengal, India. *Environ. Health Perspect.* 108:5 (2000), pp. 393–397.

Christensen, O.D., Capuano, R.A. and Moore J.N.: Trace-element distribution in an active hydrothermal system, Roosevelt hot springs thermal area, Utah. *J. Volcanol. Geotherm. Res.* 16:1–2 (1983), pp. 99–129.

Christian, J. and Hopenhayn, C.M.: Is arsenic metabolism in exposed populations influenced by selenium? *Epidemiology* 15:4 (2004), pp. S107–S107.

Círculo Médico del Rosario: Sobre la nueva enfermedad descubierta en Bell-Ville. *Rev. Médica del Rosario* VII (1917), p. 485.

Claesson, M. and Fagerberg, J.: *Arsenic in groundwater of Santiago del Estero: Sources, mobility patterns and remediation with natural materials.* MSc Thesis, Department of Land and Water Resources Engineering KTH, Stockholm, Sweden, TRITA-LWREX-03-05, 2003.

Clarke, F.W.: The data of geochemistry. USGS Bulletin 330, 1908.

Clarke, L.B. and Sloss, L.L.: Trace elements-emissions from coal combustion and gasification. *IEA Coal Research, IEACR* 49, London, UK, 1992, p. 111.

Clarke, M.B. and Helz, G.R.: Metal-thiometalate transport of biologically active trace elements in sulfidic environments. 1. Experimental evidence for copper thioarsenite complexing. *Environ. Sci. Technol.* 34 (2000), pp. 1477–1482.

Clifford, D.A.: Ion exchange and inorganic adsorption. In: F.W. Pontius (ed): *Water quality and treatment.* 4th ed, American Water Works Association, McGraw-Hill, Inc., New York, NY, 1990.

Concha, G., Nermell, B. and Vahter, M.: Metabolism of inorganic arsenic in children with chronic high arsenic exposure in northern Argentina. *Environ. Health Persp.* 106:6 (1998a), pp. 355–359.

Concha, G., Vogler, G., Lezcano, D., Nermell, B. and Vahter, M.: Exposure to inorganic arsenic metabolites during early human development. *Toxicol. Sci.* 44:2 (1998b), pp. 185–190.

Cooper, C.B., Doyle, M.E. and Kipp, K.: Risk of consumption of contaminated seafood: the Quincy Bay case study. *Environ. Health Persp.* 90 (1991), pp. 133–140.

Criaud, A. and Fouillac, C.: The distribution of arsenic (III) and arsenic (V) in geothermal waters: Examples from the Massif Central of France, the Island of Dominica in the Leeward Islands of the Caribbean, the Valles Caldera of New Mexico, U.S.A., and southwest Bulgaria. *Chem. Geology* 76:3–4 (1989), pp. 259–269.

Crump, K.S., Hoel, D.G., Langley C.H. and Peto, R.: Fundamental carcinogenic processes and their implications for low dose risk assessment. *Cancer Res.* 36 (1976), pp. 2973–2979.

CTCI Corporation: Survey and monitor of heavy metals in soil in Taiwan (survey density 100 ha/sample). Project Report, Survey and monitor of heavy metals in soil in Taiwan, Taipei, Taiwan, 2001.

Cullen, W.R.: *Is arsenic an aphrodisiac? The sociochemistry of an element.* Springer, Berlin, Germany. 2008.

Cullen, W.R. and Reimer, K.J.: Arsenic speciation in the environment. *Chem. Rev.* 89 (1989), pp. 713–764.

Cumbal, L., Bundschuh, J., Aguirre, V., Murgueitio, E., Tipán, I. and Chavez, C.: The origin of arsenic in waters and sediments from Papallacta lake in Ecuador. In: J. Bundschuh, M.A. Armienta, P. Birkle, P. Bhattacharya, J. Matschullat and A.B. Mukherjee (eds): Natural arsenic in groundwater of Latin America. In: J. Bundschuh and P. Bhattacharya (series eds): *Arsenic in the environment*, Volume 1. CRC Press/Balkema Publisher, Leiden, The Netherlands, 2009. pp. 81–90.

Cusicanqui, H., Mahon, W.A.J. and Ellis, A.J.: The geochemistry of the El Tatio geothermal field, northern Chile. *2nd United Nations Geothermal Symposium Proceedings*, Lawrence Berkeley Laboratory, Univ. of California, Berkeley, CA, 1976.

Dangic, A. and Dangic, J.: Arsenic in the soil environment of central Balkan Peninsula, southeastern Europe: occurrence, geochemistry, and impacts. In: P. Bhattacharya, A.B. Mukherjee, J. Bundschuh, R. Zevenhoven and R.H. Loeppert: *Arsenic in soil and groundwater environment; Trace metals and other contaminants in the environment*, Volume 9, Elsevier, Amsterdam, The Netherlands, 2007, pp. 207–236.

Darby, d'E.C: Arsenic and boron in the Tongonan environment. *Geotherm. Resour. Counc.* 4; Conference: Geothermal Resource Council annual meeting, Salt Lake City, UT, 1980.

Darland, J.E. and Inskeep, W.P.: Effects of pH and phosphate competition on the transport of arsenate. *J. Environ. Qual.* 26 (1997), pp. 1133–1139.

Das, H.K., Sengupta, P.K., Hossain, A., Islam, M. and Islam, F.: Diversity of environmental arsenic pollution in Bangladesh. In: M.-F. Ahmed, S.A. Tanveer and A.B.M. Badruzzaman (eds): *Bangladesh environment*, Vol. 1. Bangladesh Paribeh Andolon, Dhaka, Bangladesh, 2002, pp. 234–244.

Davies, J.A. and Kent, D.B.: Surface complexation modeling in aqueous geochemistry. In: M.F. Hochella and A.F. White (eds): Mineral-water interface geochemistry. *Rev. Mineral.* 23. Mineral Society of America, 1990, pp. 177–260.

Davies, J., Davis, R., Frank, D., Frost, F., Garland, D., Milham, S., Pierson, R.S., Raasina, R.S., Safioles, S. and Woodruff, L.: Seasonal study of arsenic in ground water: Snohomish County, Washington. Snohomish Health District and Washington State Department of Health, 1991.

De Carlo, E.H. and Thomas, D.M.: Removal of arsenic from geothermal fluids by adsorptive bubble flotation with colloidal ferric hydroxide. *Environ. Sci. Technol.* 9:6 (1985), pp. 538–544.

De Jong, S.J. and Kikietta, A.: Une particuliarité heureusement bien localisée; la présence d'arsenic en concentration toxique dans un village près de Mogtedo (Haute Volta). *Bulletin du Comité Interafricain d'Etudes Hydrauliques* (CIEH) No. 42–43 (1980).

Del Razo, L., Gonsebatt, M., Gutiérrez, M. and Ramírez, P.: Arsenite induces DNA-protein crosslink and cytokeratin expression in the WRL-68 human hepatic cell line. *Press. Carcinogen.* 21:4 (2000), pp. 701–706.

Del Razo, L., De Vizcaya, R., Izquierdo, V., Sanchez, P. and Soto, C.: Diabetogenic effects and pancreatic oxidative damage in rats subcronically exposed to arsenite. *Toxicol. Lett.* 160 (2005), pp. 135–142.

Di Benedetto, F., Costagliola, P., Benvenuti, M., Lattanzi, P., Romanelli, M. and Tanelli, G.: Arsenic incorporation in natural calcite lattice: Evidence from electron spin echo spectroscopy. *Earth Planet. Sci. Lett.* 246 (2006), pp. 458–465.

Dixit, S. and Hering, J.G.: Comparison of arsenic (V) and arsenic (III) sorption onto iron oxide minerals: Implications for arsenic mobility. *Environ. Sci. Technol.* 37 (2003), pp. 4182–4189.

Donohue, J.M. and Abernathy, C.O.: Exposure to inorganic arsenic from fish and shellfish. In: W.R. Chappell and C.O. Abernathy (eds): *Arsenic exposure and health effects.* Elsevier, Amsterdam, The Netherlands, 1999, pp. 89–98.

Drahota, P., Rohovec, J., Filippi, M., Mihaljevic, M., Rychlovsky, P., Cerveny, V. and Pertols, Z.: Mineralogical and geochemical controls of arsenic speciation and mobility under different redox conditions in soil, sediment and water at the Mokrsko-West gold deposit, Czech Republic. *Sci. Total Environ.* 407:10 (2009), pp. 3372–3384.

Drever, J.I.: *The geochemistry of natural waters.* Prentice Hall, New Jersey, NJ, 1997.

Driehaus, W.: Technologies for arsenic removal from potable water. In: J. Bundschuh, P. Bhattacharya and D. Chandrasekharam (eds): *Natural arsenic in groundwater.* Balkema, Leiden, The Netherlands, 2005, pp. 189–203.

Driehaus, W., Jekel, M. and Hildebrandt, U.: Granular ferric hydroxide: A new adsorbent for the removal of arsenic from natural water. *J. Water SRT-Aqua* 47 (1998), pp. 30–35.

Drury, S.: TOPIC3 Water and well-being: arsenic in Bangladesh. *S250 Science in Context Science*: Level 2. The Open University, printed and bound in the United Kingdom by CPI, Glasgow, UK, 2006.

Dzombak, D.A. and Morel, F.M.: *Surface complexation modeling.* John Wiley & Sons, New York, NY, 1990.

Eary, L.E.: The solubility of amorphous As_2S_3 from 25 to 90°C. *Geochim. Cosmochim. Acta* 56 (1992), pp. 2267–2280.

Eccles, L.A.: Sources of arsenic in streams tributary to Lake Crowley California. USGS Water Resources Investigation Report 76–36, 1976.

Edmonds, J.S. and Francesconi, K.A.: Arsenic in seafoods: human health aspects and regulations. *Marine Pollut. Bull.* 26:12 (1993), pp. 665–674.

Edmonds, J.S., Shibata, Y., Francesconi, K.A. and Rippington, R.J.: Morita M. : Arsenic transformations in short marine food chains studies by HPLC ICP MS. *Appl. Organomet. Chem.* 11 (1997), pp. 281–287.

Edwards, M.: Chemistry of arsenic: removal during coagulation and Fe-Mn oxidation. *J. AWWA* 86:9 (1994), pp. 64–78.

Eiche, E., Neumann, T., Berg, M., Weinman, B., van Geen, A., Norra, S., Berner, Z., Trang, P.T.K., Viet, P.H. and Stüben, D.: Geochemical processes underlying a sharp contrast in groundwater arsenic concentrations in a village on the Red River delta, Vietnam. *Appl. Geochem.* 23:11 (2008), pp. 3143–3154.

Ellis, A.J. and Mahon, W.A.J.: Natural hydrothermal systems and experimental hot-water/rock interactions. *Geochim. Cosmochim. Acta* 28:8 (1964), pp. 1323–1357.

Ellis, A.J. and Mahon, W.A.J.: *Chemistry and geothermal systems.* Academic Press, New York, NY, 1977.

Endo, G., Fukushima, S., Kinoshita, A., Kuroda, K., Morimura, K., Salim, E., Shen, J., Wanibuchi, H., Wei, M. and Yoshida, K.: Understanding arsenic carcinogenicity by the use of animal models. *Toxicol. Appl. Pharmacol.* 198 (2003), pp. 366–376.

Ewers, G.R. and Keays, R.R.: Volatile and precious metal zoning in the Broadlands geothermal field, New Zealand. *Econ. Geol.* 72:7 (1977), pp. 1337–1354.

FACOA (Fisheries Agency, Council of Agriculture): Fisheries statistical yearbook. Taipei, Taiwan, Fisheries Agency, Council of Agriculture, Executive Yuan, ROC, 2004.

Farías, S.S., Casa, V.A., Vazquez, C., Ferpozzi, L., Pucci, G.N. and Cohen, I.M.: Natural contamination with arsenic and other trace elements in ground waters of Argentine Pampean plain. *Sci. Total Environ.* 309:1–3 (2003), pp. 187–199.

Feldman, P.R. and Rosenboom, J.W.: Cambodia drinking water quality assessment, WHO in cooperation with Cambodian Ministry of Rural Development and the Ministry of Industry, Mines, and Energy. Phnom Penh, Cambodia, 2001.

Fendorf, S.E., Eick, M.J., Grossl, P. and D.L. Sparks: Arsenate and chromate retention mechanisms on goethite. 1. Surface structure. *Environ. Sci. Technol* 31 (1997), pp. 315–320.

Ferreccio, C., González, C., Milosavjlevic, V., Marshall, G., Sancha, A.M. and Smith, A.H.: Lung cancer and arsenic concentrations in drinking water in Chile. *Epidemiology* 11:6 (2000), pp. 673–679.

Ficklin, W.H., Frank, D.G., Briggs, P.K. and Tucker, R.E.: Analytical results for water, soil, and rocks collected near the vicinity of Granite Falls Washington as part of an arsenic-in-groundwater study. USGS Open-File Report 89-148, 1989.

Finkelman, R.B.: *Modes of occurrence of trace elements in coal.* PhD thesis, University of Maryland, Baltimore, MD, 1980.

Finkelman, R.B.: Potential health impacts of burning coal beds and waste banks. *Int. J. Coal Geol.* 59 (2004), pp. 19–24.

Fitz, W.J. and Wenzel, W.W.: Arsenic transformations in the soil-rhizosphere-plant system: Fundamantals and potential application to phytoremediation. *J. Biotechnol.* 99 (2002), pp. 259–278.

Fontaine, J.A.: Regulating arsenic in Nevada drinking water supplies: past problem, future challenges. In: W.R. Chappell, C.O. Abernathy and R.L. Cothern, R.L. (eds): Arsenic exposure and health effects. *Sci. Technol. Lett.* (1994), pp. 285–288.

Frost, R.F. and Griffin, R.A.: Effect of pH on adsorption of arsenic and selenium from landfill leachate by clay minerals. *Soil Sci. Soc. Amer. J.* A1 (1976), pp. 53–57.

Fujii, R. and Swain, W.C.: Areal distribution of trace elements, salinity, and major ions in shallow ground water, Tulare basin, southern San Joaquin Valley, California. USGS Water Resources Investigation Report 95-4048, 1995.

Fuller, C.C., Davis, J.A. and Waychunas, G.A.: Surface chemistry of ferrihydrite: Part 2. Kinetics of arsenate adsorption and coprecipitation. *Geochim. Cosmochim. Acta* 57 (1993), pp. 2271–2282.

Gao, S., Fujii, R., Chalmers, A.T. and Tanji, K.K.: Evaluation of adsorbed arsenic and potential contribution to shallow groundwater in Tulare Lake Bed Area, Tulare Basin, California. *Soil. Sci. Soc. Am. J.* 68 (2004), pp. 89–95.

García, M.E. and Bundschuh, J.: Control mechanisms of seasonal variation of dissolved arsenic and heavy metal concentrations in surface waters of Lake Poopó basin, Bolivia. Geological Society of America Annual Meeting, Philadelphia, 22–25 Oct. 2006, *Geological Society of America Abstracts with Programs* 38:7 (2006), p. 320.

García, M.G., Moreno, C., Galindo, M.C., Hidalgo, M. del V., Fernández, D.S. and Sracek, O.: Intermediate to high levels of arsenic and fluoride in deep geothermal aquifers from the northwestern Chaco-Pampean plain, Argentina. In: J. Bundschuh, M.A. Armienta, P. Birkle, P. Bhattacharya, J. Matschullat and A.B. Mukherjee (eds): Natural arsenic in groundwater of Latin America. In: J. Bundschuh and P. Bhattacharya (series eds): *Arsenic in the environment*, Volume 1. CRC Press/Balkema Publisher, Leiden, The Netherlands, 2009, pp. 69–79.

García-Sánchez, A., Saavedra, J. and Pellitero, E.: Distribución de As en granitoides del centro-oeste de España y sus relaciones metalogenéticas (Sn, W). *Cuad. Lab. Xeol. Laxe* 9 (1985), pp. 191–202.

Gbadebo, A.M.: Occurence and fate of arsenic in the hydrogeological systems of Nigeria. *Geological Society of America Abstracts with Programs* 37:7 (2005), p. 375.

Geucke, T., Deowan, S.A., Hoinkis, J. and Pätzold, C.: Performance of a small-scale RO desalinator for arsenic removal. *Desalination* 239 (2009), pp. 198–206.

Gieskes, J.M., You, C.-F., Lee, T., Yui, T.-F. and Chen, H.-W.: Hydrogeochemistry of mud volcanoes in Taiwan. *Acta Geol. Taiwan* 30 (1992), pp. 79–88.

Goldberg, S.: Chemical modeling of arsenate adsorption on aluminum and iron oxide minerals. *Soil Sci. Soc. Am. J.* 50 (1986), pp. 1154–1157.

Goldberg, S. and Glaubig, R.A.: Anion sorption on a calcareous montmorillonitic soil-As. *Soil Sci. Soc. Am. J.* 52 (1988), pp. 1297–1300.

Goldberg, S. and Johnston, C.T.: Mechanisms of arsenic adsorption on amorphous oxides evaluated using macroscopic measurements, vibrational spectroscopy, and surface complexation modeling. *J. Colloid Interface Sci.* 234 (2001), pp. 204–216.

Goldblatt, E.L., Van Denburgh, S.A. and Marsland, R.A.: The unusual and widespread occurrence of arsenic in well waters of Lane Country, Oregon. Lane County Health Department Report, Eugene, OR, 1963.

Goldschmidt, V.M.: The principles of distribution of chemical. elements in minerals and rocks. *J. Chem. Soc.* 1937 (1937), pp. 655–672.

Goldstein, L.: *A review of arsenic in ground water with an emphasis on Washington state.* MSc Thesis, Department of Environmental Science, Evergreen State College, Olympia, WA, 1988.

Goleva, G.A.: Metal content of hydrotherms in areas of active volcanism. In: S.I.: Naboko (ed): *Hydrothermal mineral-forming solutions in the areas of active volcanism, Nauka, Novosibirisk.* Translated in 1982 for US Dept. Interior and National Science Foundation by Amerind Publishing Co., New Delhi, India, 1974, pp. 113–115.

Gómez, A.: Chronic arsenicosis in El Zapote, Nicaragua, 1994–2002. In: A.M. Sancha (ed): *Tercer Seminario Internacional sobre Evaluación y Manejo de las Fuentes de Agua de Bebida contaminadas con Arsénico* (proceedings available as CD), Universidad de Chile, November 08–11, 2004, Santiago de Chile, Chile.

Gómez, A.: Chronic arsenicosis and respiratory effects in El Zapote, Nicaragua. In: J. Bundschuh, M.A. Armienta, P. Birkle, P. Bhattacharya, J. Matschullat and A.B. Mukherjee (eds): Geogenic arsenic in groundwater of Latin America. In: J. Bundschuh and P. Bhattacharya (series eds): *Arsenic in the environment*, Volume 1. CRC Press/Balkema Publisher, Leiden, The Netherlands, 2009, pp. 409–418.

González, E.P., Birkle, P. and Torres-Alvarado, I.: Evolution of the hydrothermal system at the geothermal field of Los Azufres, Mexico, based on fluid inclusion, isotopic and petrologic data. *J. Volcanol. Geotherm. Res.* 104:1–4 (2000), pp. 277–296.

González, E.P., Tello, E.H. and Verma, M.P.: Interacción agua geotérmica-manantiales en el campo geotérmico de Los Humeros, Puebla, México. *Ingeniería Hidráulica en México* XVI(2), 2001, pp. 185–194.

Gooch, F.A. and Whitfield, J.E.: Analyses of waters of the Yellowstone National Park, with an account of the methods of analysis employed: U.S. Geological Survey Bulletin 47, 1888.

Goyenechea, M.: Sobre la nueva enfermedad descubierta en Bell-Ville. *Rev. Med. de Rosario* 7. 1917, p. 485.

Greene, R. and Crecelius, E.: Total and inorganic arsenic in Mid-Atlantic marine fish and shellfish and implication to fish advisories. *Int. Environ. Assess. Mang.* 2:4 (2006), pp. 344–354.

Gu, Z.-M., Fang, J. and Deng, B.-L.: Preparation and evaluation of GAC-based iron-containing adsorbents for arsenic removal. *Environ. Sci. Technol.* 39 (2005), pp. 3833–3843.

Guérèquiz, A.R, Mañay, N., Goso Aguilar, C. and Bundschuh, J.: Evaluación de riesgo ambiental por presencia de arsénico en el sector oeste del acuífero Raigón, Departamento de San José, Uruguay. Presentación del proyecto. In: M. Litter (ed): Proceedings of the workshop *Arsenic distribution in Iberoamerica.* Thematic Network 406RT0282 IBEROARSEN, Centro Atómico Constituyentes, San Miguel, Province of Buenos Aires, Argentina, 2006, pp. 111–112.

Gulens, J., Champ, D.R. and Jackson, R.E.: Influence of redox environments on the mobility of arsenic in ground water. In: E.A. Jenne (ed): Chemical modeling in aqueous systems. *ACS Symposium Series* 93, Am. Chem. Soc.,Washington, DC, 1979, pp. 81–95.

Guo, H.-M., Tang, X.-H., Yang, S.-Z. and Shen, Z.-L.: Effect of indigenous bacteria on geochemical behavior of arsenic in aquifer sediments from the Hetao basin, Inner Mongolia: Evidence from sediment incubations. *Appl. Geochem.* 23 (2008), pp. 3267–3277.

Guo, H.-R.: Arsenic in drinking water and skin cancer: Comparison among studies based on cancer registry, death certificates, and physical examinations. In C.O. Abernathy, R.L. Calderón and W.R. Chappell (eds): *Arsenic: Exposure and health effects.* Chapman & Hall, London, UK, 1997, pp. 243–259.

Guo, H.-R.: Dose-response relation between arsenic in drinking water and mortality of bladder cancers. *Chin. J. Public Health* 18 (suppl.) (1999), pp. 134–139.

Guo, H.-R.: Cancer risk assessment for arsenic exposure through oyster consumption. *Environ. Health Persp.* 110 (2002), pp. 123–124.

Guo, H.-R.: The lack of a specific association between arsenic in drinking water and hepatocellular carcinoma. *J. Hepatol.* 39 (2003), pp. 383–388.

Guo, H.-R. and Tseng, Y.-C.: Arsenic in drinking water and bladder cancer: comparison between studies based on cancer registry and death certificates. *Environ. Geochem. Health 22* (2000), pp. 83–91.

Guo, H.-R. and Valberg, P.A.: Evaluation of the validity of the US EPA'S cancer risk assessment for arsenic for low-level exposures: A likelihood ratio approach. *Environ. Geochem. Health* 19 (1997), pp. 133–141.

Guo, H.-R., Chen, C.-J. and Greene, H.L.: Arsenic in drinking water and cancers: A descriptive review of Taiwan studies. *Environ. Geochem. Health* s16 (1994a), pp. 129–138.

Guo, H.-R., Chiang, H.-S., Hu, H., Lipsitz, S.R. and Monson, R.R.: Arsenic in drinking water and urinary cancers: a preliminary report. *Environ. Geochem. Health* s16 (1994b), pp. 119–128.

Guo, H.-R., Chiang, H.-S., Hu, H., Lipsitz, S.R. and Monson, R.R.: Arsenic in drinking water and incidence of urinary cancers. *Epidemiology* 8 (1997), pp. 545–550.

Guo, H.-R., Lipsitz, S.R., Hu, H. and Monson, R.R.: Using ecological data to estimate a regression model for individual data: the association between arsenic in drinking water and incidence of skin cancer. *Environ. Res.* 79 (1998), pp. 82–93.

Guo, H.-R., Yu, H.-S., Hu, H. and Monson, R.E.: Arsenic in drinking water and skin cancer: cell-type specificity. *Cancer Causes Control* 12 (2001), pp. 909–916.

Guo, H.-R., Wang, N.-S., Hu, H. and Monson, R.R.: Cell type specificity of lung cancer associated with arsenic ingestion. *Cancer Epidemiol Biomarker Prev.* 13 (2004), pp. 638–643.

Gustafsson, J.P.: Modelling competitive anion adsorption on oxide minerals and an allophone-containing soil. *Europ. J. Soil Sci.* 52 (2001), pp. 639–653.

Gustafsson, J.P. and Bhattacharya, P.: Geochemical modelling of arsenic adsorption to oxide surfaces. In: P. Bhattacharya, A.B. Mukherjee, J. Bundschuh, R. Zevenhoven and R.H. Loeppert (eds): *Arsenic in soil and groundwater environment; Trace metals and other contaminants in the environment,* Volume 9. Elsevier, 2007, pp. 153–200.

Gustafsson, J.P. and Jacks, G.: Arsenic geochemistry in forested soil profiles as revealed by solid-phase studies. *Appl. Geochem.* 10 (1995), pp. 307–315.

Hammarlund, L. and Piñones, J.: *Arsenic in geothermal waters of Costa Rica—A minor field study.* TRITA-LWR MSc Thesis, LWR-EX-09-02, Department of Land and Wayer Resources Engineering, Royal Institute of Technology, Stockholm, Sweden, 2009.

Han, B.-C., Jeng, W.-L., Jeng, M.-S., Kao, L.-T., Meng, P.-J. and Huang, Y.-L.: Rock-shells (*Thais clavigera*) as an indicator of As, Cu, and Zn contamination on the Budai Coast of the black-foot disease area in Taiwan. *Arch. Environ. Contam. Toxicol.* 32 (1997), pp. 456–461.

Han, B.-C., Jeng, W.-L., Chen, R.-Y., Fang, G.-T., Hung, T.-C. and Tseng, R.-J.: Estimation of target hazard quotients and potential health risks for metals by consumption of seafood in Taiwan. *Arch. Environ. Contam. Toxicol.* 35 (1998), pp. 711–720.

Han, F.-J.: *Cultivating industry management for smeltfish in north Taiwan.* MSc Thesis, Taiwan: Department of Fisheries Science, National Taiwan Ocean University, Teipeh, 2003 (in Chinese).

Hasan, M.A., Ahmed, K.M., Sracek, O., Bhattacharya, P., von Brömssen, M., Broms, S., Fogelstrom, J., Mazumder, M.L. and Jacks, G.: Arsenic in shallow groundwater of Bangladesh: investigations from three different physiographic settings. *Hydrogeology J.* 15:8 (2007), pp. 1507–1522.

Heinrich, C.A. and Eadington, P.J.: Thermodynamic predictions of the hydrothermal chemistry of arsenic, and their significance for the paragenetic sequence of some cassiterite-arsenopyrite-base metal sulfide deposits. *Econ. Geol.* 81:3 (1986), pp. 511–529.

Heinrichs, G.: Geogene Arsenkonzentrationen in Keupergrundwässern Frankens/Bayern. *Hydrogeologie und Umwelt* 12 (1996), pp. 1–193.

Heinrichs, G. and Udluft, P.: Natural arsenic in Triassic rocks: A source of drinking-water contamination in Bavaria, Germany. *Hydrogeol. J.* 7:5 (1999), pp. 468–476.

Helz, G.R. and Tossella, J.A.: Thermodynamic model for arsenic speciation in sulfidic waters: A novel use of ab initio computations. *Geochim. Cosmochim. Acta* 72:18 (2008), pp. 4457–4468.

Helz, G.R., Tossell, J.A., Charnock, J.M., Pattrick, R.A.D., Vaughan, D.J. and Garner, C.D.: Oligomerization in As(III) sulfide solutions: Theoretical constraints and spectroscopic evidence. *Geochim. Cosmochim. Acta* 59:22 (1995): pp. 4591–4604.

Hering, J.G., Chen, P.-Y., Wilkie, J.A., Elimelech, M. and Liang, S.: Arsenic removal by ferric chloride. *J. AWWA* 88:4 (1996), pp. 155–167.

Hiemstra, T. and van Riemsdijk, W.H.: A surface structural approach to ion adsorption: the charge distribution, CD. Model. *J. Colloid Interface Sci.* 179 (1996), pp. 488–508.

Hingston, F.J., Posner, A.M. and Quirk, J.P.: Competitive adsorption of negatively charged ligands on oxide surfaces. *Discuss. Faraday Soc.* 52 (1971), pp. 334–342.

Hinkle, S.R.: Quality of shallow ground water in alluvial aquifers of the Willamette Basin, Oregon, 1993–1995. USGS Water Resources Investigation Report 97-4082-B, 1997.

Hinkle, S.R. and Polette, D.J.: Arsenic in ground water of the Willamette Basin, Oregon. USGS Water Resources Investigation Report 98-4205, 1999.

Hitchcock, C.: *The geology of New Hampshire*: Part V: *Economic geology*. State Printer, Concord, NH, 1878.

Ho, C.-S.: An introduction to the geology of Taiwan: Explanatory text of the geologic map of Taiwan. Published by the Ministry of Economic Affairs, Taipei, Taiwan, 1975.

Ho, S.-Y., Tsai, C.-C., Tsai, Y.-C. and Guo, H.-R.: Hepatic angiosarcoma presenting as hepatic rupture in a patient with long-term ingestion of arsenic. *J. Formosan Med. Assoc.* 103 (2004), pp. 374–379.

Hollibaugh J.T., Carini, S., Gürleyük, H., Jellison, R., Joye, S.B., LeCleir, G., Meile, C., Vasquez, L. and Wallschläger, D.: Arsenic speciation in Mono Lake, California: response to seasonal stratification and anoxia. *Geochim. Cosmochim. Acta* 69 (2005), pp. 1925–1937.

Holm, T.R.: Ground-water quality in the Mahomet aquifer, McLean, Logan, and Tazewell Counties. Illinois State Water Survey Contract Report 579, Urbana-Champaign, IL, 1995.

Holm, T.R. and Curtiss, C.D.: Arsenic contamination in east-central Illinois ground waters. Illinois Department of Energy and Natural Resources, Energy and Environmental Affairs Division, Springfield, IL, 1988.

Hopenhayn-Rich, C., Biggs, M.L., Fuchs, A., Bergoglio, R., Tello, E.E., Nicolli, H. and Smith, A.H.: Bladder cancer mortality associated with arsenic in drinking water in Argentina. *Epidemiology* 7:2 (1996), pp. 117–124.

Hopenhayn-Rich, C., Biggs, M.L. and Smith, A.H.: Lung and kidney cancer mortality associated with arsenic in drinking water in Cordoba, Argentina. *Int. J. Epidemiol.* 27:4 (1998), pp. 561–569.

Hopenhayn-Rich, C., Browning, S.R., Hertz-Picciotto, I., Ferreccio, C., Peralta, C. and Gibb, H.: Chronic arsenic exposure and risk of infant mortality in two creas of Chile. *Environ. Health Persp.* 108:7 (2000), pp. 667–673.

Horng, L.-L. and Clifford, D.: The behavior of polyprotic anions in ion-exchange resins. *React. Functional Polymers* 35 (1997), pp. 41–54.

Horng, S.-F., Liaw, K.-F., Lin, L.-J., Hsueh, Y.-M., Chiou, H.-Y., Chang, M.-H. and Chen, C.-J.: A cohort study on lung cancer in the endemic area of blackfoot disease. *Chin. J. Public Health* 14 (1995), pp. 32–40.

Hristovski, K.D., Westerhoff, P.K., Möller, T. and Sylvester, P.: Effect of synthesis conditions on nano-iron (hydr)oxide impregnated granulated activated carbon. *Chem. Engin. J.* 146 (2009), pp. 237–243.

Hsieh, F.-I., Hwang, T.-S., Hsieh, Y.-C., Lo, H.-C., Su, C.-T., Hsu, H.-S., Chiou, H.-Y. and Chen, C.-J.: Risk of erectile dysfunction induced by arsenic through well water consumption in Taiwan. *Environ. Health Perspect.* 116 (2008), pp. 532–536.

Hsu, K.-H., Froines, J.R. and Chen, C.-J.: Studies of arsenic ingestion from drinking water in northeastern Taiwan: chemical speciation and urinary metabolites. In: C.O. Abernathy, R.L. Calderón and W.R, Chappell (eds): *Arsenic: Exposure and health effects.* Chapman & Hall, London, UK. 1997, pp. 190–209.

Hsu, T.-J.: Arsenic in well water. *Japan Industry Water* 199 (1975), pp. 20–22.

Hsueh, Y.-M., Cheng, G.-S., Wu, W.-M., Yu, H.-S., Kuo, T.-L. and Chen, C.-J.: Multiple risk factors associated with arsenic-induced skin cancer: effects of chronic liver disease and malnutrition status. *Br. J. Cancer* 71 (1975), pp. 109–114.

Hsueh, Y.-M., Chiou, H.-Y., Chen, S.-Y., Shyu, M.-P., Lin, L.-J., Horng, S.-F., Liaw, K.-F., Huang, Y.-L., Wu, W.L., Leu, L.C. and Chen, C.-J.: Follow-up study for the association between inorganic arsenic and skin cancer. *Chin. J. Public Health* 15 (suppl.) (1996), pp. 77–91.

Hsueh Y.-M., Chiou, H.-Y., Huang, Y.-L., Wu, W.-L., Huang, C.-C., Yang, M.-H., Lue, L.-C., Chen, G.-S. and Chen, C.-J.: Serum beta-carotene level, arsenic methylation capability, and incidence of skin cancer. *Cancer Epidemiol. Biomarkers Prev.* 6 (1997), pp. 589–596.

Hsueh, Y.-M., Wu, W.-L., Huang, Y.-L., Chiou, H.-Y., Tseng, C.-H. and Chen, C.-J.: Low serum carotene level and increased risk of ischemic heart disease related to long-term arsenic exposure. *Atherosclerosis* 141 (1998), pp. 249–257.

Hsueh, Y.-M., Ko, Y.-F., Huang, Y.-K., Chen, H.-W., Chiou, H.-Y., Huang, Y.-L.Yang, M.-H., and Chen, C.-J.: Determinants of inorganic arsenic methylation capability among residents of the Lanyang Basin, Taiwan: arsenic and selenium exposure and alcohol consumption. *Toxicol. Lett.* 137 (2003), pp. 49–63.

Huang, C.-C., Chen, J.-C., Lin, L.-P. and Liang, N.-K.: The death cause of seashells in shallow sea of southwest coast of Taiwan. Special Issue 6, Institute of Oceanography of Taiwan University, Taipei, Taiwan, 1975.

Huang, T.-H.: *Distribution of arsenic in cored sediments from Budai, Taiwan and its geological significances*. MSc Thesis, Department of Earth Sciences, National Cheng Kung University, Tainan City, Taiwan, 2009.

Huang, Y.-K., Lin, K.-H., Chen, H.-W., Chang, C.-C., Liu, C.-W., Yang, M.-H. and Yanh, M.-H.: Arsenic species contents at aquaculture farm and in farmed mouthbreeder (*Oreochromis mossambicus*) in blackfoot disease hyperendemic areas. *Food Chem. Toxicol.* 41:11 (2003), pp. 1491–1500.

Huang, Y.-K., Huang, Y.-L., Hsueh, Y.-M., Yang, M.-H., Wu, M.-M., Chen, S.-Y., Hsu, L.-I. and Chen, C.-J.: Arsenic exposure, urinary arsenic speciation, and the incidence of urothelial carcinoma: a twelve-year follow-up study. *Cancer Causes Control* 19 (2008), pp. 829–839.

Huerta-Diaz, M.A., Tessier, A. and Carignan, R.: Geochemistry of trace elements associated with reduced sulfur in freshwater sediments. *Appl. Geochem.* 13 (1998), pp. 213–233.

Hurtado-Jiménez R. and Gardea-Torresdey, J.L.: Contamination of drinking water supply with geothermal arsenic in Los Altos de Jalisco, Mexico. In: J. Bundschuh, M.A. Armienta, P. Birkle, P. Bhattacharya, J. Matschullat and A.B. Mukherjee (eds): Geogenic arsenic in groundwater of Latin America. In: J. Bundschuh and P. Bhattacharya (series eds): *Arsenic in the environment*, Volume 1. CRC Press/Balkema Publisher, Leiden, The Netherlands, 2009, pp. 179–190.

Hussam, A. and Munir, A.K.: A simple and effective arsenic filter based on composite iron matrix: Development and deployment studies for groundwater of Bangladesh. *J. Env. Sci. Health* A 42 (2007), pp. 1869–1878.

Hutchinson, J.: Arsenic cancer. *Br. Med. J.* 2 (1887), pp. 1280–1281.

Hutchinson, J.: Diseases of the skin: On some examples of arsenic-keratosis of the skin and of arsenic-cancer. *Transactions of the Pathological Society of London* 39, 1888, pp. 352–363.

Inskeep, W.P., Macur, R.E., Harrison, G., Bostick, B.C. and Fendorf, S.: Biomineralization of As(V)-hydrous ferric oxyhydroxide in microbial mats of an acid-sulfate-chloride geothermal spring, Yellowstone National Park. *Geochim. Cosmochim. Acta* 68 (2004), pp. 3141–3155.

Islam, F.S., Gault, A.G., Boothman, C., Polya, D.A., Charnock, J.M., Chartterjee, D. and Lloyd, J.R.: Role of metal-reducing bacteria in arsenic release from Bengal delta sediments. *Nature* 430 (2004), pp. 68–71.

Jacks, G. and Bhattacharya, P.: Arsenic contamination in the environment due to the use of CCA-wood preservatives. In: *Arsenic in wood preservatives, Part I*, Kemi Report 3/98, 1998, pp. 7–75.

Jackson, B.P. and Miller, W.P.: Effectiveness of phosphate and hydroxide for desorption of arsenic and selenium species from iron oxides. *J. Soil Sci. Am.* 64 (2000), pp. 1616–1622.

Jacobson, G. (1998): Arsenic poisoning from groundwater in Bengal: A vast, tragic, hydrogeological problem unfolds. Editor's Message *Hydrogeol. J.* 6, p. A4.

Jain, A. and Loeppert, R.H.: Effect of competing anions on the adsorption of arsenate and arsenite by ferrihydrite. *J. Environ. Qual.* 29 (2000), pp. 1422–1430.

Jakariya, M.: Arsenic *in tubewell water of Bangladesh and approaches for sustainable mitigation*. TRITA LWR PhD Thesis 1033, Royal Institute of Technology (KTH), Stockholm, Sweden, 2007.

Jakariya, M., Rahman, M., Chowdhury, A.M.R., Rahman, M., Yunus, M., Bhiuya, M.A., Wahed, M.A., Bhattacharya, P., Jacks, G., Vahter, M. and Persson, L-Å.: Sustainable safe water options in Bangladesh: experiences from the arsenic project at Matlab (AsMat). In: J. Bundschuh, P. Bhattacharya and D. Chandrasekharam (eds): *Natural arsenic in groundwater: occurrence, remediation and management*. Taylor and Francis/Balkema, Leiden, The Netherlands, 2005, pp. 319–330.

Jakariya, M., von Brömssen, M., Jacks, G., Chowdhury, A.M.R., Ahmed, K.M. and Bhattacharya, P.: Searching for sustainable arsenic mitigation strategy in Bangladesh: experience from two upazilas. *Int. J. Environ. Poll.* 31:3–4 (2007), pp. 415–430.

Jang, M., Min, S.H., Kim, T.H. and Park, J.K.: Removal of arsenite and arsenate using hydrous ferric oxide incorporated into naturally occurring porous diatomite. *Env. Sci. Technol.* 40 (2006), pp. 1636–1643.

Jean, J.-S.: Outbreak of enteroviruses and groundwater contamination in Taiwan: Concept of biomedical hydrogeology. Editor's Message *Hydrogeol. J.* 7 (1999), pp. 339–340.

Jean, J.-S., Nath, B., Weng, L.-O., Liu, C.-C. and Yang, Y.-W.: Mobilisation of arsenic in the groundwater of the blackfoot disease area in the Chia-Nan plain, southwestern Taiwan. *Annual Meeting of Royal Geographical Society with IBG*, 29 August 2007, London, UK, 2007.

Jekel, M. and Amy, G.L.: Arsenic removal during drinking water treatment. In: G. Newcombe and D. Dixon (eds): Interface science in drinking water treatment—Theory and application. *Interface Science and Technology*, Volume 10, 2006, pp. 193–206.

Jones, B., Renaut, R.W. and Rosen, M.R.: Biogenicity of gold- and silver-bearing siliceous sinters forming in hot (75°C) anaerobic spring-waters of Champagne Pool, Waiotapu, North Island, New Zealand. *J. Geol. Soc.* 158:6 (2001), pp. 895–911.

Joshi, A. and Chaudhuri, M.: Removal of arsenic from ground water by iron oxide coated sand. *J. Env. Eng. ASCE* 122 (1996), pp. 769–771.

Kanivetsky, R.: Arsenic in ground water of Minnesota: Hydrogeochemical modeling and characterization. Minnesota Geological Survey Report of Investigation 55, St. Paul, MN, 2000.

Kao, T.-M. and Kao, S.-J.: Investigation on causes of spontaneous gangrene. *J. Formos. Med. Assoc.* 53 (1954), p. 272.

Kao, T.-M., Kao, S.-J. and Hou, S.-T.: Spontaneous gangrene and chronic potassium poisoning. Presented at the 46th *Annual Meeting of the Formosan Medical Association*, Taipei, Taiwan, 1954.

Karcher, S., Cáceres, L., Jekel, M. and Contreras, R.: Arsenic removal from water supplies in northern Chile using ferric chloride coagulation. *J. Chart. Inst. Water Environ. Manag.* 13:3 (1999), pp. 164–169.

Karpov, G.A. and Naboko, S.I.: Metal contents of recent thermal waters, mineral precipitates and hydrothermal alteration in active geothermal fluids, Kamchatka. *J. Geochem. Exploration* 36 (1990). pp. 57–71.

Khramova, G.C.: Effects of intensified activity of the Ebeko Volcano on the composition of the water of Goryachoe Lake. In: S.I.: Naboko (ed): *Hydrothermal mineral-forming solutions in the areas of active volcanism, Nauka, Novosibirisk*. Translated in 1982 for US Dept. Interior and National Science Foundation by Amerind Publishing Co., New Delhi, India, pp. 88–94.

Kingston, R.: The Tongonan geothermal field Leyte Philippines: Report on exploration and eevelopment. Report, Kingston Reynolds Thom and Allardice Ltd., Auckland, New Zealand, 1979.

Kinniburgh, D.G., Gale, I.N., Smedley, P.L., Darling, W.G., West, J-M., Kimblin, R.T., Parker, A., Rae, J.E., Aldous, P.J. and O'Shea, M.J.: The effects of historic abstraction of groundwater from the London Basin aquifers on groundwater quality. *Appl. Geochem.* 9:2 (1994), pp. 175–196.

Kirby, J. and Maher, W.: Tissue accumulation and distribution of arsenic compounds in three marine fish species: Relationship to trophic position. *Appl. Organomet Chem.* 16 (2002), pp. 108–115.

Kirk, M.F., Holm, T.R., Park, J., Jin, Q., Sanford, R.A., Fouke, B.W. and Bethke, C.M.: Bacterial sulfate reduction limits natural arsenic contamination in groundwater. *Geology* 32 (2004), pp. 953–956.

Kirk, T. and Sarfaraz, A.: Oxidative stress as a possible mode of action for arsenic carcinogenesis. *Toxicol. Lett.* 137 (2003), pp. 3–13.

Klohn, W.: Boro y arsénico en aguas del norte de Chile. *Internat. Symp. on Hydrology of Volcanic Rocks*. Lanzarote, Spain. 1974, pp. 579–592.

Kohlmeyer, U., Kuballa, J. and Jantzen, E.: Simultaneous separation of 17 inorganic and organic arsenic compounds in marine biota by means of high-performance liquid chromatography/inductively coupled plasma mass spectrometry. *Rapid Commun. Mass Spectrom.* 16 (2002), pp. 965–74.

Kolker, A. and Nordstrom, D.K.: Occurrence and micro-distribution of arsenic in pyrite. *USGS Workshop on Arsenic in the Environment*, February 21–22, Denver, CO, 2001 (abstract available at http://wwwbrr.cr.usgs.gov/Arsenic/finalabstracts.htm).

Kolker, A., Cannon, W.F., Westjohn, D.B. and Woodruff, L.G.: Arsenic-rich pyrite in the Mississippian Marshall sandstone: Source of anomalous arsenic in southeastern Michigan ground water. *Geol. Soc. Amer. Annual Meeting Abstracts with Programs*, A-59. Boulder, CO, 1998.

Kondo, H., Ishiguro, Y., Ohno, K., Nagase, M., Toba, M. and Takagi, M.: Naturally occurring arsenic in the groundwater in the Southern Region of Fukuoka Prefecture. *Japan. Water Res.* 33 (1999), pp. 1967–1972.

Kopf, A. and Deyhle, A.: Back to the roots: boron geochemistry of mud volcanoes and its implications for mobilization depth and the global boron cycling. *Chem. Geol.* 192 (2002), pp. 195–210.

Koretsky, C.: The significance of surface complexation reactions in hydrologic systems: a geochemist's perspective. *J. Hydrol.* 230 (2000), pp. 127–171.

Korte, N.E.: Naturally occurring arsenic in groundwaters of the midwestern United States. Oak Ridge National Laboratory Environmental Technology Section Publication no. 3501. Oak Ridge, TN, 1991.

Korte, N.E. and Fernando, Q.: A review of arsenic (III) in groundwater. *Crit. Rev. Environ. Control* 21 (1991), pp. 1–39.

Kulp, R.T. and Jean, J.-S.: The role of humic substances in the biogeochemical arsenic cycle of the Chianan plain aquifer, southwestern Taiwan. Abstract *Proceedings, 2009 International Workshop on Arsenic and Humic Substances in Groundwater and Their Health Effects*, May 11–12, 2009. International Conference Hall of National Cheng Kung University, Tainan City, Taiwan, 2009.

Kuo, T.-L.: Arsenic content of artesian well water in endemic area of chronic arsenic poisoning. *Repts. Inst. Pathol. Natl. Taiwan Univ.* 20 (1968), pp. 7–13.

Lacayo, M.L., Cruz, A., Calero, S., Lacayo, J. and Fomsgaard, I.: Total arsenic in water, fish, and sediments from Lake Xolotlan, Managua, Nicaragua. *Bull. Environ. Contamin. Toxicol.* 49:3 (1992), pp. 463–470.

Lai, M.-S., Hsueh, Y.-M., Chen, C.-J., Hsu, M.-P., Chen, S.-Y., Kuo, T.-L., Wu, M.-M. and Tai, T.-Y.: Ingested inorganic arsenic and prevalence of diabetes mellitus. *Am. J. Epidemiol.* 139 (1994), pp. 484–492.

Lai, V.W.M., Cullen, W.R. and Ray, S.: Arsenic speciation in scallops. *Mar. Chem.* 66 (1999), pp. 81–89.

Lamm, S.H., Engel, A., Penn, C.A, Chen, R. and Feinleib, M.: Arsenic cancer risk confounder in southwest Taiwan data set. *Environ. Health Perspect.* 114 (2006), pp. 1077–1082.

Lamm, S.H., Luo, Z.D., Zhang, G.Y., Zhang, Y.M., Wilson, R., Byrd, D.M., Lai, S., Li, F.M., Polkanov, M., Tong, X., Loo, L. and Tucker, S.B.: An epidemiologic study of arsenic-related skin disorders and skin cancer and the consumption of arsenic-contaminated well waters in Huhhot, Inner Mongolia, China. *Health Environ. Risk Assessment* 13, 2007, pp. 713–752.

Langmuir, D.D., Mahoney, J., MacDonald, A. and Rowson, J.: Predicting arsenic concentrations in the porewaters of buried uranium mill tailings. *Geochim. Cosmochim. Acta* 63:19–20 (1999), pp. 3379–3394.

Langner, H.W., Jackson, C.R., McDermott, T.R. and Inskeep W.P.: Rapid oxidation of arsenite in a hot spring ecosystem. *Envir. Sci. Technol.* 35 (2001), pp. 3302–3309.

Larsen, E.K., Quetel, C.R., Muñoz, R., Fiala-Medioni, A. and Donard, O.F.X.: Arsenic speciation in shrimp and mussel from the Mid-Atlantic hydrothermal vents. *Mar. Chem.* 57 (1997), pp. 341–346.

Last, J.M.: *A dictionary of epidemiology.* 4th edn., Oxford University Press, Oxford, UK, 2001.

Lau, L.-S. and Mink, J.-F. Ground-water resources and development: coastal plain region, Erh Jen Chi, Kaohsiung, Taiwan. University of Hawaii Water Resources Research Center Technical Report (1971) No. 47, pp. 88–90.

Lazo, P., Cullaj, A., Arapi, A. and Deda, T.: Arsenic in soil environments in Albania. In: P. Bhattacharya, A.B. Mukherjee, J. Bundschuh, R. Zevenhoven and R.H. Loeppert (eds): *Arsenic in soil and groundwater environment; Trace metals and other contaminants in the environment,* Volume 9. Elsevier, Amsterdam, The Netherlands, 2007, pp. 237–256.

Lear, G., Polya, D.A., Song, B., Gault, A.G. and Lloyd, J.R.: Molecular analysis of arsenate-reducing bacteria within Cambodian sediments following amendment with acetate. *Appl. Environ. Microbiol.* 73 (2007), pp. 1041–1048.

Lee, C.-H., Chang, H.-R., Chen, J.-S., Chen, G.-S. and Yu, H.-S.: Defective adrenergic responses in patients with arsenic-induced peripheral vascular disease. *Angiology* 58 (2007), pp. 161–168.

Lee, J.-J.: *Geochemical characteristics of arsenic in groundwater of Lanyang plain and human heath risk assessment.* PhD Thesis, Department of Bioenvironmental Systems Engineering, National Taiwan University, Taipei, Taiwan, 2008 (in Chinese).

Lee, J.-J., Jang, C.-S., Wang, S.-W., Liang, C.-P. and Liu, C.-W.: Delineation of spatial redox zones using discriminant analysis and geochemical modeling in arsenic-affected alluvial aquifers. *Hydrol. Processes* 22 (2008a), pp. 3029–3041.

Lee, J.-J., Jang, C.-S., Liang, C.-P. and Liu, C.-W.: Assessing carcinogenic risks associated with ingesting arsenic in farmed smeltfish (ayu, *Plecoglossus altirelis*) in arseniasis-endemic area of Taiwan. *Sci. Total Environ.* 403 (2008b), pp. 68–79.

Lee, J.-J., Liu, C.-W., Jang, C.-S. and Liang, C.-P.: Zonal management of multi-purpose use of water from arsenic-affected aquifers by using a multivariable indicator Kriging approach. *J. Hydrology* 359 (2008c), pp. 260–273.

Leonard, A.: Arsenic. In: E. Merian (ed): *Metals and their compounds in the environment.* VIIC, Weinheim, Germany, 1991, pp. 751–774.

Levy, D.B., Schramke, J.A., Esposito, J.K., Erickson, T.A. and Moore, J.C.: The shallow ground water chemistry of arsenic, fluorine, and major elements: Eastern Owens Lake, California. *Appl. Geochem.* 14:1 (1999), pp. 53–65.

Lewis, C.: Seismic stratigraphy of the Chiayi area, Taiwan. Exploration and Development Research Division, Chinese Petroleum Corporation, Miaoli, Taiwan 1985.

Lewis, C., Ray, D. and Chiu, K.-K.: Primary geologic sources of arsenic in the Chainan plain (blackfoot disease area) and the Lanyang plain. *Int. Geology Rev.* 49 (2007), pp. 947–961.

Li, G.-C., Fei, W.-C. and Yen, Y.-M.: Survey of arsenical residual levels in the paddy soil and water samples from different locations of Taiwan. *Natl. Sci. Counc. Monthly ROC.* 7:8 (1979a), pp. 798–809.

Li, G.-C., Fei, W.-C. and Yen, Y.-M.: Survey of arsenical residual levels in the rice grain from various locations in Taiwan. *Natl. Sci. Counc. Monthly, ROC.* 7:7 (1979b), pp. 700–706.

Liang, C.-P., Jang, C.-S., Liu, C.-W., Lin, K.-H. and Lin, M.-C.: An integrated GIS-based approach in assessing carcinogenic risks via food-chain exposure in arsenic-affected groundwater areas. *Environ. Toxicol.* (2009). DOI: 10.1002/tox.20481.

Liao, C.-M., Chen, B.-C., Singh, S., Lin, M.-C., Liu, C.-W. and Han, B.-C.: Arsenic bioaccumulation and toxicity in tilapia (*Oreochromis mossambicus*) from blackfoot disease area in Taiwan. *Environ. Toxicol.* 18 (2003), pp. 252–259.

Liao, C.-M., Shen, H.-H., Chen, C.-L., Hsu, L.-I., Lin, T.-L., Chen, S.-C. and Chen, C.-J.: Risk assessment of arsenic-induced internal cancer at long-term low dose exposure. *J. Hazard. Mater.* 165 (2009), pp. 652–663.

Liaw, K.-F., Horng, S.-F., Lin, L.-J., Hsueh, Y.-M., Chiou, H.-Y., Chang, M.-H. and Chen, C.-J.: A cohort study on lower urinary tract cancers in the endemic area of blackfoot disease. *Chin. J. Public Health* 14 (1995), pp. 23–31.

Lien, H.-C., Tsai, T.-F., Lee, Y.-Y. and Hsiao, C.-H.: Merkel cell carcinoma and chronic arsenism. *J. Am. Acad. Dermatol.* 41 (1999), pp. 641–643.

Lillo, J.: Peligros geoquímicos: arsénico de origen natural en las aguas. University Rey Juan Carlos, Madrid, Spain, 2003.

Lim, S.-F., Zheng, Y.-M., Zou, S.-W. and Chen, J.-P.: Uptake of arsenate by an alginate-encapsulated magnetic sorbent: Process performance and characterization of adsorption chemistry. *J. Colloid Interface Sci.* 333 (2009), pp. 33–39.

Lin, C.-W. and Kao, M.-C.: Geologic map of Taiwan (scale 1:50,000, sheet 16): Suao. Central Geological Survey of Taiwan, Taipei, Taiwan, 1997.

Lin, H.-T., Wong, S.-S. and Li, G.-C.: Heavy metal content of rice and shellfish in Taiwan. *J. Food Drug Anal.* 12 (2004), pp. 167–174.

Lin, H.-T., Chen, S.-W., Shen, C.-J. and Chu, C.: Arsenic speciation in fish on the market. *J. Food Drug Anal.* 16 (2008), pp. 70–75.

Lin, J.-K. and Chiang, H.-C.: Arsenic concentration in drinking well water in Lanyang area and its preliminary health risk assessment. Taipei, Taiwan: National Science Council, ROC, NSC-90-2218-E-238-003, Taipei, Taiwan, 2002.

Lin, K.-H.: *Spatiotemporal distribution and bioaccumulation of arsenic species in the aquacultural ecosystem in the coastal areas of southwestern Taiwan.* PhD Thesis, Institute of Bioenvironmental Systems Engineering, National Taiwan University, Taipei, Taiwan, 2004.

Lin, M.-C. and Liao, C.-M.: Assessing the risks on human health associated with inorganic arsenic intake from groundwater-cultured milkfish in southwestern Taiwan. *Food Chem. Toxicol.* 46:2 (2008), pp. 701–709.

Lin, M.-C., Liao, C.-M., Liu, C.-W. and Singh, S.: Bioaccumulation of arsenic in aquacultural large-scale mullet *Liza macrolepis* from the blackfoot disease area in Taiwan. *Bull. Environ. Contamin. Toxicol.* 67 (2001), pp. 91–97.

Lin, M.-C., Cheng, H.-H., Lin, H.-Y., Chen, Y.-C., Chen, Y.-P., Chang-Chien, G.-P., Dai, C.-F., Han, B.-C. and Liu, C.-W.: Arsenic accumulation and acute toxicity in aquacultural juvenile milkfish (*Chanos chanos*) from blackfoot disease area in Taiwan. *Bull. Environ. Contam. Toxicol.* 72 (2004), pp. 248–254.

Lin, M.-H.: *Investigation of atmospheric arsenic particulates I Taichung Science Park of Central Taiwan.* MSc Thesis, Department of Environmental Engineering, Chung-Hsing University, Taichung, Taiwan, 2007 (in Chinese).

Lin, T.-F.: Development of simple methods for arsenic removal in drinking water. Project report of Taiwan Environmental Protection Agency, EPA-88-J1-02-03-403, Taipei, Taiwan, 1999 (in Chinese).

Lin, T.-F. and Wu, J.-K.: Adsorption of arsenite and arsenate within activated alumina grains: equilibrium and kinetics. *Water Res.* 35 (2001), pp. 2049–2057.

Lin, T.-F., Hsiao, H.-C., Wu, J.-K. and Hsiao, H.-C.: Removal of arsenic from groundwater using POU RO and distilling devices. *Env. Technol.* 23 (2002), pp. 781–790.

Lin, T.-F., Liu, C.-C. and Hsieh, W.-H.: Adsorption kinetics and equilibrium of arsenic onto an iron-based adsorbent and an ion exchange resin. *Water Sci. Technol.- Water Supply* 6 (2006), pp. 201–207.

Lin, T.-H., Huang, Y.-L. and Wang, M.-Y.: Arsenic species in drinking water, hair, fingernails, and urine of patients with blackfoot disease. *J. Toxicol. Environ. Health* Part A. 53 (1998), pp. 85–93.

Lin, W., Wang, S.-L., Wu, H.-J., Chang, K.-H., Yeh, P., Chen, C.-J. and Guo, H.-R.: Associations between arsenic in drinking water and pterygium in southwestern Taiwan. Environ. *Health Perspect.* 116 (2008), pp. 952–955.

Lin, Y.-B., Lin,Y.-P., Liu, C.-W. and Tan, Y.-C.: Mapping of spatial multi-scale sources of arsenic variation in groundwater on ChiaNan floodplain of Taiwan. *Sci. Total Environ.* 370 (2006), pp. 168–181.

Lin, Y.-C., Fong, S.-F., Chen, J.-Y., Wu, S.-Y., Chen, M.-L., Mao, Y.-F., Lan, C.-F., Lu, F.-J., Lee, L.-H. and Lin, C.-T.: The arsenic and fluorescent humic substances in the well waters of Yilan areas in relation to the incidence of blackfoot disease and cancer mortality of the residents. BFD Research Report, Vol. 27. Taiwan BFD Prevention and Control Center, Taiwan Provincial Government, Nantou, Taiwan, 1986. pp. 1–46 (in Chinese).

Lin, Z. and Puls, R.W.: Adsorption, desorption and oxidation of arsenic affected by clay minerals and aging process. *Environ. Geol.* 39 (2000), pp. 753–759.

Lindbäck, K. and Sjölin, A.M.: *Arsenic in groundwater in the southwestern part of the Río Dulce alluvial cone, Santiago del Estero Province, Argentina.* MSc Thesis, Department of Land and Water Resources Engineering KTH, Stockholm, Sweden, TRITA-LWREX-06-26, 2006.

Lindberg, R.D. and Runnells, D.D.: Ground water redox reactions: An analysis of equilibrium state applied to Eh measurements and geochemical modeling. *Science* 225:4665 (1984), pp. 925–927.

Ling, M.-P. and Liao, C.-M.: Risk characterization and exposure assessment in arseniasis-endemic areas of Taiwan. *Environ. Int.* 33 (2007), pp. 98–107.

Linklater, C.M., Albinsson, Y., Alexander, W.R., Casas, I., McKinley, I.G. and Sellin, S.: A natural analogue of high-pH cement pore waters from the Maqarin area of northern Jordan: Comparison of predicted and observed trace-element chemistry of uranium and selenium. *J. Contam. Hydrol.* 21:1–4 (1996), pp. 59–69.

Lippmann, M.J., Truesdell, A.H. and Frye, G.: The Cerro Prieto and Salton Sea geothermal fields—Are they really alike? *Proc. 24th Workshop on Geothermal Reservoir Engineering,* Stanford University, California, SGP-TR-162, 1999.

Litter, M., Pérez Carrera, A., Morgada, M.E., Ramos, O., Quintanilla, J. and Fernández-Cirelli, A.: Formas presentes de arsénico en agua y suelo. In: J. Bundschuh, A. Pérez-Carrera, and M.I. Litter (eds): *Distribución del arsénico en las regiones Ibérica e Iberoamericana.* Editorial Programa Iberoamericano de Ciencia y Tecnologia para el Desarrollo, Buenos Aires, Argentina, 2008, pp. 33–47. Available at: http://www.cnea.gov.ar/xxi/ambiental/iberoarsen/

Liu, C.-C., Jean, J.-S., Nath, B., Lee, M.-K., Hor, L.-I., Lin, K.-H. and Maity, J. Geochemical characteristics of the fluids and muds from two southern Taiwan mud volcanoes: Implications for water–sediment interaction and groundwater arsenic enrichment. *Appl. Geochem.* (2009) 24, pp. 1793–1802.

Liu, C.-S., Huang, I.-L. and Teng, L.-S.: Structural features off southwestern Taiwan. *Mar. Geol.* 137 (1997), pp. 305–319.

Liu, C.-W., Huang, F.-M. and Hsueh, Y.-M.: Revised cancer risk assessment of inorganic arsenic upon consumption of tilapia (Oreochromis mossambicus) from blackfoot disease hyperendemic areas. *Bull. Environ. Contam. Toxicol.* 74 (2005), pp. 1037–1044.

Liu, C.-W., Liang, C.-P., Huang, F.-M. and Hsueh, Y.-M.: Assessing the human health risks from exposure of inorganic arsenic through oyster (*Crassostrea gigas*) consumption in Taiwan. *Sci. Tot. Environ.* 361 (2006a), pp. 57–66.

Liu, C.-W., Lin, K.-H. and Jang, C.-S.: Tissue accumulation of arsenic compounds in aquacultural and wild mullet (*Mugil cephalus*). *Bull. Environ. Contam. Toxicol.* 77 (2006b), pp. 36–42.

Liu, C.-W., Liang, C.-P., Lin, K.-H., Jang, C.-S., Wang, S.-W., Huang, Y.-K. and Hsueh, Y.-M.: Bioaccumulation of arsenic compounds in aquacultural clams (*Meretrix lusoria*) and assessment of potential carcinogenic risks to human health by ingestion. *Chemosphere* 69:1 (2007), pp. 128–134.

Liu, C.-W., Huang, Y.-K., Hsueh, Y.-M., Lin, K.-H., Jang, C.-S. and Huang, L.-P.: Spatiotemporal distribution of arsenic species of oysters (*Crassostrea gigas*) in the coastal area of southwestern Taiwan. *Environ. Monitor. Assess.* 138:1–3 (2008), pp. 181–190.

Liu, C.-W., Lin, C.-C., Jang, C.-S., Sheu, G.-R. and Tsui, L.: Arsenic accumulation by rice grown in soil treated with roxarsone. *Plant Nutrit. Soil Sci.* 172 (2009), pp. 550–556.

Liu, T.-C.: Arsenic in relation to ecoenvironment. *Science Monthly* 26 (1995), pp. 134–140 (in Chinese).

Lo, M.-C. and Lin, K.-K.: Survey of arsenic levels in ground water in the Yi-Lan area. Reports on Blackfoot Disease Research 14, Taiwan Provincial Department of Health, Taichung, Taiwan, 1982, pp. 1–13.

Lo, M.-C., Hsen, Y.-C. and Lin, K.-K.: Second report on the investigation of arsenic content in underground water in Taiwan. Taiwan Provincial Institute of Environmental Sanitation, Taichung, Taiwan, 1977.

Lo, M.-J., Lin, Q.-C. and Jun, T.-C.: Survey of arsenic and other heavy metals contents in rice plant and crops in the blackfoot disease and non-blackfoot disease areas, Taiwan Province Blackfoot Disease Protection Centre Report, Nantou, Taiwan, 1983, pp. 22–35.

Loeppert, R.L. and Inskeep, W.P.: Iron. In D.L. Sparks (ed): *Methods of soil analysis. Part 3. SSSA Book Series* No. 5. ASA and SSSA, Madison, WI, 1996, pp. 639–664.

Lopéz, D., Ramson, L., Monterrosa, J., Soriano, T., Barahona, J. and Bundschuh, J.: Volcanic arsenic and boron pollution of Ilopango lake, El Salvador In: J. Bundschuh, M.A. Armienta, P. Birkle, P. Bhattacharya, J. Matschullat and A.B. Mukherjee (eds): Natural arsenic in groundwater of Latin America. In: J. Bundschuh and P. Bhattacharya (series eds): *Arsenic in the environment*, Volume 1. CRC Press/Balkema Publisher, Leiden, The Netherlands, 2009, pp. 129–143.

Lowers, H.A., Breit, G.N., Foster, A.L., Whitney, J., Yount, J., Uddin, M.N. and Muneem, A.A.: Arsenic incooporation into authigenic pyrite. Bengal basin sediment, Bangladesh. *Geochim. Cosmochim. Acta* 71 (2007), pp. 2699–2717.

Lu, F.-J.: Physiochemical characteristics of drinking water in blackfoot disease endemic areas in Chiai and Tainan Hsiens. *J. Formos. Med. Assoc.* 74 (1975), pp. 596–605.

Lu, F.-J.: Contributions of fluorescent humic substances existing in the drinking well water of Blackfoot disease endemic area in Taiwan to the environmental toxicology. *J. Formos. Med. Assoc.* 88 (1989), pp. s76–s83 (in Chinese).

Lu, F.-J., Tsai, M.-H. and Lin, K.-H.: Study on the fluorescent substances in the drinking water of blackfoot disease area: Separation and identification of fluorescent substances. *J. Formos. Med. Assoc.* 76:3 (1977) (in Chinese).

Lu, F.-J., Shieh, H.-B., Wu, S.-Y., Sun, J.-T., Guo, J.-L., Lee, J.-J. and Hu, H.-D.: The fluorescent substances in well waters of blackfoot disease area of Taiwan: The interrelationship among fluorescence intensity, arsenic, pH, and TDS in relation to epidemical degrees of BFD. BFD Research Report, Vol. 33, Taiwan BFD Prevention and Control Center, Taiwan Provincial Government, Nantou, Taiwan, 1985 (in Chinese).

Lu, F.-J., Guo, H.-R., Chiang, H.-S. and Hong, C.-L.: Relationships between the fluorescent intensity of well water and the incidence rate of bladder cancer. *J. Chin. Oncol. Soc.* 2 (1986), pp. 14–23.

Lu, J.-N. and Chen, C.-J.: Prevalence of hepatitis B surface antigen carrier status among residents in the endemic area of chronic arsenicism in Taiwan. *Anticancer Res.* 11 (1991), pp. 229–233.

Lu, S.-N., Chow, N.-H., Wu, W.-C., Chang, T.-T., Huang, W.-S., Chen, S.-C., Lin, C.-H. and Carr, B.I.: Characteristics of hepatocellular carcinoma in a high arsenicism area in Taiwan: a case-control study. *J. Occup. Environ. Med.* 46 (2004), pp. 437–441.

Luján, J.C.: Desarsenicación del agua utilizando hidrogel activado de hidroxido de aluminio. *Ciencia, Tecnologia y Medio Ambiente* 1:1 (2001), Unversidad Tecnológia Nacional, Tucumán, pp. 9–13.

Luján, J.C. and Graieb, O.J.: Eliminación del arsénico en agua por destilación a escala doméstica en zonas rurales. *Revista Médica* 1 (1994), Tucumán, pp. 247–255.

Luján, J.C. and Graieb, O.J.: Eliminación del arsénico en agua por destilación a escala doméstica en zonas rurales. *Revista Ciencia y Tecnología* 3:7 (1995), Universidad Tecnológia Nacional, Tucumán, p. 13.

Luo, Z.D et al.: Epidemiological survey on chronic arsenic poisoning in Inner Mongolia. *Journal of Endemic Disease Inner Mongolia* 18:1 (1993a), pp. 4–6 (in Chinese).

Luo, Z.D., Zhang, Y. and Ma, L.: Epidemiological survey of chronic arsenic poisoning at Tie Mengeng and Zhi Jiliang villages in Inner Mongolia. *Chinese Public Health* 9:8 (1993b), pp. 347–348 (in Chinese).

Luo, Z.D., Zhang, Y.M., Ma, L., Zhang, G.Y., He, X., Wilson, R., Byrd, M., Griffiths, J.G., Lai, S., He, L., Grumski, K. and Lamm, S.H.: Chronic arsenicism and cancer in Inner Mongolia: consequences of well-water arsenic levels greater than 50 µg/l. In: C.O. Abernathy, R.L. Calderon and W.R Chappell: *Arsenic: Exposure and health effects*. Chapman and Hall, London, UK, 1997, pp. 55–68.

Macy, J.M., Nunan, K., Hagen, K.D., Dixon, D.R., Harbour, P.J., Cahill, M. and Sly, L.I.: *Chrysiogenes arsenatis*, gen. Nov., sp. Nov., a new arsenate-respiring bacterium isolated from gold mine wastewater. *Int. T. Syst. Bacteriol.* 46 (1996), pp. 1153–1157.

Maher, W., Goessler, W., Kirby, J. and Raber, G.: Arsenic concentrations and speciation in the tissues and blood of sea mullet (Mugil cephalus) from Lake Macquarie NSW. *Austral. Marine Chem.* 68 (1999), pp. 169–182.

Malik, A.H., Khan, Z.M., Mahmood, Q., Nasreen, S. and Bhatti, Z.A.: Perspectives of low cost arsenic remediation of drinking water in Pakistan and other countries. *J. Haz. Mat.* 168 (2009), pp. 1–12.

Mandal, B.K., Roychowdhury, T., Samanta, G., Basu, G.K., Chowdhury, P.P., Chandra, C.R., Lodh, D., Karan, N.K., Dhar, R.K., Tamili, D.K., Das, D., Saha, K.C. and Chakraborti, D.: Arsenic in groundwater in seven districts of West Bengal, India: the biggest arsenic calamity in the world. *Curr. Sci.* 70:11 (1996), pp. 976–986.

Manning, B.A. and Goldberg, S.: Modeling competitive adsorption of arsenate with phosphate and molybdate on oxide minerals. *Soil Sci. Soc. Am. J.* 60 (1996), pp. 121–131.

Manning, B.A. and Goldberg, S.: Adsorption and stability of arsenic (III) at the clay mineral-water interface. *Environ. Sci. Technol.* 31 (1997), pp. 2005–2011.

Manning, B.A., Fendorf, S.E. and Goldberg, S.: Surface structures and stability of As(III) on goethite: spectroscopic evidence for inner-sphere complexes. *Environ. Sci. Technol.* 32 (1998), pp. 2383–2388.

Marin, A.R., Masscheleyn, P.H. and Partick, W.H. Jr.: The influence of chemical form and concentration of arsenic on rice growth and tissue arsenic concentration. *Plant Soil.* 139 (1992), pp. 175–183.

Mariner, R.H. and Willey, L.M.: Geochemistry of thermal waters in Long Valley, Mono County, California. *J. Geophys. Res.* 81:5 (1976), pp. 792–800.

Marvinney, R.G., Loiselle, M.C., Hopeck, J.T., Braley, D. and Krueger, J.A.: Arsenic in Maine groundwater: An example From Buxton, Maine. *Proceedings of the 1994 Focus Conference on Eastern Regional Ground Water Issues*, Dublin, Ohio: National Ground Water Association, 1994, pp. 701–714.

Mason, R.P., Laporte, J. and Andres, S.: Factors controlling the bioaccumulation of mercury, methylmercury, arsenic, selenium, and cadmium by freshwater invertebrates and fish. *Arch. Environ. Contam. Toxicol.* 38 (2000), pp. 283–297.

Matisoff, G., Khourey, C.J., Hall, J.F., Varnes, A.W. and Strain, W.H.: The nature and source of arsenic in northeastern Ohio ground water. *Ground Water* 20:4 (1982), pp. 446–456.

Matschullat, J.: Arsenic in the geosphere—A review. *Sci. Total Environ.* 249 (2000), pp. 297–312.

McArthur, J.M., Ravenscroft, P., Safiulla, S. and Thirlwall, M.F.: Arsenic in groundwater: testing pollution mechanisms for sedimentadry aquifers in Bangladesh. *Water Resour. Res.* 37 (2001), pp. 109–117.

McCarthy, K.T., Pichler, T. and Price, R.E.: Geochemistry of Champagne hot springs shallow hydrothermal vent field and associated sediments, Dominica, Lesser Antilles. *Chem. Geol.* 224:1–3 (2005), pp. 55–68.

McCleskeya, B.R., Nordstrom, D.K. and Maest, A.S.: Preservation of water samples for arsenic(III/V) determinations: an evaluation of the literature and new analytical results. *Appl. Geochem.* 19:7 (2004), pp. 995–1009.

McGeer, J.C., Brix, K.V., Skeaff, J.M., Deforest, D.K., Brigham, S.I., Adams, W.J. and Green, A.: Inverse relationship between bioconcentration factor and exposure concentration for metals: implications for hazard assessment of metals in the aquatic environment. *Environ. Toxicol. Chem.* 22 (2003), pp. 1017–1037.

McKenzie, E.J., Brown, K.L., Cady, S.L. and Campbell, K.A.: Trace metal chemistry and silicification of microorganisms in geothermal sinter, Taupo volcanic zone, New Zealand. *Geothermics* 30:4 (2001), pp. 483–502.

McNeill, L.S. and Edwards, M.: Predicting As removal during metal hydroxide precipitation. *J. AWWA* 88 (1997a), pp. 75–86.

McNeill, L.S. and Edwards, M.: Arsenic removal during precipitative softening. *J. Env. Eng. ASCE* 123 (1997b), pp. 453–460.

McRae, C.M.: *Geochemistry and origin of arsenic-rich pyrite in the Uphapee Creek, Macon County, Alabama.* MSc Thesis, Department of Geology, Auburn University, AL, 1995.

Meharg, A.A. and Rahman, M.M.: Arsenic contamination of Bangladesh paddy field soils: Implication for rice contribution to arsenic consumption. *Environ. Sci. Technol.* 37 (2003), pp. 229–234.

Melamed, R., Jurinak, J.J. and Dudley, L.M.: Effect of adsorbed phosphate on transport of arsenate through an Oxisol. *Soil . Sci. Soc. Am. J.* 59 (1995), pp. 1289–1294.

Mellano, M.F. and Ramirez, A.E.: *Groundwater arsenic in the area around Maria Elena in Santiago del Estero Province, North-western Argentina: Hydrogeochemical characteristics, arsenic mobilization and experimental studies on arsenic removal using natural clays.* Department of Land and Water Resources Engineering KTH, Stockholm, Sweden. TRITA-LWR-EX-04-40, 2004.

Meng, X.G., Bang, S. and Korfiatis, G.P.: Effects of silicate, sulfate, and carbonate on arsenic removal by ferric chloride. *Water Res.* 34 (2000), pp. 1255–1261.

Meng, X.G., Korfiatis, G.P., Bang, S.B. and Bang, K.W.: Combined effects of anions on arsenic removal by iron hydroxides. *Toxicol. Lett.* 133 (2002), pp. 103–111.

Mercado, S., Bermejo, F., Hurtado, R., Terrazas, B. and Hernández, L.: Scale incidence on production pipes of Cerro Prieto geothermal wells. *Geothermics* 18:1/2 (1989), pp. 225–232.

Ministerio de Salud: Resolución (MS) 153/2001. Del 22/2/2001. B.O.: 1/3/2001. Impleméntase el Programa de Minimización de Riesgos por Exposición a Arsénico en Agua de Consumo del Departamento de Salud Ambiental de la Dirección de Promoción y Protección de la Salud. MRA-LIGAN (Minería de la República Argentina-Unidad de Gestión Ambiental Nacional), 2005. Estudios Ambientales de Base—Inventario de Recursos Naturales, 2001, Buenos Aires, Argentina; http://www.mineria.qov.arlambiente/estudios/inicio.asp (accessed June 2009).

Minkkinen, P. and Yliruokanen, I.: The arsenic distribution in Finnish peat bogs. *Kemia-Kemi*, Finland 7–8 (1978), pp. 331–335.

Mohana, D. and Pittman, C.U. Jr.: Arsenic removal from water/wastewater using adsorbents—A critical review. *J. Hazardous Materials* 142: 1–2 (2007), pp. 1–53.

Möller, T., Sylvester, P., Shepard, D. and Morassi, E.: Arsenic in groundwater in New England: point-of-entry and point-of-use treatment of private wells. *Desalination* 243 (2009), pp. 293–304.

Moore, J.N., Ficklin, W.H. and Johns, C.: Partitioning of arsenic and metals in reducing sulfidic sediments. *Environ. Sci. Technol.* 22 (1988), pp. 432–437.

Morales, K.H., Ryan, L., Kuo, T.-L., Wu, M.-M. and Chen, C.-J.: Risk of internal cancers from arsenic in drinking water. Environ. *Health Perspect.* 108 (2000), pp. 655–661.

Morgada, M.E., Mateu, M., Bundschuh, J. and Litter, M.I.: Arsenic in the Iberoamerican region. The IBEROARSEN Network and a possible economic solution for arsenic removal in isolated rural zones. *Journal e-Terra* 5:5 (2008); e-terra.geopor.pt.

Morgada de Boggio, K.E., Levy, I.K., Mateu, M., Litter, M.I., Bhattacharya, P. and Bundschuh, J.: Low-cost technologies for arsenic removal in the Chaco-Pampean plain, Argentina. In: J. Bundschuh, M.A. Armienta, P. Birkle, P. Bhattacharya, J. Matschullat and A.B. Mukherjee (eds): Natural arsenic in groundwater of Latin America. In: J. Bundschuh and P. Bhattacharya (series eds): *Arsenic in the environment*, Volume 1. CRC Press/Balkema Publisher, Leiden, The Netherlands, 2009, pp. 677–683.

Moses, C.O., Nordstrom, D.K., Herman, J.S. and Mills, A.L.: Aqueous pyrite oxidation by dissolved oxygen and by ferric iron. *Geochim. Cosmochim. Acta* 51:6 (1987). pp. 1561–1571.

Mukherjee, A.: *Deeper groundwater chemistry and flow in the arsenic affected western Bengal basin, West Bengal, India.* PhD Thesis, University of Kentucky, Lexington, KY, 2006.

Mukherjee, A., Fryar, A.E. and Howell, P.: Regional hydrostratigraphy and groundwater flow modeling of the arsenic contaminated aquifers of the western Bengal basin, West Bengal, India, Hydrogeol. J. 15 (2007), pp. 1397–1418.

Muñoz, O., Devesa, V., Suner, M.A., Velez, D., Montoro, R., Urieta, I., Macho, M.L. and Jalon, M.: Total and inorganic arsenic in fresh and processed fish products. *J. Agricul. Food Chem.* 48 (2000), pp. 4369–4376.

Nadakavukaren, J.J., Ingermann, R.L. and Jeddeloh, G.: Seasonal variation of arsenic concentration in well water in Lane County, Oregon. *Bull. Environ. Contam. Toxicol.* 33:3 (1984): pp. 264–269.

Nath, B., Jean, J.-S., Lee, M.-K., Yang, H.-J. and Liu, C.-C.: Geochemistry of high arsenic groundwater in Chia-Nan plain, southwestern Taiwan: Possible sources and reactive transport of arsenic. *J. Contam. Hydrol.* 99 (2008), pp. 85–96.

Navarro, M., Sánchez, M., López, H. and López, M.: Arsenic contamination levels in waters, soils, and sludges in southeast Spain. *Bull. Environ. Contam. Toxicol.* 50 (1993), pp. 356–362.

Newcombe, R.L. and Möller, G.: Arsenic removal from drinking water: A review. Blue Water Technologies, Inc., Hayden, ID, 2006; http://www.blueh2o.net/docs/asreview%20080305.pdf (accessed July 2009).

Ngai, T.K.K., Shrestha, R.R., Dangol, B., Maharjan, M. and Murcott, S.E.: Design for sustainable development—Household drinking water filter for arsenic and pathogen treatment in Nepal. *J. Env. Sci. and Health* A 42 (2007), pp. 1879–1888.

Nicholson, R.V.: Iron-sulfide oxidation mechanisms—Laboratory studies. In: J.L. Jambor and D.W. Blowes (eds): *The environmental geochemistry of sulfide mine-wastes.* Short Course Handbook 22, Waterloo, Mineralogical Association of Canada, Ontario, Canada, 1994, pp. 163–183.

Nicholson, R.V., Gillham, R.W. and Reardon, E.J.: Pyrite oxidation in carbonate-buffered solutions. 1. Experimental kinetics. *Geochim. Cosmochim. Acta* 52 (1988), pp. 1077–1085.

Nickson, R., McArthur, J., Burgess, W., Ahmed, K.M., Ravenscroft, P. and Rahman, M.: Arsenic poisoning of Bangladesh groundwater. *Nature* 395 (1998), p. 338.

Nickson, R., McArthur, J.M., Ravenscroft, P., Burgess, W.G. and Ahmed, K.M.: Mechanism of arsenic release to groundwater, Bangladesh and West Bengal. *Appl. Geochem.* 15 (2000), pp. 403–413.

Nickson, R.T., McArthur, J.M., Shrestha, B., Kyaw-Myint, T.O. and Lowry, D.: Arsenic and other drinking water quality issues, Muzaffargarh District, Pakistan. *Appl. Geochem.* 20 (2005), pp. 55–68.

Nicolli, H.B., O'Connor, T.E., Suriano, J.M., Koukharsky, M.L., Gómez-Peral, M.A., Bertini, L.M., Cohen, I.M., Corradi, L., Balean, O.A. and Abril, E.G.: Geoquímica del arsénico y de otros oligoelementos en aguas subterráneas de la llanura sudoriental de la Provincia de Córdoba. Academia Nacional de Ciencias, Córdoba. *Miscelánea* 71 (1985), pp. 1–112.

Nicolli H.B., Suriano, J.M., Gómez Peral, M.A., Ferpozzi, L.H. and Baleani, O.M.: Groundwater contamination with arsenic and other trace elements in an area of the Pampa, Province of Córdoba, Argentina. *Environ. Geol. Water Sci.* 14 (1989), pp. 3–16.

Nicolli, H.B., Tineo, A., García, J.W., Falcón, C.M. and Merino, M.H.: Trace-element quality problems in groundwater from Tucumán, Argentina. In: R. Cidu (Ed.), *Water-Rock Interaction* 2. Balkema, Lisse, The Netherlands, 2001, pp. 993–966.

Nicolli, H.B., Tineo, A., Falcón, C.M., García, J.W., Merino, M.H., Etchichury, M.C. Alonso, M.S. and Tofalo, O.R.: Arsenic hydrogeochemistry in groundwater from the Burruyacú basin, Tucumán province, Argentina. In: J. Bundschuh, M.A. Armienta, P. Birkle, P. Bhattacharya, J. Matschullat and A.B. Mukherjee (eds): Natural arsenic in groundwater of Latin America. In: J. Bundschuh and P. Bhattacharya (series eds): *Arsenic in the environment*, Volume 1. CRC Press/Balkema Publisher, Leiden, The Netherlands, 2009, pp. 47–59.

Nimick, D.A., Moore, J.N., Dalby, C.E. and Savka, M.W.: The fate of geothermal arsenic in the Madison and Missouri rivers, Montana and Wyoming. *Water Resour. Res.* 34 (1998), pp. 3051–3067.

Niu, S., Cao, S. and Shen, E.: The status of arsenic poisoning in China. In: C.O. Abernathy, R.K. Calderon and W.R. Chappell (eds): *Arsenic exposure and health effects*. Chapman and Hall Press, London, UK, 1997, pp. 78–83.

Noguchi, K. and Nakagawa, R.: Arsenic and arsenic-lead sulfide sediments from Tamagawa Hot Springs, Akita Prefecture. *Proc. Japan Acad.* 45 (1969), pp. 45–50.

Nordstrom, D.K.: Aqueous pyrite oxidation and the consequent formation of secondary iron minerals. In: J.A. Kittrick, D.S. Fanning and L.R. Hossner (eds): *Acid sulfate weathering*. Soil Sci. Soc. Am. Publ., 1982, pp. 37–56.

Nordstrom, D.K.: An overview of arsenic mass poisoning in Bangladesh and West Bengal. In: C. Young (ed): *Minor elements 2000, As, Sb, Se, Te, and Bi*. Soc. Mining Metallurgy and Exploration, 2000, pp. 21–30.

Nordstrom, D.K.: Worldwide occurrences of arsenic in groundwater. *Science* 296 (2002), pp. 2143–2145.

Nordstrom, D.K. and Alpers, C.N.: Geochemistry of acid mine waters. In: G.S. Plumlee and M.J. Logsdon (eds): *Reviews in economic geology*, Vol. 6A: *The environmental geochemistry of mineral deposits*, Part A: *Processes, methods and health issues*. Soc. Econ. Geol., Littleton, CO, 1999, pp. 133–160.

Nordstrom, D.K. and Muñoz, J.L.: *Geochemical thermodynamics*. Blackwell, Cambridge, UK, 1994.

Nordstrom, D.K. and Southam, G.: Geomicrobiology of sulfide mineral oxidation. Chap. 11, in: J.F. Banfield and K.H. Nealson: Interactions between microbes and minerals. *Reviews in Mineralogy* 35, Min. Soc. Am., Washington, DC, 1997, pp. 361–390.

Nordstrom, D.K., Ball, J.W. and McCleskey, R.B.: Ground water to surface water: Chemistry of thermal outflows in Yellowstone National Park. In: W. Inskeep and T.T. McDermott (eds): *Geothermal biology and geochemistry in Yellowstone National Park*. Thermal Biology Institute, Montana State University, Bozeman, MT, 2005, pp. 73–94.

Nordstrom, D.K., McCleskey, R.B. and Ball, J.W.: Sulfur geochemistry of hydrothermal waters in Yellowstone National Park, Wyoming, USA. IV. Acid-sulfate waters. *Appl. Geochem.* 24 (2009), pp. 191–207.

Norvell, J.L.S.: *Distribution of, sources of, and processes mobilizing arsenic, chromium, selenium, and uranium in the central Oklahoma aquifer*. MSc Thesis, Department of Geology, Colorado School of Mines, Golden, CO, 1995.

Nriagu, J.O.: A global assessment of natural sources of atmospheric trace metals. *Nature* 338 (1989), pp. 47–49.

Nriagu, J.O. and Pacyna, J.M.: Quantitative assessment of worldwide contamination of air, water, and soils by trace metals. *Nature* 333 (1988), pp. 134–139.

Nriagu, J.O., Bhattacharya, P., Mukherjee, A.B., Bundschuh, J., Zevenhoven, R. and Loeppert, R.H.: Arsenic in soil and groundwater: an overview. In: P. Bhattacharya, A.B. Mukherjee, J. Bundschuh, R. Zevenhoven and R.H. Loeppert (eds): *Arsenic in soil and groundwater environment; Trace metals and other contaminants in the environment*, Volume 9. Elsevier, 2007, pp. 3–60.

OECD: Harmonized integrated hazard classification system for human health and environmental hazards of chemical substances and mixtures. Annex 2. Guidance Document 27, ENV/JM/MONO 8.OECD, Paris, France, 2001.

Oremland, R.S. and Stolz, J.F.: The ecology of arsenic. *Science* 300 (2003), pp. 393–944.

Owen-Joyce, S.J.: Hydrology of a stream-aquifer system in the Camp Verde area, Yavapai County, Arizona. *Arizona Department of Water Resources Bulletin* 3, Phoenix, AZ, 1984.

Owen-Joyce, S.J. and Bell, C.K.: Appraisal of water resources in the Upper Verde River area, Yavapai and Coconino Counties, Arizona. *Arizona Department of Water Resources Bulletin* 2, Phoenix, AZ, 1983.

Pacyna, J.O.: Atmospheric trace elements from natural and anthropogenic sources. In: J.O. Niragu and C.I. Davidson (eds): Toxic metals in the atmosphere. *Adv. Environ. Sci. Technol.* 17, John Wiley & Sons, New York, NY, 1986, pp. 33–52.

Panno, S.V., Hackley, K.C., Cartwright, K. and Liu, C.L.: Hydrochemistry of the Mahomet bedrock valley aquifer, east-central Illinois: Indicators of recharge and ground-water flow. *Ground Water* 32:4 (1994), pp. 591–604.

Papke, R.T., Ramsing, N.B., Bateson, M.M. and Ward, D.M.: Geographical isolation in hot spring cyanobacteria. *Environ. Microbiol.* 5:8 (2004), pp. 650–659.

Pearce, F.: Arsenic's fatal legacy grows worldwide. *New Scientist* 2407 (2003), pp. 1–3.

Pederick, R.L., Gault, A.G., Charnock, J.M., Polya, D.A. and Llyod, J.R: Probing the biogeochemistry of arsenic: Response of two contrasting aquifer seiments from Cambodia to stimulation by arsenate and ferric iron. *J. Environ. Sci. Health* Part A 42 (2007), pp. 1753–1774.

Peryea, F.J. and Kammereck, R.: Phosphate enhanced movement of arsenic out of lead arsenate-contaminated topsoil. *Water Air Soil Pollut.* 93 (1997), pp. 243–254.

Peters, S.C.: Arsenic in groundwaters in the Northern Appalachian Mountain belt: A review of patterns and processes. *J. Contam. Hydrol.* 99:1–4 (2008), pp. 8–21.

Peters, S.C. Blum, J.D., Klaue, B. and Karagas, M.R.: Arsenic occurrrence in New Hampshire ground water. *Environ. Sci. Technol.* 33:9 (1999), pp. 1328–1333.

Petrusevski, B., Sharma, S., Schippers, J.C. and Shordt, K.: *Arsenic in drinking water*. IRC International Water and Sanitation Centre, The Hague, The Netherlands, 2007.

Petrusevski, B., Sharma, S., van der Meer, W.G., Kruis, F., Khan, M., Barua, M. and Schippers, J.C.: Four years of development and field-testing of IHE arsenic removal family filter in rural Burlangdadesh. *Water Sci. Technol.* 58 (2008), pp. 53–58.

Phuong, T.D., Van Chuong, P., Khiem, D.T. and Kokot, S.: Elemental content of Vietnamese rice-Part 1: Sampling, analysis and comparison with previous studies. *Analyst* 124 (1999), pp. 553–560.

Pichler, T. and Veizer, J.: Precipitation of Fe(III) oxyhydroxide deposits from shallow-water hydrothermal fluids in Tutum Bay, Ambitle Island, Papua New Guinea. *Chem. Geol.* 162:1 (1999), pp. 15–31.

Pichler, T., Veizer, J. and Hall, G.: Natural input of arsenic into a coral reef ecosystem by hydrothermal fluids and its removal by Fe(III) oxyhidroxides. *Environ. Sci. Technol.* 33 (1999), pp. 1373–1378.

Pierce, M.I. and Moore, C.B.: Adsorption of arsenite and arsenate on amorphous iron hydroxide. *Water Res.* 16 (1982), pp. 1247–1253.

Pirnie, M.: Technologies and costs for removal of arsenic from drinking water. US EPA Report 815-R-00-028, 2000.

Pizarro, G. and Balabanoff, L.: Estudio de eliminación del arsénico en el agua de Antofagasta. *Bol. Soc. Chil. Quím.* 12 (1973), pp. 1–8.

Planer-Friedrich, B., London, J., McCleskey, R.B., Nordstrom, D.K. and Wallschläger, D.: Thioarsenates in geothermal waters of Yellowstone National Park: determination, preservation and geochemical importance. *Environ. Sci. Technol.* 41 (2007), pp. 5245–5251.

Plant, J.A., Kinniburgh, D.G., Smedley, P.L., Fordyce, F.M. and Klinck, B.A.: Arsenic and selenium. In: B.S. Lollar, H.D. Holland and K.K. Turekian (eds): *Treatise on geochemistry*. Elsevier Ltd., San Diego, CA, 2004, pp. 17–66.

Pokrovski, G.S., Kara, S. and Roux, J.: Stability and solubility of arsenopyrite, FeAsS, in crustal fluids. *Geochim. Cosmochim. Acta* 66:13 (2002), pp. 2361–2378.

Postma, D., Boesen, C., Kristiansen, H. and Larsen, F.: Nitrate reduction in an unconfined sandy aquifer: water chemistry, reduction processes, and geochemical modeling. *Water Resour. Res.* 27:8 (1991), pp. 2027–2045.

Puga, F., Olivos, P., Greyber, R., González, I., Heras, E., Barrera, S. and González, E.: Hidroarsenicismo crónico: Intoxicación arsenical crónica en Antofgasta: Estudio epidemiológico y clínico. *Rev. Chilena Pediatría* 44:3 (1973), pp. 215–223.

Queirolo, F., Stegen, S., Mondaca, J., Cortes, R., Rojas, R., Contreras, C., Muñoz, L., Schwuger, M. and Ostapczuk, P.: Total arsenic, lead, cadmium, copper, and zinc in some salt rivers in northern Andes of Antofagasta, Chile. *Sci. Total Environ.* 255 (2000), pp. 85–95.

Quintanilla, J., Ramos, O., Ormachea, M., García, M.E., Medina, H., Thunvik, R., Bhattacharya, P. and Bundschuh, J.: Arsenic contamination, speciation and environmental consequences in the Bolivian plateau. In: J. Bundschuh, M.A. Armienta, P. Birkle, P. Bhattacharya, J. Matschullat and A.B. Mukherjee (eds): Natural arsenic in groundwater of Latin America. In: J. Bundschuh and P. Bhattacharya (series eds): *Arsenic in the environment*, Volume 1. CRC Press/Balkema Publisher, Leiden, The Netherlands, 2009, pp. 91–99.

Randall, S.R., Sherman, D.M. and Ragnarsdottir, K.V.: Sorption of As(V) on green rust (Fe$_4$(II)Fe$_2$(III)(OH)$_{12}$SO$_4$.3H$_2$O) and lepidocrocite (γ-FeOOH): surface complexes from EXAFS spectroscopy. *Geochim. Cosmochim Acta* 65 (2001), pp. 1015–1023.

Ravenscroft, P.: Predicting the global extent of arsenic pollution of groundwater and its potential impact on human health. Report prepared for UNICEF, New York, NY, 2007.

Ravenscroft, P., Brammer, H. and Richards, K.: *Arsenic pollution —A global synthesis*. Wiley-Blackwell, 2009.

Redman, A.D., Macalady, D.L. and Ahmann, D.: Natural organic matter affects arsenic speciation and sorption onto hematite. *Environ. Sci. Technol.* 36 (2002), pp. 2889–2896.

Reichert, F. and Trelles, R.: Yodo y arsénico en las aguas subterráneas. *Anal. Asoc. Quim. Argent.* 9 (1921), pp. 89–95.

Reid, K.D., Goff, F. and Counce, D.A.: Arsenic concentration and mass flow rate in natural waters of the Valles caldera and Jemez Mountains region, New Mexico. *New Mexico Geology* 25:3 (2003), pp. 75–82.

Reynolds, J.G., Naylor, D.V. and Fendorf, S.E.: Arsenic sorption in phosphate-amended soils during flooding and subsequent aeration. *Soil Sci. Soc. Am. J.* 63 (1999), pp. 1149–1156.

Reza, A.H.M.S., Jean, J.-S., Yang, H.-J., Lee, M.-K., Hsu, H.-F., Liu, C.-C., Lee, Y.-C., Bundschuh, J., Lin, K.-H. and Lee, C.-Y.: Arsenic and humic substances in alluvial aquifers of Bangladesh and Taiwan: A comparative study. J. Contam. Hydrology 2009 (under review).

Rietra, R.P.J.J., Hiemstra, T. and van Riemsdijk, W.H.: Interaction between calcium and phosphate adsorption on goethite. Environ. Sci. Technol. 35 (2001), pp. 3369–3374.

Rimstidt, J.D., Chermak, J.A. and Gagen, P.M.: Rates of reaction of galena, sphalerite, chalcopyrite, and arsenopyrite with Fe(III) in acidic solutions. In: C.N. Alpers and D.W. Blowes (eds): Environmental geochemistry of sulfide oxidation. American Chemical Society, Symposium Series 550, Washington, DC, 1994, pp. 2–13.

Risk Assessment Forum: Special report on ingested inorganic arsenic: skin cancer; nutritional essentiality. US Environmental Protection Agency, EPA publication EPA\625\3-87/103, Washington, DC, 1988.

Ritchie, J.A.: Arsenic and antimony in some New Zealand thermal waters. *New Zealand J. Sci.* 4 (1961), pp. 218–229.

Rittle, K.A., Drever, J.I. and Colberg, P.J.S.: Precipitation of arsenic during bacterial sulfate reduction. *Geomicrobiol. J.* 13:1 (1995), pp. 1–11.

Rivara, M., Cebrián, M., Corey, G., Hernandez, M. and Romieu, I.: 1997. Cancer risk in an arsenic-contaminated area of Chile. *Toxicol. Ind. Health* 13:2–3 (1997), pp. 321–338.

Roberts, K., Stearns, B. and Francis, R.L.: Investigation of arsenic in southeastern North Dakota ground water. A Superfund Remedial Investigation Report. North Dakota State Department of Health, Bismarck, ND, 1985.

Roberts, L.C., Hug, S.J., Ruettimann, T., Billah, M., Khan, A.W. and Rahman, M.T.: Arsenic removal with iron(II) and iron(III) in waters with high silicate and phosphate concentrations. *Environ. Sci. Technol.* 38 (2004), pp. 307–315.

Robertson, F.N.: Arsenic in ground-water under oxidizing conditions, south-west United States. *Environ. Geochem. Health* 11:3–4 (1989), pp. 171–186.

Romero, F.M., Armienta M.A. and Carrillo-Chavez A.: Arsenic sorption by carbonate-rich aquifer material, a control on arsenic mobility at Zimapán, Mexico. *Archiv. Environ. Contamin. Toxicol.* 47 (2004), pp. 1–13.

Romero, L., Alonso, H., Campano, P., Fanfani, L., Cidub, R., Dadea, C., Keegan, T., Thornton, I. and Farago, M.: Arsenic enrichment in waters and sediments of the Rio Loa (Second Region, Chile). *Appl. Geochem.* 18 (2003), pp. 1399–1416.

Rossman, T.: Mechanism of arsenic carcinogenesis: an integrated approach: fundamental and molecular mechanisms of mutagenesis. *Mutation Res.* 533 (2003), pp. 37–65.

Rowe, G.L., Ohsawa, S., Takano, B., Brantley, S.L., Fernández, J.F. and Barquero, J.: Using crater lake chemistry to predict volcanic activity at Poás Volcano, Costa Rica. *Bull. Volcanol.* 54 (1992a), pp. 494–503.

Rowe, G.L., Brantley, S.L., Fernández, M., Fernández, J.F., Borgia, A. and Barquero, J.: Fluidvolcano interaction in an active stratovolcano: the crater lake system of Poás Volcano, Costa Rica. *J. Volcanol. Geother. Res.* 49 (1992b), p. 2351.

Roychowdhury, T., Uchino, T., Tokunaga, H. and Ando, M.: Survey of arsenic in food composites from an arsenic affected area of West Bengal, India. *Food Chem. Toxicol.* 40 (2002), pp. 1611–1621.

Rubel, F.: Design manual: Removal of arsenic from drinking water by adsorptive media. EPA-600-R-03-019, US Environmental Protection Agency, 2003.

Rudnick, R.L. and Gao, S.: Composition of the continental crust. In: R.L. Rudnick (ed): The crust. In: H.D. Holland and K.K. Turekian (eds): *Treatise on geochemistry*, Volume 3. Elsevier, Amsterdam, The Netherlands, 2005, pp. 1–64.

Sadiq, M.: Arsenic chemistry in soils: An overview of thermodynamic predictions and field observations. *Water Air Soil Pollut.* 93 (1995), pp. 117–136.

Saha, K.C.: Chronic arsenical dermatosis from tubewell water in West Bengal during 1983–87. *Indian J. Dermatol.* 40:1 (1995), pp. 1–12.

Salcedo, J.C., Portales, A., Landecho, E. and Diaz, R.: Transverse study of a group of patients with vasculpathy from chronic arsenic poisoning in communities of the Francisco de Madero and San Pedro Districts, Coahuila, Mexico. *Rev. Fac. Med. (Torreón)* 12 (1984). p. 16.

Sancha, A.M. and Frenz, P.: Estimate of the current exposure of the urban population of northern Chile to arsenic. In: *Interdisciplinary perspectives on drinking water risk assessment and management.* IAHS Publ. 260, 2000, pp. 3–8.

Sancha, A.M., O'Ryan, R. and Pérez, O.: The removal of arsenic from drinking water and associated costs: the Chilean case. *Interdisciplinary Perspectives on Drinking Water Risk Assessment and Management.* IAHS Publ. 260, 2000, pp. 17–25.

Sánchez-Rodas, D., Geiszinger, A., Gomez-Ariza, J.L. and Francesconi, K.A.: Determination of an arsenosugar in oyster extracts by liquid chromatography-electrospray mass spectrometry and liquid chromatography-ultraviolet photo-oxidation-hydride generation atomic fluorescence spectrometry. *Analyst* 127 (2002), pp. 60–65.

Sarkar, S., Blaney, L.M., Gupta, A., Ghosh, D. and SenGupta, A.K.: Use of ArsenXnp, a hybrid anion exchanger, for arsenic removal in remote villages in the Indian subcontinent. *React. Funct. Polym.* 67 (2007), 1599–1611.

Schlottmann, J.L. and Breit, G.N.: Mobilization of As and U in the central Oklahoma aquifer. In: Y.K. Kharaka and A.S. Maest (eds): *Water-rock interaction.* Balkema, Rotterdam, The Netherlands, 1992, pp. 835–838.

Schoof, R.A., Yost, L.J., Crecelius, E., Irgolic, K., Goessler, W., Guo, H.-R. and Greene, H.: Dietary arsenic intake in Taiwanese districts with elevated arsenic in drinking water. *Hum. Ecol. Risk Assess.* 4 (1998), pp. 117–135.

Schoof, R.A., Yost, L.J., Eickhoff, J., Crecelius, E.A., Cragin, D.W., Meacher, D.M. and Menzel, D.B.: A market basket survey of inorganic arsenic in food. *Food Chem. Toxicol.* 37 (1999), pp. 839–846.

Schreiber, M.A., Simo, J.A. and Freiberg, P.G.: Stratigraphic and geochemical controls on naturally occurring arsenic in groundwater, eastern Wisconsin, USA. *Hydrogeol. J.* 8:2 (2000), pp. 161–176.

Scott, S.D.: Chemical behaviour of sphalerite and arsenopyrite in hydrothermal and metamorphic environments. *Mineral. Mag.* 47:4 (1983), pp. 427–435.

See, L.-C., Chiou, H.-Y., Lee, J.-S., Hsueh, Y.-M., Lin, S.-M., Tu, M.-C., Yang, M.-L. and Chen, C.-J.: Dose-response relationship between ingested arsenic and cataracts among residents in southwestern Taiwan. *J. Environ. Sci. Health* A 42 (2007), pp. 1843–1851.

Selinus, O., Finkelman, R.B., Centeno, J.A. and Lax, K.: Medical geology: a new future for geoscience. *European Geologist* 2005, pp. 27–30, http://www.medicalgeology.org/PDF/EuropeanGeologist.pdf (accessed November 2009).

Shen, J.J.S. and Wang, H.-J.: Sources and genesis of the Chinkuashi Au-Cu deposits in northern Taiwan; Constraints from Os and Sr isotopic composition of sulfides *Earth Planet. Sci. Lett.* 222 (2004), pp. 71–83.

Sherman, D.M. and Randall, S.R.: Surface complexation of arsenic(V) to iron(III) hydroxides: structural mechanism from ab initio molecular geometries and EXAFS spectroscopy. *Geochim. Cosmochim. Acta* 67 (2003), pp. 4223–4230.

Shih, T.T.: A survey of the active mud volcanoes in Taiwan and a study of their types and the character of the mud. *Petrol. Geol. Taiwan* 5 (1967), pp. 259–311.

Shrestha, R.R., Shrestha, M.P., Upadhyay, N.P., Pradhan, R., Khadka, R., Maskey, A., Maharjan, M., Tuladhar, S., Dahal, B.M. and Shrestha, K.: Groundwater arsenic contamination, its health impact and mitigation program in Nepal. *J. Environ. Sci. Health* A38 (2003), pp. 185–200.

Shyu, G.-S. and Lin, G.-C.: Survey of heavy metal content in soil in Beitou Guandu area. Environmental Protectection Bureau, Taipei, Taiwan, 2006.

Shyu, G.-S., Chang, T.-K. and Lin, S.-C.: Monitoring and investigation of arsenic in soil, Beitou in Taiwan. *Workshop on Arsenic Contamination in Beitou Guandu Area*, Taipei, 2009, pp. 20–47.

Sigg, L. and Stumm, W: The interaction of anions and weak acids with the hydrous goethite (α-FeOOH) surface. *Colloids Surfaces* 2 (1981), pp. 101–117.

Simeoni, M.A., Batts, B.D. and McRae, C.: Effect of groundwater fulvic acid on the adsorption of arsenate by ferrihydrite and gibbsite. *Appl. Geochem.* 18 (2003), pp. 1507–1515.

Simo, J.A., Freiberg, P.G. and Freiburg, K.S.: Geologic constraints on arsenic in groundwater with applications to groundwater modeling. Groundwater Research Report WRC GRR 96-01, University of Wisconsin, Madison, WI, 1996.

Simon, G., Huang, H., Penner-Hahn, J.E., Kesler, S.E. and Kao, L.S.: Oxidation state of gold and arsenic in gold-bearing arsenian pyrite. *Amer. Mineralogist* 84:7–8 (1999), pp. 1071–1079.

Singer, P.C. and Stumm, W.: Acidic mine drainage: The rate-determining step. *Science* 167:3921 (1970), pp. 1121–1123.

Smedley, P.L.: Arsenic in rural groundwater in Ghana: part special issue: hydrogeochemical studies in sub-Saharan Africa. *J. Afr. Earth Sci.* 22:4 (1996), pp. 459–470.

Smedley, P.L.: Arsenic occurence in groundwater in South and East Asia —scale, causes and mitigation. The World Bank, Washington, DC, 2005.

Smedley, P.L.: Sources and distribution of arsenic in groundwater and aquifers. *Proceedings of the Symposium Arsenic in groundwater—A World problem*, 29 November 2006, Utrecht, The Netherlands, 2006, pp. 4–32, http://www.iah.org/downloads/occpub/arsenic_gw.pdf (accessed October 2009).

Smedley, P.L. and Kinniburgh, D.G.: Source and behaviour of arsenic in natural waters. U.N. Synthesis Report on Arsenic in Drinking Waters. World Health Organization, Geneve, Switzerland, 2001, pp. 1–61.

Smedley, P.L. and Kinniburgh, D.G.: A review of the source, behaviour and distribution of arsenic in natural waters. *Appl. Geochem.* 17 (2002), pp. 517–568.

Smedley, P.L., Zhang, M., Zhang, G.Y. and Luo, Z.D.: Arsenic and other redox-sensitive elements in groundwater from the Huhhot Basin, Inner Mongolia. In: R. Cidu (ed): *Water Rock Interaction*, Vol. 1. Swets & Zeitlinger, Lisse, The Netherlands, 2001, pp. 581–584.

Smedley, P.L, Kinniburgh, D.G., Macdonald, D.M.J., Nicolli, H.B., Barros, A.J. and Tullio, J.O.: Arsenic associations in sediments from the loess aquifer of La Pampa, Argentina. *Appl. Geochem.* 20 (2005), pp. 989–1016.

Smedley, P.L., Nicolli, H.B., Macdonald, D.M.J. and Kinniburgh, D.G.: Arsenic in groundwater and sediments from La Pampa province, Argentina. In: J. Bundschuh, M.A. Armienta, P. Birkle, P. Bhattacharya, J. Matschullat and A.B. Mukherjee (eds): Natural arsenic in groundwater of Latin America. In: J. Bundschuh and P. Bhattacharya (series eds): *Arsenic in the environment*, Volume 1. CRC Press/Balkema Publisher, Leiden, The Netherlands, 2009, pp. 35–45.

Smith, A.D. and Lewis, C.: Geochemistry of metabasalt and associated metasedimentary rocks from the Lushan Formation of the Upthrust Slate Belt, South-Central Taiwan: *Int. Geology Rev.* 49 (2007), pp. 1–13.

Smith, A.H., Goycolea, M., Haque, R. and Biggs, M.L.: Marked increase in bladder and lung cancer mortality in a region of Northern Chile due to arsenic in drinking water. *Am. J. Epidemiol.* 147:7 (1998), pp. 660–669.

Smith, A.H., Arroyo, A.P., Mazumder, D.N.G., Kosnett, M.J., Hernandez, A.L., Beeris, M., Smith, M.M. and Moore, L.E.: Arsenic-induced skin lesions among Atacameño people in northern Chile despite good nutrition and centuries of exposure. *Environ. Health Persp.* 108:7 (2000), pp. 617–620.

Smith, E., Naidu, R. and Alston, A.M.: Chemistry of inorganic arsenic in soils: II. Effect of phosphorus, sodium, and calcium on arsenic sorption. *J. Environ. Quality* 31 (2002), pp. 557–563.

Smith, J.V.S., Jankowski, J. and Sammut, J.: Vertical distribution of As(III) and As(V) in a coastal sandy aquifer: factors controlling the concentration and speciation of arsenic in the Stuarts Point groundwater system, northern New South Wales, Australia. *Appl. Geochem.* 18:9 (2003), pp. 1479–1496.

Sommers, S.C. and McManus, R.G.: Multiple Asal cancers of skin and internal organs. *Cancer* 6 (1952), pp. 347–59.

Sorg, T.: Arsenic: USEPA demonstration program and latest research results. Presentation, USEPA Workshop on Inorganic Contaminant Issues, 21–23 August 2007, Cincinnati, OH, 2007.

Spliethoff, H.M., Mason, R.P. and Hemond, H.F. Interannual variability in the speciation and mobility of arsenic in a dimictic lake. *Environ. Sci. Technol.* 29:8 (1995), pp. 2157–2161.

Spycher, N.F. and Reed, M.H.: As(III) and Sb(III) sulfide complexes: An evaluation of stoichiometry and stability from existing experimental data. *Geochim. Cosmochim. Acta* 53 (1989), pp. 2185–2194.

Sracek, O., Bhattacharya, P., Jacks, G., Gustafsson, J. and Brömssen, M.: Behavior of arsenic and geochemical modeling of arsenic enrichment in aqueous environments. *Appl. Geochem.* 19 (2004), pp. 169–180.

Sracek, O., Novák, M., Sulovský, P., Martin, R., Bundschuh, J. and Bhattacharya, P.: Mineralogical study of arsenic-enriched aquifer sediments at Santiago del Estero, Northwest Argentina. In: J. Bundschuh, M.A. Armienta, P. Birkle, P. Bhattacharya, J. Matschullat and A.B. Mukherjee (eds): Geogenic arsenic in groundwater of Latin America. In: J. Bundschuh and P. Bhattacharya (series eds): *Arsenic in the environment*, Volume 1. CRC Press/Balkema Publisher, Leiden, The Netherlands, 2009, pp. 61–67.

Stach, L.W.: Stratigraphic subdivision and correlation of the upper Cenozoic sequence in the foothills region east of Chiayi and Hsinying, Taiwan. *Symposium Petroleum Geology Taiwan*, Taipei, Taiwan, 1957, pp. 177–230.

Stauder, S., Raue, B. and Sacher, F.: Thioarsenates in sulfidic waters. *Environ. Sci. Technol.* 39 (2005), pp. 5933–5939.

Stauffer, R.E. and Thompson, J.M.: Arsenic and antimony in geothermal waters of Yellowstone National Park, Wyoming, USA. *Geochim. Cosmochim. Acta* 48:12 (1984), pp. 2547–2561.

Stauffer, R.E., Jenne, E.A. and Ball, J.W.: Chemical studies of selected trace elements in hot-spring drainages of Yellowstone National Park. U.S. Geological Survey Professional Paper 1044-4, 1980.

Stollenwerk, K.G.: Geochemical processes controlling transport of arsenic in groundwater: A review of adsorption. In: A.H. Welch and K.G. Stollenwerk (eds): *Arsenic in ground water: Geochemistry and occurrence*. Kluwer, Boston, MA, 2003, pp. 67–100.

Storniolo, A., Martín, R., Thir, M., Cortes, J., Ramirez, A., Mellano, F., Bundschuh, J. and Bhattacharya, P.: Disminución del contenido de arsénico en el agua mediante el uso de material geológico natural. In: G. Galindo, J.L. Fernández-Turiel, M.A. Parada and D. Gimeno Torrente (eds): Arsénico en aguas: origen, movilidad y tratamiento. Taller. II Seminario Hispano-Latinoamericano sobre temas actuales de hidrología subterránea—IV Congreso Hidrogeológico Argentino. 25–28 Octobere 2005, Río Cuarto, Argentina, 2005, pp. 173–182.

Stumm, W.: *Chemistry of the solid-water interface*. John Wiley & Sons, New York, NY, 1992.

Su, S.-W. and Chen, Z.-S.: Impacts of arsenic contaminated soils on agroecosystems in Guandu plain, Taipei: Assessment by As fractionation: *Proceedings of 14th International Conference on Heavy Metals in the Environment*, 16–23 Noverber 2008. Taipei, Taiwan 2008. pp. 93–96.

Su, S.-W., Lin, S.-C., Huang, W.-D., Chang, T.-K. and Chen, Z.-S.: Rice safety in arsenic contaminated rice-growing soils of Guandu plain in Taipei. In: R.S. Chung, D.Y. Lee and Y.S. Shih (eds): *Newsletter of Soil and Fertiliers* 2007, pp. 58–59.

Su, S.-W., Lin, T.-L. and Chen, Z.-S.: Arsenic-contaminated soil and food safety of crops: Recent studies in Guandu plain of Taiwan and literature review. *Workshop on Arsenic Contamination in Beitou Guandu Area*, Taipei, Taiwan, 2009, pp. 68–85.

Suhendrayatna, O.A., Nakajima, T. and Maeda, S.: Studies on the accumulation and transformation of arsenic in freshwater organisms I. Accumulation, transformation and toxicity of arsenic compounds on the Japanese Medaka, *Oryzias latipes*. *Chemosphere* 46 (2002a), pp. 319–324.

Suhendrayatna, O.A., Nakajima, T. and Maeda, S.: Studies on the accumulation and transformation of arsenic in freshwater organisms II. Accumulation and transformation of arsenic compounds by Tilapia mossambica. *Chemosphere* 46 (2002b), pp. 325–331.

Sverjensky, D. and Fukushi, K.: A predictive model (ETLM) for As(III) adsorption and surface speciation on oxides consistent with spectroscopic data. *Geochim. Cosmochim. Acta 70* (2006), pp. 3778–3802.

Swartz, R.J.: *A study of the occurrence of arsenic on the Kern Fan element of the Kern Water Bank, southern San Joaquin Valley, California.* MSc Thesis, Department of Geology, California State University, Bakersfield, CA, 1995.

Swartz, R.J., Thyne, G.D. and Gillespie, J.M.: Dissolved arsenic in the Kern Fan San Joaquin Valley, California: Naturally occurring or anthropogenic. *Environ. Geosci.* 3:3 (1996), pp. 143–153.

Swedlund, P.J. and Webster, J.G.: Adsorption and polymerization of silicic acid on ferrihydrite, and its effect on arsenic adsorption. *Water Res.* 33 (1999), pp. 3413–3422.

Taipei EPB: Survey and validation of As concentration in farm-land soil in Taipei city. Environmental Protection Bureau, Taipei, Taiwan, 2006.

Taipei Water Department: http://www.taipei.gov.tw/cgi-bin/Message/MM_msg_control?mode=viewnews&ts=4907caed:a5f&theme=/3790000000/379410000A/NG-message (accessed November 2009).

Taiwan COA: Taiwanese Food Supply and Demand. Annual Report. Council of Agriculture, Taiwan, 2004.

Taiwan COA: Taiwanese Food Supply and Demand. Annual Report. Council of Agriculture, Taiwan, 2005.

Taiwan COA: Taiwanese Food Supply and Demand. Annual Report. Council of Agriculture, Executive Yuan: Taiwan, 2006.

Taiwan COA: Taiwan Fisheries Yearbook, Council of Agriculture, Executive Yuan: Taiwan, 2007. Available at http://www.fa.gov.tw/eng/index.php (accessed November 2009).

Taiwan EPA: National drinking water quality standards. Environmental Protection Agency, Taipei, Taiwan, 1998.

Taiwan EPA: Soil quality survey database. Taiwan Environmental Protection Agency, Taipei, Taiwan, 2002, http://edb.epa.gov.tw/new/soil_Quality/soil_Qry.asp (accessed November 2008).

Taiwan EPA: Soil and Groundwater Remediation Act. Taiwan Environmental Protection Agency (EPA), Taipei, Taiwan, 2003.

Taiwan EPA: Data integration and establishment program of soil and groundwater contamination remediation: Annual report on soil and groundwater contamination remediation in 2005 (21 July 2005–21 July 2006), Taiwan Environmental Protection Agency, Taipei, Taiwan, 2006.

Taiwan FACOA: Fisheries Statistical Yearbook. Taipei: Fisheries Agency, Council of Agriculture, Taiwan, ROC, 2005 (in Chinese).

Taiwan Fishery Agency: Annual report of fishery agency—establishment of geographical information systems in aquacultural fish. Taiwan Fishery Agency, Council of Agricuture, Executive Yuan, Tainan, 2003.

Taiwan Provincial Department of Health Blackfoot Disease Control and Prevention in Taiwan, Republic of China. Taichung, Taiwan Provincial Department of Health, Taichung, Taiwan, 1993.

Taiwan Provincial Government: Survey on the groundwater resources in the Chianan coastal alluvial plain. Groundwater Engineering Division of Taiwan Provincial Government, Research Report, December 1959, Nantou, Taiwan (in Chinese).

Takamatsu, T., Kawashima, M. and Koyama, M.: The role of Mn^{2+}-rich hydrous manganese oxide in the accumulation of arsenic in lake sediments. *Water Res.* 19 (1985), pp. 1029–1032.

Tan, L.-B., Chen, K.-T. and Guo, H.-R.: Clinical and epidemiological features of patients with genitourinary tract tumour in a blackfoot disease endemic area of Taiwan. *Br. J. Urol. Int.* 102 (2008), pp. 48–54.

Tanaka, T.: Distribution of arsenic in the natural environment with an emphasis on rocks and soils. *Appl. Organomet. Chem.* 2 (1988), pp. 283–295.

Tandukar, N., Bhattacharya, P. and Mukherjee, A.B.: Preliminary assesment of arsenic contamination in groundwater in Nepal. In: *Managing arsenic for our future; proceedings of the International Conference on Arsenic in the Asia-Pacific region*, Adelaide, South Australia, November, 21–23, 2001. Adelaide: Aris Pty. Ltd., 2001, pp. 103–105.

Tandukar, N., Bhattacharya, P., Jacks, G. and Valero, A.A.: Naturally occurring arsenic in groundwater of Terai region in Nepal and mitigation options. In: J. Bundschuh, P. Bhattacharya and D. Chandrashekharam (eds): *Natural arsenic in groundwater: occurrence, remediation and management.* Balkema/Taylor and Francis Group, Leiden, The Netherlands, 2005, pp. 41–48.

Tandukar, N., Bhattacharya, P., Neku, A. and Mukherjee, A.B.: Extent and severity of arsenic poisoning in Nepal. In: Groundwater arsenic contamination in India: Extent and severity. In: R. Naidu, E. Smith, G. Owens, P. Bhattacharya and P. Nadebaum (eds): *Managing arsenic in the environment: from soil to human health*. CSIRO Publishing, Melbourne, Australia, 2006, pp. 595–604.

Tarbuck, E.J. and Lutgens, F.K.: Earth: *An introduction to physical geology*. 7th ed, Prentice Hall, New Jersey, NJ, 2002.

Tello, E.E.: *Hidroarsenisismo Crónico Regional Endémico (HACRE), sus manifestaciones clínicas*. Imprenta de la Universidad Nacional de Córdoba, Cordoba, Argentina, 1951.

Tello, E.E.: Arsenicismos hídricos: Que es el hidroarsenicismo crónico regional endémico argentino (HACREA)? *Arc. Argent. Dermat.* T XXXVI 4 (1986), pp. 197–214.

Tello, E.E.: Carcinomas of internal organs and their relationship to arsenical drinking water in the Republic of Argentina. *Med. Cutan. Ibero. Lat. Am.* 16 (1988), pp. 497–501.

Thomson, B.M., Jeffrey Cotter, T.J. and Chwirka, J.D.: Design and operation of point-of-use treatment system for arsenic removal. *J. Env. Eng. ASCE* 129 (2003), pp. 561–564.

Thompson, J.M.: Chemistry of thermal and nonthermal springs in the vicinity of Lassen Volcanic National Park. *J. Volcanol. Geoth. Res.* 25:1–2 (1985), pp. 81–104.

Trelles, R., Larghi, A. and Paez, J.P.: *El problema sanitario de las aguas destinadas a la bebida humana, con contenidos elevados de arsénico, vanadio y flúor*. Universidad de Buenos Aires, Facultad de Ingeniería, Instituto de Ingeniería Sanitaria, Publ. 4, Buenos Aires, Argentina, 1970.

Tsai, S.-M., Wang, T.-N. and Ko, Y.-C.: Cancer mortality trends in a blackfoot disease endemic community of Taiwan following water source replacement. *J. Toxicol. Environ. Health* A 55 (1998), pp. 389–404.

Tsai, S.-M., Wang, T.-N. and Ko, Y.-C.: Mortality for certain diseases in areas with high levels of arsenic in drinking water. *Arch. Environ. Health* 54 (1999), pp. 186–193.

Tsai, S.-Y., Chou, H.-Y., The, H.-W., Chen, C.-M. and Chen, C.-J.: The effects of chronic arsenic exposure from drinking water on the neurobehavioral development in adolescence. *Neurotoxicology* 24 (2003), pp. 747–753.

Tseng, C.-H.: Abnormal current perception thresholds measured by neurometer among residents in blackfoot disease-hyperendemic villages in Taiwan. *Toxicol. Lett.* 146 (2003), pp. 27–36.

Tseng, C.-H., Chong, C.-K., Chen, C.-J. and Tai, T.-Y.: Dose-response relationship between peripheral vascular disease and ingested inorganic arsenic among residents in blackfoot disease endemic villages in Taiwan. *Atherosclerosis* 120 (1996). pp. 125–133.

Tseng, C.-H., Tai, T.-Y., Chong, C.-K., Lai, M.-S., Lin, B.-J., Chiou, H.-Y., Hsueh, Y.-M., Hsu, K.-H. and Chen, C.-J.: Long-term arsenic exposure and incidence of non-insulin-dependent diabetes mellitus: a cohort study in arseniasis-hyperendemic villages in Taiwan. *Environ. Health Perspect.* 108 (2000), pp. 847–851.

Tseng, C.-H., Chong, C.-K., Tseng, C.-P., Hsueh, Y.-M., Chiou, H.-Y., Tseng, C.-C. and Chen, C.-J.: Long-term arsenic exposure and ischemic heart disease in arseniasis-hyperendemic villages in Taiwan. *Toxicol. Lett.* 137 (2003), pp. 15–21.

Tseng, W.-B.: Mechanism of blackfoot disease. *J. Formos. Med. Assoc.* 72 (1973), pp. 11–24 (in Chinese).

Tseng, W.-P.: Effects and dose-response relationships of skin cancer and blackfoot disease with arsenic. *Environ. Health Perspect.* 19 (1977), pp. 109–119.

Tseng, W.-P.: Diagnosis of blackfoot disease by Doppler ultrasonography. Blackfoot Disease Research Report 7, 1978, pp. 11–17. Taiwan Provincial Department of Health, Taichung.

Tseng, W.-P.: Blackfoot disease in Taiwan: a 30-year follow-up study. *Angiology* 40 (1988), pp. 547–558.

Tseng, W.-P., Chen, W.-Y., Sung, J.-L. and Chen, J.-S.: A clinical study of blackfoot disease in Taiwan: An endemic peripheral vascular disease. *Mem. College Med. Natl. Taiwan Univ.* 7 (1961), pp. 1–18.

Tseng, W.-P., Chu, H.-M., How, S.-H., Fong, J.-M., Lin, C.-S. and Yeh, S.: Prevalence of skin cancer in an endemic area of chronic arsenicism in Taiwan. *J. Nat. Cancer Inst.* 40 (1968), pp. 453–463.

Tuutijarvi, T., Lu, J., Sillanpaa, M. and Chen, G.: As(V) adsorption on maghemite nanoparticles. *J. Hazard. Mat.* 166 (2009), pp. 1415–1420.

Ure, A. and Berrow, M.: The elemental constituents of soils. In H.J.M. Bowen (ed): *Environmental chemistry*. Royal Society of Chemistry, London, UK, 1982, pp. 94–203.

Uriarte, M.G., Paoloni, J.D., Navarro, E., Fiorentino, C.E. and Sequeira, M.: Landscape, surface runoff, and groundwater quality in the district of Puan, Province of Buenos Aires, Argentina. *J. Soil Water Conserv.* 57:3 (2002), pp. 192–195.

US EPA: Health Assessment Document for Inorganic Arsenic: Final Report. Washington, D.C.: United States Environmental Protection Agency. (EPA publication EPA-600/8-83-021F), 1984.

US EPA: Guidance manual for assessing human health risks from chemically contaminated fish and shellfish, EPA-503/8-89-002, 1989.

US EPA: Risk-based concentration table. US EPA, Philadelphia, PA, 1998.

US EPA: National primary drinking water regulations; Arsenic and clarifications to compliance and new source contaminants monitoring; Final rule. Fed. Reg. 66:6076–7066, 2001.

US EPA: Treatment technologies for arsenic removal. EPA-600-S-05-006, US Environmental Protection Agency, 2005.

US EPA: National primary drinking water regulations. EPA-816-F-09-004, Environmental Protection Agency, 2009.

US FDA: Guidance document for arsenic in shellfish. US Food and Drug Administration, Washington, DC, 1993. pp. 25–27.

US FDA: Fish and fisheries products hazards and controls guidance, third edition, appendix 5: FDA and EPA Safety Levels in Regulations and Guidance. Washington, DC, 2001.

Van Cappellen, P., Charlet, L., Stumm, W. and Wersin, P.: A surface complexation model of carbonate mineral-aqueous solution interface. *Geochim. Cosmochim. Acta* 57 (1993), pp. 3505–3518.

Van Geen, A., Rose, J., Thoral, S., Garnier, J.M., Zheng, Y. and Bottero, J.Y.: Decoupling of As and Fe releases to Bangladesh groundwater under reducing conditions. Part II: Evidence from sediment incubations. *Geochim. Coschim. Acta* 68 (2994), pp. 3475–3486.

Varsányi, I., Fodré, Z. and Bartha, A.: Arsenic in drinking water and mortality in the southern Great Plain, Hungary. *Environ. Geochem. Health* 13 (1991), pp. 14–22.

Vermeer, A.W.P. and Koopal, L.K.: Adsorption of humic acids to mineral particles. 2. Polydispersity effects with polyelectrolyte adsorption. *Langmuir* 14 (1998), pp. 4210–4216.

Vilano, M. and Rubio, R.: Determination of arsenic species in oyster tissue by microwave-assisted extraction and liquid chromatography-atomic fluorescence detection. *Appl. Organomet. Chem.* 15 (2001), pp. 658–666.

Violante, A. and Pigna, M.: Competitive sorption of arsenate and phosphate on different clay minerals and soils. *Soil Sci. Soc. Am. J.* 66 (2002), pp. 1788–1796.

Viraraghavan, T., Subramanian, K.S. and Aruldoss, J.A.: Arsenic in drinking water: Problems and solutions. *Water Sci. Technol.* 40 (1999), pp. 69–76.

Voelker, D.C.: Observation-well network in Illinois, 1984. USGS Open-File Report 86-416, 1986.

von Brömssen, M., Jakariya, M., Bhattacharya, P., Ahmed, K.M., Hasan, M.A., Sracek, O., Jonsson, L., Lundell, L. and Jacks, G.: Targeting low-arsenic aquifers in Matlab Upazila, Southeastern Bangladesh. *Sci. Total Environ.* 379 (2007), pp. 121–132.

Wagner, B., Töpfner, C., Lischeid, G., Scholz, M., Klinger, R. and Klaas, P.: Hydrogeochemische Hintergrundwerte der Grundwässer Bayerns. *GLA Fachberichte* 21, Bayerisches Geologisches Landesamt, Munich, Germany, 2003.

Wang, C.-H., Jeng, J.-S., Yip, P.-K., Chen, C.-L., Hsu, L.-I., Hsueh, Y.-M., Chiou, H.-Y., Wu, M.-M. and Chen, C.-J.: Biological gradient between long-term arsenic exposure and carotid atherosclerosis. *Circulation* 105 (2002), pp. 1804–1809.

Wang, C.-H., Hsiao, C.-K., Chen, C.-L., Hsu, L.-I., Chiou, H.-Y., Chen, S.-Y., Hsueh, Y.-M., Wu, M.-M. and Chen, C.-J.: A review of the epidemiological literature on the role of environmental Arsenic exposure and cardiovascular diseases. *Toxicol. Appl. Pharmacol.* 222 (2007), pp. 315–326.

Wang, C.-H., Chen, C.-L., Hsiao, C.-K., Chiang, F.-T., Hsu, L.-I., Chiou, H.-Y., Hsueh, Y.-M., Wu, M.-M. and Chen, C.-J.: Increased risk of QT prolongation associated with atherosclerotic diseases in arseniasis endemic area of southwestern coast Taiwan. *Toxicol. Appl. Pharmacol.* 239 (2009), pp. 320–324.

Wang, G.: Arsenic poisoning from drinking water in Xinjiang. *Chinese J. Prevent. Med.* 18 (1984), pp. 105–107.

Wang, L. and Huang, J.: Chronic arsenism from drinking water in some areas of Xinjiang, China. In: J.O. Nriagu (ed): *Arsenic in the environment*, Part II: *Human health and ecosystem effects*. Wiley, New York, NY, 1994, pp. 159–172.

Wang, S., Shu, M. and Yang, C.: Morphological study of mud volcanoes on land in Taiwan. *J. Nat. Taiwan Museum* 31 (1988), pp. 31–49.

Wang, S.-L., Chiou, J.-M., Chen, C.-J., Tseng, C.-H., Chou, W.-L., Wang, C.-C., Wu, T.-N. and Chang, L.W.: Prevalence of non-insulin-dependent diabetes mellitus and related vascular diseases in southwestern arseniasis-endemic and nonendemic areas in Taiwan. *Environ. Health Perspect.* 111 (2003), pp. 155–159.

Wang, S.-W.: *Effect of sediments on the distribution of arsenic species in aquacultural ecological system of blackfoot disease areas*. MSc Thesis, Dept. Bioenviron. Sys. Eng., National Taiwan Univ., Taipei, Taiwan, 2003.

Wang, S.-W., Lin, K.-H., Hsueh, Y.-M. and Liu, C.-W.: Arsenic distribution in a tilapia (*Oreochromis mossambicus*)-water-sediment aquacultural ecosystem in blackfoot disease hyperendemic areas. *Bull. Environ. Contam. Toxicol.* 78:2 (2007), pp. 137–141.

Wasserman, G., Liu, X., Parvez, F., Ahsan, H., Factor-Litvak, P., van Geen, A., Slavkobich, V., Lolacono, N., Cheng, Z., Hussain, I., Momotaj, H. and Graziano, J.: Water arsenic exposure and children's intellectual function in Araihazar Bangladesh. *Environ. Health Perspect.* 112:13 (2004), pp. 1329–1333.

Water Resources Bureau: Groundwater resources map of Taiwan. Water Resources Bureau, Ministry of Economic Affairs, Taipei, Taiwan, 2003.

Wauchope, R.D. and Mc Dowell, L.L.: Adsorption of phosphate, arsenate, methanearsonate, and cacodylate by lake and stream sediments: Comparisons with soils. *J. Environ. Qual.* 13 (1984), pp. 499–504.

Waychunas, G.A., Rea, B.A., Fuller, C.C. and Davis, J.A.: Surface chemistry of ferrihydrite. 1. EXAFS study of the geometry of coprecipitated and adsorbed arsenate. *Geochim. Cosmochim. Acta* 57:10 (1973), pp. 2251–2269.

Waychunas, G.A., Davis, J.A. and Fuller, C.C.: Geometry of sorbed arsenate on ferrihydrite and crystalline FeOOH—reevaluation of EXAFS results and topological factors in predicting sorbate geometry, and evidence for monodentate complexes. *Geochim. Cosmochim. Acta* 59 (1995), pp. 3655–3661.

Waychunas, G.A., Fuller, C.C., Rea, B.A. and Davis, J.A.: Wide angle X-ray scattering (WAXS) study of "two-line" ferrihydrite structure: effect of arsenate sorption and comparison with EXAFS results. *Geochim. Cosmochim. Acta* 60:10 (1996), pp. 1765–1781.

Webster, J.G.: The solubility of As_2S_3 and sulphide-bearing fluids at 25 and 90°C. *Geochim. Cosmochim. Acta* 54:4 (1990), pp. 1009–1017.

Webster, J.G.: Arsenic. In: C.P. Marshall and R.W. Fairbridge (eds): *Encyclopaedia of geochemistry*. Chapman Hall, London, UK, 1999, pp. 21–22.

Webster, J.G. and Nordstrom, D.K.: Geothermal arsenic. In: A.H. Welch and K.G. Stollenwerk (eds): *Arsenic in ground water: Geochemistry and occurrence*. Springer, New York, NY, 2003, pp. 101–125.

Welch, A.H. and Lico M.S.: Factors controlling As and U in shallow ground water, southern Carson Desert, Nevada. *Appl. Geochem.* 13:4 (1998), pp. 521–539.

Welch, A.H., Lico, M.S. and Hughes, J.L.: Arsenic in ground water of the western United States. *Ground Water* 26:3 (1988), pp. 333–347.

Welch, A.H., Westjohn, D.B., Helsel, D.R. and Wanty, R.B.: Arsenic in ground water of the United States: occurrence and geochemistry. *Ground Water* 38 (2000), pp. 589–604.

Wenzel, W.W., Kirchbaumer, N., Prohaska, T., Stingeder, G., Lombi, E. and Adriano, D.C.: Arsenic fraction in soils using an improved sequential extraction procedure. *Anal. Acta.* 436 (2001), pp. 309–323.

Westerhoff, P.K., Benn, T.M., Chen, A.S.C., Wang, L. and Cumming, L.J.: Assessing arsenic removal by metal (hydr)oxide adsorptive media using rapid small scale column tests. US EPA Report, EPA-600-R-08-051, Environmental Protection Agency, Cincinnati, OH, USA, 2008.

Westjohn, D.B., Kolker, A., Cannon, W.F. and Sibley, D.F.: Arsenic in ground water in the "thumb area" of Michigan. In: *The Mississippian Marshall Sandstone revisited, Michigan: Its geology and geologic resources*, 5th Symposium, Michigan Department of Environmental Quality, East Lansing, MI, 1998, pp. 24–25.

White, D.E.: Environments of generation of some base-metal ore deposits. *Econ. Geol.* 63 (1968), pp. 301–335.

White, D.E.: Active geothermal systems and hydrothermal ore deposits. *Econ. Geology*, Seventy-fifth Anniversary Volume (1905–1980), 1981, pp. 392–423.

WHO: *Guidelines for drinking-water quality*, 2nd ed, *Recommendations*, Vol. 1., World Health Organization, Geneva, Switzerland, 1993.

WHO: *Guidelines for drinking-water quality: Arsenic in drinking water*. Fact Sheet No. 210. World Health Organization, Geneva, Switzerland, 2001, http://www.who.int/mediacenter/factsheets/fs210/en/print.html (accessed May 2009).

WHO: *Guidelines for drinking water quality*, 3rd ed *Recommendations*. Vol. 1., World Health Organization, Geneva, Switzerland, 2008.

Wilkie, J.A. and Hering, J.G.: Rapid oxidation of geothermal arsenic(III) in steamwaters of the eastern Sierra Nevada. *Environ. Sci. Technol.* 32:5 (1998), pp. 657–662.

Wilkin, R.T., Wallschläger, D. and Ford, R.G.: Speciation of arsenic in sulfidic waters. *Geochem. Trans.* 4 (2003), pp. 1–7.

Williams, M., Fordyce, F., Paijitprapanon, A. and Charoenchaisri, P.: Arsenic contamination in surface drainage and groundwater in part of the Southeast Asian Tin belt, Nakhon Si Thammarat Province, southern Thailand. *Environ. Geol.* 27 (1996): pp. 16–33.

Williams, P.N., Price, A.H., Raab, A., Hossain, S.A., Feldmann, J. and Meharg, A.A.: Variation in arsenic speciation and concentration in paddy rice related to dietary exposure. *Environ. Sci. Technol.* 39 (2005), pp. 5531–5540.

Williams, P.N., Islam, M.R., Adomako, E.E., Raab, A., Hossain, S.A., Zhu, Y.G. and Meharg, A.A.: Increase in rice grain arsenic for regions of Bangladesh irrigating paddies with elevated arsenic in groundwater. *Environ Sci Technol.* 40 (2006), pp. 4903–4908.

Wilson, R: The largest man made environmental catastrophe. Report to annual meeting of the Royal Geographical Society, London, UK, 2006, http://physic.harvard.edu/%7Ewilson/arsenic/conferences/2007_RGS/S1.1%20R%20Wilson%20ALT.pdf (accessed October 2009).

Wood, S.A., Tait, C.D. and Janecky, D.R.: A Raman spectroscopy study of arsenite and thioarsenite species in aqueous solutions at 25°C. *Geochem. Trans.* 3 (2002), pp. 31–39.

WRUD: Preliminary study on arsenic contamination in selected areas of Myanmar. Water Resources Utilization Department (WRUD) in the Ministry of Agriculture and Irrigation of Myanmar, Rangun, Myanmar, 2001.

Wu, C.-M.: The groundwater geology characteristic of an endemic blackfoot disease area. Blackfoot Disease Research Report Vol. 5, Prevention and Control Center of Blackfoot Disease, Division of Health of Taiwan Provincial Government, Nantou County, Taiwan, 1978, pp. 1–24 (in Chinese).

Wu, C.-M.: Hydrogeological characteristics of groundwater in the blackfoot disease area. Hydrogeology Conference organized by Central Geological Survey, Taipei, Taiwan, 1989.

Wu, H.-Y., Chen, K.-P., Tseng, W.-P. and Hsu, C.-L.: Epidemiologic studies on blackfoot disease: I. Prevalence and incidence of the disease by age, sex, year, occupation and geographical distribution. *Mem College Med. Natl. Taiwan Univ.* (1961), pp. 733–750.

Wu, M.-M., Kuo, T.-L., Hwang, Y.-H. and Chen, C.-J.: Dose-response relation between arsenic concentration in well water and mortality from cancers and vascular diseases. *Am. J. Epidemiol.* 130 (1989), pp. 1123–1132.

Wu, M.-M., Chiou, H.-Y., Wang, T.-W., Hsueh, Y.-M., Wang, I.-H., Chen, C.-J. and Lee, T.-C.: Association of blood arsenic levels with increased reactive oxidants and decreased antioxidant capacity in a human population of northeastern Taiwan. *Environ. Health Perspect.* 109 (2001), pp. 1011–1017.

Wu, M.-M., Chiou, H.-Y., Ho, I.-C., Chen, C.-J. and Lee, T.-C.: Gene expression of inflammatory molecules in circulating lymphocytes from arsenic-exposed human subjects. *Environ. Health Perspect.* 111 (2003), pp. 1429–1438.

Wu, W.-J., Huang, C.-T., Liao, Y.-H., Huang, J.-H. and Wang, S.-L.: Source and fate of arsenic in soil at Guandu, Beitou. *Workshop on Arsenic Contamination in Beitou Guandu Area*, Taipei, Taiwan, 2009, pp. 48–67.

Xia, S., Dong, B., Zhang, Q., Xu, B., Gao, N. and Causserand, C.: Study of arsenic removal by nanofiltration and its application in China. *Desalination* 204 (2007), pp. 374–379.

Xu, H., Allard, B. and Grimvall, A.: Influence of pH and organic substance on the adsorption of As(V) on geologic materials. *Water Air Soil Pollut.* 40 (1988), pp. 293–305.

Xu, H., Allard, B. and Grimvall, A.: Effect of acidification and natural organic materials on the mobility of arsenic in the environment. *Water Air Soil Pollut.* 57–58 (1991), pp. 269–278.

Yan, X., Kerrich, R. and Hendry, M.: Distribution of arsenic(III), arsenic(V) and total inorganic arsenic in pore-waters from a thick till and clay-rich aquitard sequence, Saskatchewan, Canada. *Geochim. Cosmochim. Acta* 64 (2000), pp. 2637–2648.

Yang, C.-Y.: Does arsenic exposure increase the risk of development of peripheral vascular diseases in humans? *J. Toxicol. Environ. Health* A 69 (2006), pp. 1797–1804.

Yang, C.-Y., Chang, C.-C., Tsai, S.-S., Chuang, H.-Y., Ho, C.-K. and Wu, T.-N.: Arsenic in drinking water and adverse pregnancy outcome in an arseniasis-endemic area in northeastern Taiwan. *Environ. Res.* 91 (2003), pp. 29–34.

Yang, C.-Y., Chiu, H.-F., Wu, T.-N., Chuang, H.-Y. and Ho, S.-C.: Reduction in kidney cancer mortality following installation of a tap water supply system in an arsenic-endemic area of Taiwan. *Arch. Environ. Health* 59 (2004), pp. 484–488.

Yang, C.-Y., Chiu, H.-F., Chang, C.-C., Ho, S.-C. and Wu, T.-N.: Bladder cancer mortality reduction after installation of a tap-water supply system in an arsenious-endemic area in southwestern Taiwan. *Environ. Res.* 98 (2005), pp. 127–132.

Yang, C.-Y., Chang, C.-C., Ho, S.-C. and Chiu, H.-F.: Is colon cancer mortality related to arsenic exposure? *J. Toxicol. Environ. Health* A 71 (2008a), pp. 533–538.

Yang, C.-Y., Chang, C.-C. and Chiu, H.-F.: Does arsenic exposure increase the risk for prostate cancer? *J. Toxicol. Environ. Health* A. 71 (2008b), pp. 1559–1563.

Yang, T.-H. and Tsai, J.: Preliminary report on selenium levels in water, soil, and food from the endemic blackfoot disease area. *J. Formos. Med. Assoc.* 60 (1961), pp. 80–81.

Yang, Y.-W.: *The redox reaction of arsenic in the groundwater of the blackfoot disease area in the Chia-Nan plain, southwestern Taiwan.* MSc thesis, Department of Earth Sciences, National Cheng Kung University, Tainan, Taiwan, 2006 (in Chinese).

Yau, P.-S.: *Relationship between arsenic content in rice plant and paddy soil in the arsenic contaminated area.* MSc Thesis, Department of Bioenvironmental Systems Engineering, National Taiwan University, Taipei, Taiwan, 2008.

Yavuz, C.T., Mayo, J.T., Yu, W.W., Prakash, A., Falkner, J.C., Yean, S., Cong, L., Shipley, H.J., Kan, A., Tomson, M., Natelson, D. and Colvin, V.L.: Low-field magnetic separation of monodisperse Fe_3O_4 nanocrystals. *Science* 314 (2006), pp. 964–967.

Yeh, G.-H., Yang, T.-F., Chen, J.-C., Chen, Y.-G. and Song, S.-R.: Fluid geochemistry of mud volcanoes in Taiwan. In: G. Martinelli and B. Panahi (eds): *NATO Science Series*: IV: *Earth and Environmental Sciences* 51 (2004), pp. 227–237.

Yeh, S.: Relative incidence of skin cancer in Chinese in Taiwan: with special reference to Asal cancer. *Natl. Cancer Inst. Monogr.* 10 (1963), pp. 81–107.

Yeh, S.: Skin cancer in chronic arsenicism. *Hum. Pathol.* 4 (1973), pp. 469–485.

Yeh, S. and How, S.-W.: A pathological study on the blackfoot disease in Taiwan. *Repts. Inst. Pathol. Natl. Taiwan Univ.* 14, 1963, pp. 25–73.

Yeh, S., How, S.-W. and Lin, C.-S.: Asal cancer of skin: Histologic study with special reference to Bowen's disease. *Cancer* 21 (1968), pp. 312–339.

Yu, R.-C., Hsu, K.-H., Chen, C.-J. and Froines, J.R.: Arsenic methylation capacity and skin cancer. *Cancer Epidemiol Biomarker Prev.* 9 (2000), pp. 1259–1262.

Yudovich, Y.E. and Ketris, M.P.: Arsenic in coal: a review. *Int. J. Coal Geol.* 61 (2005), pp. 141–196.

Zachara, J.M., Cowan, C.E. and Resch, C.T.: Metal cation/anion adsorption on calcium carbonate: Implications to metal ion concentrations in groundwater. In: H.E. Allen, E.M. Perdue and D.S. Brown (eds): *Metals in groundwater.* Lewis, Chelsea, UK, 1993, pp. 37–71.

Zavala, Y.J. and Duxbury, J.M.: Arsenic in rice: I. Estimating normal levels of total arsenic in rice grain. *Environ. Sci. Technol.* 42 (2008), pp. 3856–3860.

Zeng, H., Fisher, B. and Giammar, D.E.: Individual and competitive adsorption of arsenate and phosphate to a high-surface-area iron oxide-based sorbent. *Erviron. Science Technol.* 42 (2008): pp. 147–152.

Zhang, Y., Ma, L. and Luo, Z.D.: Water quality analysis of arsenic-enriched groundwater in the large area of western Huhhot Basin. *Rural Eco-Environment* 10:1 (1994), pp. 59–61 (in Chinese).

Zheng, B., Ding, Z., Huang, R., Zhu, J., Yu, X., Wang, A., Zhou, D., Mao, D. and Su, H.: Issues of health and disease relating to coal use in southwestern China. *Int. J. Coal Geol.* 40 (1999), pp. 119–132.

Ziegler, A.C., Wallace, W.C., Blevins, D.W. and Maley, R.D.: Occurrence of pesticides, nitrite plus nitrate, arsenic, and iron in water from two reaches of the Missouri River alluvium, northwestern Missouri—July 1988 and June–July 1989. USGS Open-File Report 93–101, 1993.

Zuena, A.J. and Keane, P.E.: Arsenic contamination of private potable wells. In: *EPA National Conference on Environmental Engineering Proceedings*, US EPA, Boston, MA, 1985, pp. 717–725.

Subject index

accumulation of arsenic (*see* arsenic accumulation *and* arsenic bioaccumulation)

adsorbents of arsenic (*see also* arsenic sorbents) 43, 137, 138, 149, 150

adsorption of arsenic (*see* arsenic adsorption *and* arsenic sorbents)

alkaline lake 25

alkalinity 103, 105, 107, 110–112

aluminum oxyhydroxides 20, 28, 30, 40–46

alluvial
  aquifer (*see* aquifer)
  deposit 3, 16, 23, 86–88, 90, 95
  fan 88

amphibole 21

amphibolite 23

anaerobic bacteria 40, 42, 109

andesite 38, 45, 96, 98

anemia (*see* arsenic health effects)

anhydrite 21, 22

anion exchange (*see also* arsenic removal from drinking water) 44, 135, 138

annabergite 21

anthropogenic arsenic (*see* arsenic sources)

antimony 33

antioxidant 76

apatite 21, 22, 107

aquifer
  artesian 59, 61–67, 69, 73, 75, 76, 78, 80, 82, 104,
  confined 32, 39, 90, 97, 102–104, 107
  deep 26, 30, 39, 87, 90, 97, 103, 108–110, 115
  multilayer 39, 88
  phreatic (unconfined) 39, 88, 104, 115
  oxidizing 1–3, 25, 27, 28, 30, 32, 36, 37, 39, 41, 45, 92, 107, 115
  recharge 39, 88, 89–91, 107
  reducing 2, 3, 25, 28, 29, 30, 31, 36, 40–43, 88, 90, 92, 96, 99, 103, 107–110, 113, 115, 146, 147
  sedimentary 2, 6, 27, 29, 31, 32, 40, 45, 104–109, 113, 146, 147

argillic alteration 22

argillite 46, 90

arsenate (*see* arsenic species)

arsenate minerals 20, 42

arsenate-phosphate competition for sorption sites (*see also* arsenic mobilization) 29, 44, 137, 138, 140

arsenian pyrite 20, 21

arseniasis-endemic area 77, 79, 80, 123, 134, 148, 149
  Chianan plain
    arsenic in sediments 104–107
    water chemistry 101–104
  Lanyang (Yilan) plain
    arsenic in sediments 104, 106, 107
    water chemistry 101–104

arseniasis-endemic areas (*see also* arsenic health effects) 77, 79, 80, 134, 148, 149

arsenic accumulation (*see also* arsenic bioaccumulation)
  in soil 115–119

arsenic (ad)sorption (*see also* arsenic mobilization *and* arsenic sorbents) 2–4, 20, 24, 26–28, 30, 31, 40–44, 46, 47, 106–108, 129, 135–138, 140, 141, 142, 144–146, 149, 150
  by calcite 46, 47
  by clay minerals 46
  by Fe, Al and Mn oxides/hyroxides 42, 43
  capacity 43, 46, 107, 138, 150
  ions competing for adsorption sites 24, 28, 29, 44, 137, 138, 140
  pH dependence 43, 44
  redox potential dependence 43, 44

arsenic bioaccessibility (*see* arsenic bioavailability)

arsenic bioaccumulation 125–131
  factor (BF) 101
  in crops 119–122
  in aquacultural organisms
    clam 128, 129
    milkfish 127, 128

arsenic bioaccumulation (*Continue*)
    mullet 128
    oyster 129, 130
    tilapia 125–127
  in human body 5
  in plants 119–122
arsenic bioavailability 5, 20
  for humans 18
  from soils for plants 115, 118–121, 147
arsenic biomarkers (*see* arsenic metabolites
    and arsenic accumulation)
arsenic biotransformation 76
arsenic carcinogenesis (*see* arsenic health
    effects)
arsenic chemistry 20
arsenic coagulants/flocculants (*see also*
    arsenic removal from water) 139, 143
  aluminum salts 136
  iron salts 136
    ferric chloride 136, 139, 141, 149
arsenic contamination/contents 1, 2, 8, 19
  atmosphere 19
  drinking water 1–18, 23, 29, 32, 33, 61,
    62, 65, 67–71, 73–84, 135, 136, 139,
    145, 148–150
  earth's crust 19, 20, 22
  economic impact 4, 16
  food (*see* arsenic bioavailability)
  flood and delta plains 2, 3, 16, 18, 24
  global problem 9–14, 16
  groundwater 1, 2–5, 7–13, 16–20, 24, 29,
    67–71, 73–84, 99, 113
  human uptake through food chain 5, 18,
    101, 120, 123, 145, 147, 148
  impact on
    aquaculture 101, 115, 123–134, 145
    irrigation 5, 7, 16, 20, 37, 45,
      115, 118, 119, 123, 147
    livestock production 5, 16
    public health 1, 5–8, 18, 74, 118, 134
  lithosphere 19, 20
  plants/crops (edible) 5, 119–122, 145, 147
  rainwater 20
  seawater 20, 24, 39, 96, 99, 109,
    110, 113, 146
  social impact 4, 16, 18
  soils 5, 20, 91, 99, 113, 115–117, 120
  surface water 1, 2, 5, 7, 9, 19, 20, 24–27,
    29, 30, 31, 33, 34, 37, 38, 101
  worldwide distribution 9–13
arsenic contents in rocks (*see also* arsenic
    sources, rocks)
  overview 2–24

arsenic coprecipitation 2, 24, 27, 28, 31, 42,
    46, 136, 137
arsenic desorption (*see also* arsenic
    mobilization) 2, 20, 23, 24, 26–30, 40,
    42, 44, 45, 47, 92, 105, 108, 113, 129,
    146, 147
arsenic discoveries in groundwater 6–8
arsenic dissolution (*see* arsenic
    mobilization)
arsenic distribution in aquifers (*see also*
    arsenic contamination)
  worldwide 9–14, 16, 17
  Taiwan 101–109, 130, 146, 147
  patchy distribution in aquifer 41, 45
  control (*see* arsenic mobilization controls)
arsenic distribution in soils (*see also* arsenic
    contamination)116
  Taiwan 115–119
arsenic epidemiology
  studies 8, 67, 73, 74, 84
    of blackfoot disease 59, 65
  characteristics of blackfoot disease 57–59
arsenic exposure to humans 5–9, 14–16, 18
  acute (short-term) 5
  children 7
  cummulative 66, 67, 69, 70, 75, 82
  drinking water 1, 5–8, 63, 64, 67–71,
    73–84, 130, 132, 134, 145, 148
  edible plants/crops 5, 119–122, 145, 147
  exposure time 6, 7, 37
  fish and shellfish 133, 134, 145, 147, 148
  food chain 5, 18, 101, 120, 123,
    145, 147, 148
  ingestion 5
  inhalation 5
  in-utero 149
  long-term 69, 70, 82, 148
  population 15
  postnatal 149
arsenic forms (*see* arsenic species)
arsenic health effects 5–7, 9, 14
  anemia 5
  arsenicism 6
  arsenicosis 85
  Bellville disease 6
  blackfoot disease (BFD) 1, 7, 49–59, 73,
    85, 86, 93, 95, 96, 102–104, 113, 119,
    120, 123, 125, 128, 132, 135, 145–147
  cause 61–71, 135
  co-factors 69
  endemic area 55–59, 61–63, 73, 81, 85,
    93, 95, 96, 101, 104, 107, 109, 113,
    115, 119, 120, 123–134, 146

history 49–59
bronchitis 82
cancer 5, 9
  internal 7, 15, 61, 77–82, 148
    brain 79, 84
    cervix 77, 84
    colon 77, 79, 80, 84,
    esophagus 77, 79, 84
    kidney 55, 75, 77, 79, 80, 83
    leukemia 77, 79, 84
    liver 55, 56, 75, 77–81, 83, 84, 103
    lung 5, 55, 56, 75, 77, 78–81, 83, 84,
      103, 148
    nasal cavity 75, 79, 80, 83
    ovary 79
    prostate 77, 79, 80, 83
    small intestine 79, 84
    stomach cancer 77,79,80, 84
    ureter 77, 79,
    urethra 77, 79
    urinary bladder 5, 55, 56, 75, 77–80,
      83, 84, 103
  skin 5, 7, 55, 56, 61, 66, 67, 68, 73–76,
    83, 84, 103, 148
carcinogenic 5, 7, 14, 15, 76, 79, 80, 83,
  84, 133, 134
cataract 81, 82, 148
cerebral infarction 70
developmental retardation 5, 148, 149
diabetes 61, 69, 70, 82, 148, 149
  mellitus 5, 70, 73, 82
erectile dysfunction 61, 148
eye diseases 82
fatality rate 7
gastrointestinal tract 5
hypertension 61, 69, 70, 148, 149
ischemic heart disease 69, 70, 71, 73, 82
nervous system 5
neurological disorders 148
pterygium 148
respiratory system 5
skin lessions 5–7, 9, 55, 56, 61, 67–69,
  73–76, 83,84, 103, 148
  Bowen disease 67, 73
  cancer (*see* cancer)
  hyperkeratosis 5, 59, 67, 68, 73, 103
  hyperpigmentation 5, 7, 59, 67,
    68, 73
  hypopigmentation 67
vascular diseases 5, 148
  atherosclerosis 56, 68–70, 76
    carotid 61, 69, 70
  cardiobrain vascular disease 103

cardiovascular disease 6, 56, 69,
  70, 73, 149
cerebrovascular disease 55, 70, 73
macrovascular disease 61, 69
microvascular disease 61, 69
peripheral vascular disease (blackfoot
  disease (BFD) (*see* arsenic health
  effects, blackfoot disease)
arsenic health risk 4, 115, 122, 129, 148
  assessment 5, 132–134, 147, 149
arsenic immobilization 2, 3, 24–47, 147
  adsorption on mineral phases 2, 26–28,
    46, 107
  coprecipitation with oxy(hydr)oxides 2,
    24, 26–28, 31, 42, 46, 136, 137
  formation of secondary arsenic minerals
    21, 22, 23, 26, 27, 40
    by oxidation of arsenopyrite 21
    by oxidation of realgar 21
  sulfide mineral precipitation 28, 91,
    146, 147
arsenic in
  groundwater (*see* arsenic contamination)
  food (human diet) (*see* arsenic exposure
    and arsenic bioaccumulation)
  crops/plants (*see* arsenic
    bioaccumulation, crops/plants)
  fish/shellfish (*see* arsenic
    bioaccumulation, fish/shellfish)
  geothermal water 2, 11–13, 19–22,
    24–26, 28–30, 33–39, 87, 91, 93, 96,
    99, 113, 115, 117, 118, 145, 146
  hot spring (*see* geothermal spring)
  mud volcano 87, 95, 109–113
    fluid 95, 110–113
    mud 95, 110
  rock (*see* arsenic sources)
  rock-forming minerals (*see also* arsenic
    minerals) 20–22
  soil (*see* arsenic contamination)
  water (*see* arsenic contamination)
arsenic ingestion/intake (*see* arsenic
  exposure)
arsenic ion exchangers (*see* arsenic removal
  from water)
arsenic metabolites 131, 132
  dimethylarsenate (DMA) 122, 131, 134
  dimethylarsinic acid [DMA(V)]
    76, 81, 82, 123, 125
  monomethylarsenate (MMA)
    122, 131, 134
  monomethylarsonic acid [MMA(V)]
    76, 81, 82, 123, 125

arsenic metabolites (*Continue*)
  monomethylarsonous acid [MMA(III)]
    123, 125
arsenic methylation capacity
  fish and shellfish 131, 132
arsenic minerals (*see also* arsenic in rock
    forming minerals)1, 20–22
  annabergite 21
  arsenate minerals 20, 42
  arsenian pyrite 20, 21
  arsenite minerals 20
  arsenolite 21
  arsenopyrite 20–22, 31, 33, 34, 94, 95, 109
  claudetite 21
  cobaltite 21
  conicalcite 21
  elemental arsenic 20, 21
  enargite 21, 98
  hematolite 21
  hoernesite 21
  luzonite 98
  native arsenic 20, 21
  niccolite 21
  orpiment 21, 25, 36
  pharmacosiderite 21
  realgar 21, 36
  scorodite 21, 31
  tennantite 21
arsenic mobilization into groundwater/
    surface water 1–3, 17–19, 23–47, 96,
    107–109, 145–147
  mechanisms 2, 24–48, 113, 145
    competing ion effect 24, 28, 44, 137
      bicarbonate 44, 138, 140
      phosphate 29, 44, 137, 138
      silicate 44, 137, 138, 140
    desorption at high pH 2, 3, 26,
      28, 40, 42–47
    dissolution in geothermal reservoirs 26,
      33–39
    dissolution of metal oxy(hydr)oxides
      at about neutral pH (*see* reductive
      dissolution)
    dissolution of metal oxy(hydr)oxides at
      low pH 40–42
    formation of complexes between
      arsenic and humic acids 27, 28, 47
    ion exchange 4, 26, 29, 43–45
    microbiological processes
      bacterial reduction of arsenic(V) 41,
        96, 109, 147
      bacterial reduction of iron(III)
        41, 147

bacterial reduction of
  manganese(IV) 41, 147
bacterial sulfate reduction 41,
  96, 147
reductive dissolution 2, 3, 26–28,
  40–42, 96, 107, 108
sulfide oxidation at low pH 2, 26, 28,
  30–32, 92
  kinetics 30–32
  mechanism 30–32
water-rock interactions 24, 26, 27,
  29, 30, 110
  anion exchange 44
  cation-exchange 29, 43, 44
mobility controls 1–3, 19, 45, 46,
  101–109, 145–147
available arsenic sources (*see also*
  arsenic sources) 1, 3, 20–23, 26,
  92–101, 145–147
  atmospheric deposition 19, 20
  erosion 19, 23, 41, 86
  input from geothermal reservoirs
    2, 11–13, 19–22, 24–26, 28–30,
    33–40, 45, 87, 91, 93, 96, 99, 113,
    115, 117, 118, 145, 146
  input from mud volcanoes 109–113
  minerals (*see* arsenic sources)
  release due to weathering 1, 2, 7, 16,
    17, 19, 20, 22, 26, 29, 31, 38, 45,
    94, 96, 119
  rocks (*see* arsenic sources)
biogeochemical conditions (*see also*
  microbial activity) 1, 19, 26
  microbes 26,
  microbial activity 28, 109, 110
climatic conditions 1, 3, 19, 27
evaporative arsenic concentration
  increase 3, 27, 29, 34, 40
geochemical conditions 1–3, 18,
  19, 22, 23, 26, 29–31, 37, 40,
  42, 43, 45
  competing ions 44, 137, 138, 140
  ionic strength of water 24, 28, 43
  organic matter content 20, 23–26,
    28–30, 40, 41, 88, 106, 107, 109,
    119, 146, 147
  pH value 1–3, 24, 25–31,34, 38–45,
    47, 62–64, 91, 92, 103, 104, 105,
    111, 112, 119
    increase by cation exchange 29,
      43, 44
presence of humic acid 27, 28, 47,
  67, 88, 145

redox conditions 1–3, 24–28, 41, 107, 108, 119, 145, 146
geological setting 1, 3, 17, 19, 20, 26, 29, 33
geomorphology 3, 19, 24, 26, 27, 29
groundwater exploitation 3, 17, 18, 24, 27, 30, 32
groundwater-surface water interactions 24, 27, 29
  irrigation 20, 29, 119, 123, 147
hydrogeological conditions 1, 3, 7, 19, 26, 85–92
land use pattern 3, 18, 132
  farming habits 147
sedimentology 85–92
species of arsenic present 24–30, 40–44, 107, 108
arsenic occurrence (*see* arsenic sources *and* arsenic contamination)
arsenic problem mitigation (*see also* arsenic removal from water) 16–18, 144
  economic constraints 3, 62, 64, 69, 136, 145
  lack of interest 4, 16
  lack of public pressure 17
arsenic regulatory limits 5, 7, 16, 84, 148
  drinking water 7, 14–16, 145
    arsenic guideline value 14
    maximum contaminant level (MCL) 14, 17, 74, 136, 138
  food 15
arsenic release (*see* arsenic mobilization)
arsenic removal from water (*see also* arsenic problem mitigation) 3, 17
  ex-situ treatment 3
    adsorption methods 3, 135–138, 149, 150
      activated alumina (AA) 137
      iron (hydr)oxide-based adsorbents
        Aqua-Bind MP 137
        Bayoxide E33 ferric oxide 137
        granular activated carbon 139
        Granular Ferric Hydroxide (GFH) 137
        iron oxide modified activated AA 137
        iron-coated sand 137
        MEDIA G2 137
      titania-based adsorbents 137
    adsorption/coprecipitation (coagulation) 4, 135–141, 143, 144, 149
      aluminum sulfate 136

iron-based coagulants 136, 137
      ferric chloride 136, 139, 141, 149
    lime softening 136, 137
  arsenic removal plant 136, 137, 139, 141–143, 149, 150
  coagulation/flocculation 4, 135, 136, 138, 139, 141, 143, 144
  ion exchange 4, 135–138, 144, 149, 150
    anion exchange resins 138
    hybride sorbents 138
      ArsenX^{np} 138
      FO 36 138
      READ-As 138
  membrane techniques 3, 138, 139
    microfiltration (MF) 138, 139
    nanofiltration (NF) 138, 139, 144
    reverse osmosis (RO) 3, 138–140, 144
    ultrafiltration (UF) 138, 139
  point of entry (POE) 133, 138–140, 143–144
  point of use (POU) 133, 138–140, 143–144
  in-situ treatment
    reactive barriers 3
    precipitation methods (*see* coagulation/flocculation)
  methods for centralized water supply 2, 4, 136–139, 143, 144, 149
  methods for decentralized water supply 17, 139, 140, 144, 149, 150
  point of use and entry devices 139–141
arsenic sorbents (*see* arsenic removal from water)
arsenic sources 1, 2, 17, 19–24, 29, 33, 37, 39–47, 92–101
  anthropogenic activities 19, 99–101, 115
    agriculture
      fertilizers 19
      pesticides 19
    industry 19, 99
    mining 5, 9, 10, 19, 25, 28, 30–32, 115
      gold 98–101, 113
      smelting and ore processing 19, 21
    wood preservation agents 19
  atmospheric deposition 19, 20
  geogenic 1, 5, 7–9, 15, 19–24, 26, 29, 40–46, 92–99, 113, 115, 118, 145–147
    arsenic minerals (*see* arsenic minerals)
    geothermal manifestations 19, 33–39, 87, 91, 93, 115–118, 145
      fossil geotermal systems 22

arsenic sources (*Continue*)
    fumaroles 2, 19, 25, 33, 38–40
    geotermal fluids 2, 12, 13, 19, 20,
        24, 26, 33–39, 87, 91, 93, 96, 99,
        113, 115, 117, 118, 146
    solfatares 19, 33
    geothermal spring deposits
        21, 91, 96
    minerals (*for individual minerals see*
        arsenic minerals *and* arsenic in
        rock-forming minerals) 20–22
    rocks with typically high arsenic
        content (*see* also arsenic in rocks)
        22–24
        bitumen deposits 31
        black shale 22, 24, 32, 93, 94, 95
        coal deposits 19, 22–24, 32, 95, 96,
            98, 99
        epithermal gold deposits 35, 36
        geothermal spring deposits 21, 96
        granites 2, 22, 35, 45
        hydrothermal deposits, 17, 20, 21–24,
            26, 29, 31, 33, 95, 96, 98
        hydrothermally altered rocks 24
        lignite 24
        marine sediments 2, 23, 94, 107,
            108, 110
        norites 21
        peat deposits 22, 24, 25, 31, 32, 95,
            96, 109
        pegmatites 22
        phosphorites 22–24, 32
        porphyry deposits 22
        secondary source minerals 1, 2, 19,
            21–24, 26, 27, 32, 40, 42
        sulfide ore deposits 2, 7, 23, 24,
            26–32, 95, 98, 99, 109, 146, 147
        volcanic rocks 2, 7, 16, 17, 19
            volcanic ash 20, 22, 29, 45
                glass 22, 45
    volcanic emissions 2, 19, 20
    weathering/weathering products 1, 2,
        7, 16, 17, 19, 20, 22, 29, 31, 33, 34,
        36–39, 45, 94, 96, 98, 119
    seaspray 19
arsenic species
    bioavailability for humans 147
    in food
        in fish/shellfish
            clam 128, 129
            mullet 128
            oyster 129, 130
            tilapia 125, 126
        in rice 115, 122
        inorganic
            arsenate [As(V)] 122, 123, 125,
                126, 131
            arsenite [As(III)] 123, 125, 126,
                131, 134
            methylation 131, 132
        organic (*see also* arsenic metabolites)
            122–134
            arsenobetaine (AsB) 123, 134
        health risk assessment 132–134, 147
    in urine 76, 81
    in water 3, 24–28, 40, 43, 107, 108, 123,
        135–140, 146, 147
        in alkaline lake 25
        in geothermal water 25, 36–38, 118
        in mining water 25
        in soda lake 25
        in sulfidic waters 25, 26
        inorganic
            arsenate [As(V)] 4. 25, 26, 27, 30,
                36–38, 42–46, 96, 106–109, 118,
                123, 135–137, 139, 140, 147, 149,
                150
            arsenite [As(III)] 4. 25, 26, 30, 33,
                36–38, 41–44, 96, 106–109, 118,
                122, 123, 135, 136, 138, 139, 140,
                143, 144, 146, 149
            thioanions 25
        reaction kinetics 17, 25, 26, 31, 37
        redox equilibrium 25, 26, 28
arsenic transport 19, 113, 119
    aeolic 2, 18, 20, 29
    fluvial 2, 3, 7, 16, 18, 29, 94, 95, 98, 99,
        102, 110
    in water 24, 26
        groundwater 7, 17, 18, 24, 30, 96,
            107–109
            reaction with humic acid 27, 28, 47,
                67, 81, 82, 88, 90, 102, 103, 104,
                107, 109, 113, 145–147
        retardation 30
        surface water 3, 7, 16, 18, 30, 91, 94,
            95, 98, 99, 102, 147
    in soil 18
arsenic treatment plant (*see* arsenic removal
    from water)
arsenic problem mitigation 1, 3, 4, 8, 16–18,
    136, 137, 141, 143, 144
    alternative water resources 3, 17, 102,
        123, 145

arsenic removal plant 7, 136, 137, 139, 141–143, 149, 150
  awareness creation 18
  for single households 4
  global solutions 16
  in isolated urban areas 2
  in rural areas 2
  international cooperation 1, 4, 17
    needs 17, 18
  water treatment (*see* arsenic removal from water)
arsenic treatment (*see* arsenic removal from water)
arsenite (*see* arsenic species)
arsenite minerals 20
arsenolite 21
arsenopyrite 20–22, 31, 33, 34, 94, 95, 109
authigenic pyrite 24, 26, 32
authigenic sulfides 24, 26, 32

bacteria 31, 40, 42, 107, 109, 147
bacterial
  oxidation of
    iron(II) 31
    pyrite 31
  reduction of
    arsenic(V) 41, 96, 109, 147
    iron(III) 41, 147
    manganese(IV) 41, 147
    sulfate 41, 96, 147
barite 21
basalt 34, 45, 95, 96, 98
batch experiments 138
Bellville disease (*see* arsenic health effects)
bicarbonate 44, 62, 63, 64, 90, 138, 147
bioaccessibility of arsenic (*see* arsenic bioaccessibility)
bioaccumulation of arsenic (*see* arsenic bioaccumulation)
bioavailability (*see* arsenic bioavailability)
bioavailable arsenic (*see* arsenic bioavailability)
biomarkers (*see* arsenic biomarkers)
biotite 21, 22, 107
biotransformation of arsenic (*see* arsenic biotransformation)
bitumen 31
black shale 22–24, 32, 93–95
blackfoot disease (BFD) (*see* arsenic health effects)

bladder cancer (*see* arsenic health effects)
boron 33
Bowen disease (*see* arsenic health effects)
breccia 98
bronchitis disease (*see* arsenic health effects)

calcite 21, 22, 27, 46, 47
calcium 44, 62, 63, 64, 90
cancer (*see* arsenic health effects)
carbon dioxide 62–64
carbonate minerals 21, 22, 24, 41, 44, 45, 47, 137
carbonate rocks 22, 39
carboxylic acid 47
carcinogenesis of arsenic (*see* arsenic health effects)
cardiovascular disease (*see* arsenic health effects)
cation exchange 4, 20, 29, 44
cerebrovascular disease (*see* arsenic health effects)
chalcopyrite 20, 21
chloride 3, 34, 36, 63, 64, 91, 110, 113, 136, 139, 141, 149
chlorite 46, 94, 108
chronic exposure to arsenic (*see* arsenic exposure)
cinnabar 96, 98
claudetite 21
clay 3, 21, 23, 24, 32, 39, 46, 87, 88, 94, 95, 107, 109, 119
clay minerals 2, 23, 27, 30, 46
coagulant for arsenic removal (*see* arsenic coagulants)
coagulation (*see* arsenic removal from water)
coagulation-flocculation (*see* arsenic removal from water)
coal 19, 22, 23, 24, 31, 32, 95, 96, 98, 99
cobaltite 21
column experiments 30, 138, 140
confined aquifer (*see* aquifer)
conicalcite 21
controls on arsenic mobilization (*see* arsenic mobilization)
cutaneous manifestations (*see* arsenic health effects)

daily arsenic intake (*see* arsenic exposure)
delta plain 2, 3, 8, 10, 11, 15, 16, 18, 24, 26, 28, 40, 41, 45, 95, 140

diabetes (*see* arsenic health efects)
diagenetic sulfide formation 22, 24,
    26, 31, 32
dimethylarsenate (DMA) (*see* arsenic
    metabolites)
dimethylarsinous acid [DMA(III)]
    (*see* arsenic metabolites)
dimethylarsinic acid [DMA(V)] (*see* arsenic
    metabolites)
dolomite 21–23, 47
drinking water (*see also* arsenic
    contamination, drinking water)
    1–18, 23, 29, 32, 33, 61, 62,
    65, 67–71, 73–84, 135, 136,
    139, 145, 148–150
  analysis standards 9, 15–17, 91, 101, 131,
      136, 139, 140, 141, 143
  quality standard 140, 143
    Taiwan 131, 139
  resources 5, 33, 59
  sources 59, 62, 65
  supply 7, 8, 14, 29, 31, 80, 82, 115
    surface water 85, 148
  system 143, 144

Eh (*see* redox potential)
electrical conductivity 45, 110
elemental arsenic (*see* arsenic minerals)
enargite 21, 98
epidemiology (*see* arsenic epidemiology)
erectile dysfunction (*see* arsenic health
    effects)
esfalerite 21
ethanoic acid 109
evaporative arsenic concentration increase
    3, 27, 29, 34, 40
evaporites 23, 27

feldspar 21, 23, 107
feldspatic alteration 22
ferric chloride (FeCl₃) (*see* arsenic
    coagulants)
flocculation (*see* arsenic removal from
    water)
flood plain 2, 3, 8, 16, 24
fluid-rock interaction
    (*see* water-rock interactions)
fluoride 33, 62, 63, 64
Fowler's solution 5

galena 20, 21, 27
geogenic arsenic (*see* arsenic sources,
    geogenic)

geothermal arsenic (*see also* arsenic sources)
    2, 11–13, 19–22, 24–26, 28–30, 33–39,
    87, 91, 93, 96, 99, 113, 115, 117, 118,
    145, 146
geothermal manifestations
    (*see also* arsenic sources) 19, 33–39,
    87, 91, 93, 115–118, 145
geothermal spring (*see also* arsenic sources)
    13, 19, 21, 25, 28, 29, 33–40,
    87, 91, 93, 96, 99, 109–113, 117,
    118, 131, 134
gneiss 23, 45, 88
goethite (α-FeOOH) 41, 43, 44, 137
gold deposit 36, 42, 96, 98, 99
gold mines (*see also* arsenic sources)
    4, 85, 87, 98, 99, 100, 113, 147
gold-copper ores 96, 98
granular ferric hydroxide (GFH)
    (*see* arsenic removal from water)
greenstone 23
groundwater (*see also* aquifer)
  arsenic (*see* arsenic contamination,
      aquifer/groundwater)
  classification
    acid sulfate-chloride waters 34, 36, 91
    Ca-HCO₃ type 45
    Ca-Na-HCO₃ type 90
    Na-Cl type 38
    Na-Cl-HCO₃ type 38
    Na-HCO₃ type 45
    sulfidic 25, 26, 28
  deep 26, 30, 39, 87, 90, 97, 103,
      108–110, 115
  evaporative concentration increase
      3, 27, 29, 34, 40
  geothermal 2, 11–13, 19–22, 24–26,
      28–30, 33–39, 87, 91, 93, 96, 99,
      113, 115, 117, 118, 145, 146
  level 89, 97
    fluctuations 96, 97
  oxidizing conditions 1–3, 25, 27, 28, 30,
      32, 36, 37, 39, 41, 45,
      92, 107, 115
  recharge 39, 88, 89–91, 107
  reducing conditions 2, 3, 25, 28, 29, 30,
      31, 36, 40–43, 88, 90, 92, 96, 99, 103,
      107–110, 113, 115, 146, 147
  residence time 29, 30, 34, 37, 44, 45
  shallow 39, 88, 104, 115
gypsum 21, 23

haloisite 46
health (*see* arsenic health effects)

health effects (*see* arsenic health effects)
health risk assessment
    (*see* arsenic health risk)
hematite (α-Fe$_2$O$_3$) 21, 43, 44, 96
hematite 21, 43, 44, 96
hematolite 21
hoernesite 21
hornfels 23
hot spring (*see* geothermal spring)
humic substances 27, 47, 67, 81, 82,
    88, 90, 102, 103, 104, 107, 109,
    113, 145–147
human
  exposure to As (*see* arsenic exposure)
  health (*see* arsenic health effects)
hyperkeratosis (*see* arsenic health effects)
hyperpigmentation (*see* arsenic health
    effects)
hypertension (*see* arsenic health effects)
hypopigmentation (*see* arsenic health
    effects)

igneous intrusion 95
igneous rocks 22–24, 95
illite 46, 94, 108
ilmenite 21
iron formations 23
iron oxides 21, 26, 27, 30, 31, 34,
    40–46, 94, 96, 118, 119, 137,
    140, 144, 149
iron oxyhydroxides 2, 3, 19, 20, 21, 26,
    28, 30, 32, 36, 40–47, 96, 99, 105,
    106–109, 109, 113, 137, 147
inorganic arsenic
    (*see* arsenic species)
intake of arsenic (*see* arsenic exposure)
interactions between arsenic and humic
    substances 27, 47, 67, 81, 82,
    88, 90, 102, 103, 104, 107,
    109, 113, 145–147
iron-coated adsorbents (*see* arsenic removal
    from water)
iron-rich sediments 23
irrigation 5, 7, 16, 20, 29, 37, 45, 115, 118,
    119, 123, 147
ischemic heart disease (*see* arsenic health
    effects)

jarosite 21, 22, 96, 98

kaolinite 46
keratosis (*see* arsenic health effects)
kidney cancer (*see* arsenic health effects)

lacustrine deposit (*see* sediment, lacustrine)
leukemia (*see* arsenic health effects)
limestone 3, 23, 46, 47
limonite 96
lithium 33
loess sediments 6, 20, 23, 24, 39, 45
long-term exposure (*see* arsenic exposure)
luzonite 98

maghemite 150
magnesium 63, 64, 137
magnetite 21, 22, 150
manganese 62–64, 108, 140
  oxide 26, 30, 40–46
  oxyhydroxides 3, 20, 28, 30,
    37, 40–47, 96, 107–109, 113, 147
marcasite 20, 21, 32
mechanisms of As mobilization (*see* arsenic
    mobilization)
metabolite (*see* arsenic metabolites)
metal oxides 21, 24–27, 30, 31, 40–47,
    92, 94, 96, 118, 119, 135–138,
    140, 144, 149
metal oxyhydroxides 2, 3, 19–21, 24–32, 36,
    37, 40–47, 92, 96, 99, 105, 106–108,
    109, 113, 119, 137
metamorphic rocks 21–23, 45,
    88, 94, 96, 98
metasandstone 96, 98
methane emission 108, 110
methylation (*see* arsenic metabolites)
microbe-mediated arsenic release
    (*see* arsenic mobilization, microbes)
mineralized areas (*see* arsenic sources)
mining (*see* arsenic sources,
    anthropogenic)
mitigation (*see* arsenic problem mitigation
    and arsenic removal from water)
monomethylarsenate (MMA)
    (*see* arsenic metabolites)
monomethylarsonous acid [MMA(III)]
    (*see* arsenic metabolites)
monomethylarsonic acid [MMA(V)]
    (*see* arsenic metabolites)
montmorillonite 46
mud volcanoes 109–113
mudstone 22, 32, 88, 108, 109, 113
muscovite 107

nanotechnology 149, 150
native arsenic 20, 21
natural arsenic contamination
    (*see* arsenic sources)

neurotoxicity of arsenic
        (*see* arsenic health effects)
niccolite 21
non-vascular diseases
        (*see* arsenic health effects)

olivine 21
organic acid 25, 28, 47, 109
organic matter 20, 23–26, 29, 30, 40–42, 88,
        106, 107, 109, 110, 146, 147
orpiment 21, 25, 36
oxide minerals 20, 25–31,
        40–47, 94, 96, 146

peat 22, 24, 25, 31, 32, 95, 109
pharmacosiderite 21
phosphate minerals 21, 22, 29, 44
phosphate-exchangeable arsenic 29, 44,
        137, 138
phosphorite 23, 24, 32
phreatic aquifer (*see* aquifer)
phyllite 22, 23, 94
plasma antioxidant capacity 75
propylitic alteration 22
pterygium (*see* arsenic health effects)
public health 1, 5–8, 18, 74, 118, 134
    risk assessment 5, 18, 74, 79,
        132–134, 147
        authorities 8
pyrite 20, 21, 24, 26, 31–34, 95, 96, 98, 99
    precipitation 24
pyroxene 21

quartz 21–23, 88, 96, 107, 108
quartzite 23

realgar 21, 36
redox potential (Eh, ORP) 2, 24, 25,
        30, 38, 39, 41–44, 103, 105, 197,
        111, 112, 146
redox-mediated arsenic release (*see* arsenic
        mobilization)
reductive dissolution (*see* arsenic
        mobilization)
remediation (*see* arsenic problem mitigation
        and arsenic removal from water)
removal of arsenic (*see* arsenic removal
        from water)
reverse osmosis (*see* arsenic removal from
        water)
risk of arsenic exposure
        (*see* arsenic exposure)

salinity 27, 39, 90, 105, 110
sandstone 23, 24, 30, 32, 33, 39, 88, 90, 95,
        98, 108
schist 22, 23, 45, 88
scorodite 21, 31
sediment 2, 5, 6, 9, 16, 17, 20, 22, 24, 26,
        31, 32, 34, 39–42, 44, 46, 88, 90, 95,
        104–109, 113, 146, 147
    aeolian 2, 6, 20, 23, 29, 37, 45
    alluvial 9, 16, 23, 87, 95, 104–107
    fish ponds 126–129
    fluvial 2, 16, 18, 23, 29, 32, 45, 86, 87
    lacustrine 23, 30
    marine 2, 23, 32, 94, 95, 107, 110
        deep sea 88, 108
        shallow sea 108
    river 3, 23, 30, 118
        sediment load 3, 96
    unconsolidated 20, 23, 40, 42
sedimentary aquifer (*see* aquifer)
sedimentary rocks 22, 23, 34, 95, 98
shale 2, 22, 23, 24, 32, 88, 93–96, 98
siderite 21, 22, 41, 47, 108
silica; dissolved 34, 44, 45, 51, 62–64, 67,
        137, 138, 140
silicate minerals 20–22, 107
silt 23, 39, 43, 87, 88, 94, 108
siltstone 23, 88
slate 22, 23, 88, 90, 94, 96, 98, 99, 109
soda lake 24
sodium 90, 113
soil (*see* arsenic contamination, soil)
sorbents for arsenic removal (*see* arsenic
        removal from water)
species dependent arsenic mobility 24–30,
        40–44, 107, 108
species of arsenic (*see* arsenic species)
standard of arsenic in food 120, 122
sulfate minerals 20–22, 31, 147
sulfide minerals 2, 7, 20–23, 26, 27, 31, 32,
        41, 94, 99, 110, 147
sulfosalts 20
surface water 2, 5, 7–13, 18–20,
        24–27, 29–31, 33, 34, 37,
        38, 43, 65, 71, 85, 87, 88,
        101, 117, 119, 123, 135, 136,
        141, 143, 148
    arsenic content (*see* arsenic
        contaminaton, surface water)

tennantite 21
thermal fluids (*see* arsenic sources)

thioanions (*see* arsenic species, water)
toxicological effects of arsenic (*see* arsenic
    health effects)
transport of arsenic (*see* arsenic transport)
tube well 8, 9, 140

unconfined aquifer (*see* aquifer)
unconsolidated sediments (*see* sediments,
    unconsolidated)
urinary arsenic (*see* arsenic metabolites)

vascular diseases (*see* arsenic health effects)
volcanic ash 20, 22, 29, 45
volcanic emissions (*see* arsenic sources)
volcanic glass 22, 45

water (*see* groundwater, surface water)
    quality (*see* arsenic contamination,
        surface water, groundwater)
    resources 5–7, 9, 16, 18, 19, 37, 123, 128

supply
    centralized water supply 2, 4, 136–139,
        143, 144, 149, 150
    change of source 135, 136
    decentralized water supply 17, 139,
        140, 144, 149, 150
    rural water supply 4, 6, 8, 16, 17, 102,
        140, 141, 150
    treatment plant (*see also* arsenic removal
        from water) 136, 137, 139, 141–143,
        149, 150
    techniques (*see* arsenic removal from
        water)
water-rock interactions 24, 26, 27,
    29, 30, 110
weathering 1, 2, 7, 16, 17, 19, 20, 22, 29, 31,
    38, 45, 94, 96, 98, 119
wood preservation (*see* arsenic sources,
    anthropogenic)

Locality index

Acámbaro 13
Aceh 11
Acoculco 13
Afyon 11
Akutan island 12
Alaska 12
Aleutian islands 12
Altos de Jalisco 13, 39
Ambitle island 12
Andean highlands 2, 8, 13, 27
Anding 94
Anei 86, 105
Anhui 11
Ankobra basin 10
Anle 98
Anping 129
Appalachian Highlands 12, 28
Aquitaine 10
Arad county 10
Argentine Andean highland 13
Arizona 12, 27, 28, 44, 139
Arkhangai 11
Assam 11, 14
Asturias 10
Atacama desert 3, 13, 27

Baihe 86, 88–89, 94, 97
Baikal lake 11
Baja California state 12, 14, 39
Bali (Taiwan) 93
Banská Bystrica 10
Bavaria 10, 32
Bay of Plenty 12
Beimen 50, 56–57, 59, 61, 65, 82, 85–86, 89,
        94–96, 101–102, 104–105, 107–108,
        123–124, 133
Beitou 91, 93, 113
Bengal basin 11
Bihar 11, 14
Bihor 10
Black Forest 2, 10
Bowen island 12
Brahmaputra 2–3, 24, 28, 41

Bramaputra valley 11
Bridgwater 10
British Columbia 12, 23
Budai 50, 56–57, 59, 61, 65, 82, 86, 88, 89,
        94–95, 97, 101–102, 104, 107, 110,
        123–124, 128–129, 132, 135

Caborca Magdalena 12
Cáceres province 10
Cactus-Sitio Grande 13, 35, 39
Caldas 13
Caldes de Malavella 10
Camargo 12
Cameroon 10
Carlisle basin 10
Celina-Mokrsko 10
Central Mountain Range 85, 88, 90, 92
Cerro Mina de Agua 13
Cerro Prieto geothermal area 12
Chaco plain 13, 45
Chalkidiki 10
Champagne hot springs 13
Chandigarh 11
Changhua 97, 99, 116, 129
Chennai 11
Chianan plain 2, 7, 11, 41, 85–89, 92–97,
        101–110, 113, 123, 145–147
Chiayi 50, 56–58, 61, 86, 89, 94, 97, 102,
        116, 124, 129, 135
Chiayi Hill 88–89, 93–94
Chihuahua state 6, 12
Chinautla 13
Chinkuashi gold mine 85, 87, 96, 98–100,
        113, 147
Chittagong coastal plain 11
Chocosuela-Platanar area 13
Choushuei river 97
Chukou 86, 89, 94–95
Chunglun hot spring 86, 87, 110
Cienfuegos 13
Citarum river 11
Cidu 98
Coahuila state 6, 12

Coatepeque lake 13
Cobalt (Ontario) 12
Colombo area 11
Comarca Lagunera 12
Copahue 13
Coquimbo 13
Croatia 10

Dakota 12, 22, 28, 44
Dakunshuei mud volcano 87
Danshuei river 93
Daton 93
Datong 11
Delicias 12
Dingshan 94
Dongshan 61, 64, 90, 92, 101, 130
Dornod steppe 11
Duero basin 10
Durango city 12
Durango state 6, 12

Ekondo Titi 10
El Alto (La Paz) 13
El Charco 13
Emet-Hisarcik area 11
Emilia-Romagna 10
Erjen river 85, 86, 89, 94
Etchojoa 12
Extremadura 10

Fairbanks 12
Fensmark 10
Finland 10
Finnish Lapland 10
Flanders 10
Florida 12
Fort Myers 12
Fraser valley 12
Fukui 11–12

Ganges 2–3, 24, 28, 41, 95
Ganges-Brahmaputra-
    Meghna plain 2–3, 24, 28
Geothermal Spring Valley
    36, 91, 93, 113
Ghazni 11
Girona province 10
Gisborne 12
Gongliao 98
Gouda 10
Govi Altai 11
Guanajuato state 7, 13

Guandu plain 11, 85, 87, 91, 93, 101, 113,
    115, 117–121, 145–147
Guanzihling mud volcano 86, 87, 89, 110
Guanzihling 110
Guayllabamba 13
Gueishan island 59, 87, 90, 98
Gujarat 11, 33

Hat Yai 11
Hawai 12, 34–35
Hawkes Bay 12
Hetao basin 11
Hermosillo 12
Heybeli spa 11
Hidalgo state 13
Himachal Pradesh 11
Houshayu (Beijing) 11
Hsiaokunshuei mud volcano 86–87,
    110–112
Hsiaying 50, 56–57, 59, 65, 86, 89, 94, 102
Hsinchu 89, 94, 116, 117
Hsinshi 89, 94
Hsinwen 86, 89, 94
Hsinying 86, 89, 94, 96
Hsuechia 50, 56–57, 59, 61, 65, 86, 89,
    94–95, 97, 101–102, 104–105,
    107–108, 123–124
Hualien 87, 116
Huang river 93, 101
Huanggang creek 91, 93, 117
Huelva province 10
Huhot basin 11
Hungary 10

Ilo 8, 13
Ilopango lake 13
Indus plain 2–3, 11, 18, 24, 28, 40
Interior Plains (USA) 12, 22,
    28, 32, 40, 44
Intermontane Plateaus (USA) 12, 27, 40, 44
Irrawaddy 11
Ischia 10
Isla de la Juventud 13
Izmir 14

Jalisco state 13
Jhonghi 88
Jhongshan 98
Jhuangwei 61, 64, 90, 92, 130
Jiangjiun river 87, 89, 96
Jiangjiun 86, 94, 96, 102, 104–105, 107–108
Jiaosi 61, 64, 90–92, 130

Jilin province 11
Jimenes 12
Jinhu 94
Jishuei river 87, 89, 94, 96–97, 101, 104
Jujo-Tecominoacán 13, 35, 39
Julimes 12

Kaduna 10
Kamchatka 12, 35, 36
Kaňk 10
Kaohsiung 58–59, 86–87, 89, 97, 113, 116
Kaoping river 87
Kathmandu valley 11
Kawerau geothermal field 12
Keelung 91, 93, 98, 116–117
Kilahuea 12, 35
Kinuma 13
Kuantien 89
Kuanyin High 86
Kuitun area 11
Kunshueiping mud volcano 86, 87, 110, 113
Kurdistan 11
Kyushu island 11–12

Labrador 12
Lanyang 11, 61, 64–65, 75–76, 82, 84–85,
    87–88, 90, 92–93, 95–96, 98–99,
    101–104, 106–109, 113, 115, 123–124,
    130–131, 133, 145–148
Lazio 10
León province 10
Liaoning province 11
Lientan 89
Linbei 11
Lipéz 13
Lishan fault 87, 98
Lithuania 10
Liujia 86
Liujiao 86, 89, 105, 107–108
Liuying 94
Liverpool-Rufford 10
Llano La Tejera 13
Loa river basin 13, 37
Los Azufres geothermal area 13
Los Humeros 13, 34–35, 39
Lucao 86, 105, 108
Luna-Sen 13, 35, 39
Luodong 90, 92, 130

Madoc 12
Madrid basin 10
Madrid Tertiary detrital aquifer 9–10

Malcantone watershed 10
Manchester 10
Mangyshlak 11
Manipur 11
Manzanillo 13
Manzano Mountains 12
Margajita river 13
Massif Central 10
Mekong river 2–3, 11, 18, 24, 28
Meoqui 12
Miaoli 116
Michigan 12, 28, 32
Michoacán state 13
Minas Gerais 13
Miravalles 13, 35, 38
Mixco 13
Moa 13
Morales 12, 74
Mount Apo 11
Mount Etna 10
Mount Vesuvius 10

Naba 94
Nagaland 11
Nakorn Chaisi 11
Nantou 86, 89, 94, 97, 116
Napo province 13, 38
Nariño 13
NE Caucasus foothills 11
New Foundland 12
New Mexico 12, 139
Niigata plain 11–12
Ningxia province 11
Nitra 10
North Humberside 10
Nova Scotia 12
Nuannuan 98

Odiel river basin 10
Offin basin 10
Ogcheon belt 11
Ogun state 10
Okavango delta 10
Okinawa basin 87, 90, 98
Oklahoma 12, 44
Olomega lake 13
Oruro 13
Osaka 11

Pachang river 86–87, 89, 94–96, 101–102, 104
Pacific Mountain System 12, 28, 40–41, 44
Pacific Ocean 88

Paderborn 10
Pampa plain 13
Pannonian basin 9–10
Papallacta lake 13, 38
Peikang High 87, 96
Peikang 85–87, 94, 96–97
Penghu islands 87, 96
Perth 11
Pezinok 10
Pingsi 98
Pingtung 58–59, 86, 97, 116, 119
Pol-Chuc-Abkatún 13, 35, 39
Poopó basin 13
Potosí dep 13
Potosí 7, 12, 13
Punjab 11
Puno 13
Putzu river 86, 87, 94, 96–97, 104
Pyrenees 10

Quebrada de Camarones 13
Quijos county 13, 38

Radovljica 10
Red River 2–3, 11, 16, 18, 24,
        28, 40, 140
Renai 98
Rhine valley 10
Rhon Phibun district 11
Riaño-Valdeburón 10
Rift valley 10
Rincón de la Vieja 13, 35, 38
Río Verde 12
Rivers state 10
Rocky Mountains 12
Romania 10
Rueifang 98

Saint Elizabeth 13
Sakhalin 12
Salamanca 10, 13
San Antonio-El Triunfo 12
San Ignacio-Mulegé 12
San Joaquin valley 3, 12, 27, 40
San José department 13
San Luis Potosí 7, 12
Sansing 90, 92, 130
Santa Cruz de la India 13
Santa Ma. de la Paz 12
Santa Rosa del Peñón 13
Saskatchewan 12
Sendai 11–12

Serbia 10
Serengeti 10
Severn Trent 10
Shanxi 9, 11
Shinji plain 11–12
Shuangsi 98
Siena 10
Sierra El Mechudo 12
Sierra Guadalupe 12
Sindh 11
Sinyi 98
Skellefte orefield 10
Slovakia 2, 10
Snow Mountain Range 88, 90, 92, 130
Soda Dam 12, 35
Songshan 93
Sonora state 12
Stromboli 10
Stuarts Point 12
Sudetes mountains 10
Sueihuotongyuan mud volcano 87

Tabasco state 13
Taichung 99, 116
Tainan 7, 50, 56–57, 58–59, 61, 76,
        85–86, 89, 94, 97, 102, 108, 116,
        123–124, 129, 133, 135, 146
Taipei 49, 61, 85, 87, 91, 93, 113,
        116–118, 121, 125, 145
Taitung 86, 87, 89, 94, 97, 116
Taiwan Strait 50, 85–87, 89, 94,
        96–97, 104, 113
Taiwan 1–3, 5–9, 11, 14–17, 19, 27,
        33–34, 36, 41, 49–50, 56–59,
        61–66, 69–71, 73–91, 93–96,
        98–99, 102–103, 110, 113,
        115–117, 119–123,
        125, 127–129, 131–136,
        139, 141–143, 145–150
Takatsuki 11–12
Tambo river 13, 29, 35, 38
Tampa 12
Tanshuei river 117
Taoyuan 116, 119
Tatio 13, 29, 35, 37
Taupo volcanic zone 12
Terai 11
Thessaloniki 10
Thessaly 10
Thoubal 11
Tiber valley 10
Tin belt 11

Tipitapa geothermal area 13
Tlamacazapa 13
Tolima 13
Tongonan geothermal area 11
Tongxiang 11
Toucheng 90, 92, 130
Transylvania 10
Trás-os-Montes 10
Tripura 11
Tsengwen river 86–87, 89, 96–97
Tsochen 89
Tumbaco 13
Tumon Bay 12
Tutum Bay 12, 36

Ulan-Ude 11
Uppsala 10
Uttar Pradesh 11, 14

Vale of York 10
Valle de Siria 13
Valle del Elqui 13
Valle del Guadiana 12
Valles Caldera 12, 35
Vapi 11
Västerbotten county 10
Veneto 10
Vojvodina province 10
Vosges mountains 10

Waikato river 12
Waiotapu valley 12
Wairau plain 12
Wanganui 12

Washington 12, 28, 44
Western Foothill Belt 89, 94
Wiesbaden 10
Wisconsin 12, 24, 30, 32
Wugu 93
Wuhe 11
Wujie 61, 64, 90, 92, 108, 130
Wushanting mud volcano 86–87,
        110–112

Yatenga province 10
Yenshueikeng mud volcano 86, 87
Yenshuei 86–87, 89, 94, 97, 102,
        104–105, 107–108
Yichu 50, 56–57, 59, 61, 65, 86,
        89, 94–95, 97, 101–102, 105,
        107–108, 123, 128
Yilan 64, 75, 83, 87–88, 90–92,
        96, 99, 116, 130, 135,
        145, 149
Yinyang sea 98–100
Yongkang 94
Yuanshan 90, 92, 130
Yunlin 59, 119
Yunnan province 11

Zapote 13, 15
Zhejiang 11
Zhengzhou 11
Zhongmou 11
Zimapán 13, 46
Zrenjanin 10

Arsenic in the Environment

Book Series Editor: Jochen Bundschuh & Prosun Bhattacharya

ISSN: 1876-6218

Publisher: CRC Press / Balkema, Taylor & Francis Group

1. Natural Arsenic in Groundwaters of Latin America
   Jochen Bundschuh, M.A. Armienta, Peter Birkle, Prosun Bhattacharya,
   Jörg Matschullat, & A.B. Mukherjee
   2009
   ISBN: 978-0-415-40771-7 (Hbk)

2. The Global Arsenic Problem: Challenges for Safe Water Production
   Nalan Kabay, Jochen Bundschuh, Bruce Hendry, Marek Bryjak, Kazuharu Yoshizuka,
   Prosun Bhattacharya & Süer Anaç
   2010
   ISBN: 978-0-415-57521-8 (Hbk)